J. Glicenstein J. Ohana C. Leclercq

Tumours of the Hand

With 130 Figures, Mostly in Colour

Springer-Verlag
Berlin Heidelberg New York
London Paris Tokyo

Julien Glicenstein, M.D.
63, Boulevard des Invalides, F-75007 Paris

Jacques Ohana, M.D.
14 bis, Avenue Bosquet, F-75007 Paris

Caroline Leclercq, M.D.
15, Rue Franklin, F-75016 Paris

Translator: David Le Vay, M.D.
Dewhurst Lodge, Wadhurst, East Sussex, TN5 6QB, Great Britain

---

Title of the original French edition: *Tumeurs de la main*

ISBN-13: 978-3-642-71836-6 e- ISBN-13: 978-3-642-71834-2
DOI: 10.1007/978-3-642-71834-2

Library of Congress Cataloging-in-Publication Data. Glicenstein, J. (Julien), 1934–. (Tumeurs de la main. English) Tumours of the hand / J. Glicenstein, J. Ohana, C. Leclercq; (translator, David Le Vay). p. cm. Translation of: Tumeurs de la main. Includes bibliographies and index.
1. Hand-Tumors. I. Ohana, J. (Jacques), 1952–. II. Leclercq, C. (Caroline) 1952–. III. Title. (DNLM: 1. Hand. 2. Neoplasms. WE 830 G559t) RC280. H35G4513 1988 616.99'297-dc19 DNLM/DLC for Library of Congress 87-27382

Reproduction of the figures: Gustav Dreher, Württembergische Graphische Kunstanstalt GmbH, Stuttgart
Typesetting, Printing and Binding: Appl, Wemding. 2124/3140-543210

Foreword

Tumours of the hand are highly varied, their only common feature being their location in an organ in which preservation of sensation and mobility must be our primary concern. Although the subject of numerous studies, they are not well known to the majority of practitioners, as most of these studies deal with only one type of tumour. Now, however, we have a book devoted entirely to these tumours as a whole, with a clear and logical approach to the clinical features, histology, differential diagnosis and treatment of each type, together with an extensive bibliography. The need for such a work - one I believe to be unique of its kind - has long been felt, for tumours of the hand often raise difficult problems of diagnosis, prognosis and treatment.

As regards differential diagnosis, this book is plainly very useful; it can be consulted like a classification. In recent years, diagnosis has been improved by new methods of investigation, such as thermography, scintigraphy, arteriography and serial angiography, etc. Though very expensive, the use of these sophisticated techniques sometimes proves valuable for the diagnosis of certain bony or vascular tumours that would otherwise be difficult to approach. Such methods should not be overused; when the lesion is easily accessible, biopsy is the correct procedure. In every case, biopsy provides the only certain method of diagnosis, even though histological interpretation may be difficult and calls for great experience; the penalties of error can be very serious.

The prognosis of tumours of such diversity is very variable, since they may arise in any of the tissues of the hand, be benign or malignant, be specific to the hand or nail, occur also at other sites or be metastatic. The varieties are so numerous and some are so rare that it is difficult to collect enough cases to enable one to make a definitive prognosis. This book may provide the basis for different groups of specialists - dermatologists, surgeons, oncologists - to collect precise statistics permitting the assessment of the prognosis of each type of tumour and the value of different treatments on a rational basis.

Treatment has also advanced. Simple excision of even the commonest type, of synovial cysts, the tumoral nature of which is in any case debatable, has now been challenged in view of the frequency of recurrences. Use of the operating microscope and histological monitoring during surgery of lesions lacking definite boundaries allow better control of the extent of excision, which must be as economical as safety allows with an organ as important as the hand. Reconstruction after mutilating operations is facilitated by the entire range of techniques of plastic surgery, from simple skin grafts to composite flaps whose transfer is made possible by microsurgical methods. Major reconstruction of the hand conforms to rules aimed primarily at re-establishing its sensory and prehensile functions, which means that it must be performed by hand surgeons only. However, the value of a multidisciplinary approach to those tumours for which diagnosis is difficult and prognosis uncertain is self-evident, especially when the question of amputation arises. All available experts should then be called on - histologists, surgeons, oncologists, sometimes rheumatologists or dermatologists.

The team that has carried out the enormous amount of work required for the preparation of this book is composed of plastic surgeons and hand surgeons. The very useful volume now presented to us is the fruit of exceptional clinical experience allied with exhaustive research, as the quality of this documentation bears witness.

Raoul Tubiana

Contents

III. Tumours of Bone

IV. Ungual Tumours

Introduction

Any of the tissues making up the hand - the skin, soft parts, vessels, nerves, bone, nails - may become the site of tumoral proliferation; however, the incidence, interpretation and even the treatment of these tumours are not completely understood.
The absolute and relative incidences of hand tumours are difficult to determine. Haber et al. (1965) collected 2321 lesions seen over 25 years at the Chicago Memorial Hospital, i.e. a little less than 100 tumours a year. Detailed study of this series reveals, apart from 156 verrucas and papillomata, benign tumours most of which do not require specialist advice, 507 cases of Dupuytren's contracture - not normally regarded as tumoral in nature - and 697 neurilemmomata consisting for the most part of neuromata following traumatic nerve section. On the other hand, the authors report only 3 melanotic whitlows over the entire period. Posch (1966) reports a more limited but more consistent series of 679 cases in which synovial cysts (ganglia) form 59% of cases, giant-cell tumours 10% and glomus tumours 3.5%. Boyes (1970) finds a similar relationship: 394 cases with 52% synovial cysts, 13% giant-cell tumours and 3.5% mucous cysts. Bogumill et al. (1975) collected 129 lesions over 3.5 years: synovial cysts were by far the most common tumours (55%), while other lesions were very rare except for giant-cell tumours of the tendon sheaths (10%) and mucous cysts (8%). The same predominance of synovial cysts is noted by Stack (1960), Butler et al. (1960) and Leung (1981).
Our own series, amounting to 471 cases seen over 10 years, gives a rather different picture (Table 1). Synovial cysts constitute only 35% of cases and mucous cysts 12.5%, whereas, as in other series, inclusion cysts (epidermal cysts), giant-cell tumours and glomus tumours are the most common of the "rare" tumours. Tumours of bone make up only 2%, and malignant tumours mainly affect the skin.
To assess the true incidence of hand tumours, several factors must be taken into account. Those involved in examining affected patients include hand surgeons, plastic surgeons, orthopaedists, dermatologists, rheumatologists, surgeons and general practitioners. Surgeons are consulted only when an operation is contemplated. Plastic surgeons operate on many of the skin tumours, while the orthopaedists examine most of the bony lesions. The result is that some authors incorporate into their statistics lesions not considered as tumoral by others. Allowing for these different factors, it may be estimated that the most frequently found tumours of the hand are synovial cysts, lesions that represent over 50% of cases, but whose tumoral nature is in any case disputed.

Table 1. Classification of 471 cases

Cysts:	
synovial	165 (35%)
epidermal	35 (7.5%)
mucous	58 (12.5%)
interdigital sinuses	2
Giant-cell tumours of tendon sheaths	22 (4.75%)
Lipomata	12
Fibromata	6
Koenen's tumour (periungual fibroma)	3
Phalangeal pads (solitary) (knuckle pads)	4
Angiomata	13
Vein tumours	2
Lymphangioma	2
Arteriovenous fistula	1
Aneurysm	4
Neural tumours:	16 (3.4%)
fibrolipoma	7
schwannoma	6
neurofibroma	3
Glomus tumours	20 (4.5%)
Naevi	28 (6%)
Melanomata	9
Keratoacanthomata	1
Bowen's disease	6
Epitheliomata:	
basal-cell	2
squamous-cell	14 (3%)
Verrucas and keratoses	20 (4.2%)
Radiodermatitis	5
Botryomycoma	9
Chondroma	4
Exostosis	3
Osteoid osteoma	2
Synovial tumours	1
Malignant tumours of the soft parts:	2
fibrosarcoma	1
epithelioid sarcoma	1

The others include giant-cell lesions, inclusion cysts and mucous cysts (5%-10%), and glomus tumours and angiomata (less than 5%). The bone tumours, particularly the chondromata, are more common than the statistics would suggest.
Finally, squamous-cell skin cancers and malignant melanomata are the most widespread malignant tumours, while cancers of the soft parts are exceptional (Clifford and Kelly 1955; Kendall et al. 1969).
The very notion of a tumour of the hand is debatable, for many of the lesions affecting the hand are difficult to classify, whether as dystrophies, hamartomata or true tumours, and every series indicates variability in this matter. We ourselves have excluded all post-traumatic lesions (neuromata and scarring) and non-tumoral proliferations (Dupuytren's contracture and metabolic, calcified, protein or gouty deposits); we have retained all the lesions that present clinically as tumours (synovial cysts, fibrolipomata of the median nerve, arteriovenous fistulae, etc.), together with skin lesions that may undergo malignant change (e.g. naevi).
The classification we have adopted is as follows:

Skin tumours: verrucas, naevi, cysts, precancerous keratoacanthomata (keratoses, radiodermatitis, melanosis), cancerous (incipient and invasive epithelioma, skin sarcoma, melanoma)
Tumours of the soft parts: benign and malignant tumoral lesions of synovial origin; lipomatous tumours; fibrous, cartilaginous, muscular, neural and vascular tumours
Bone tumours: whether of cartilaginous, bony or fibrous origin; malignant tumours
Ungual tumours: benign and malignant

The clinical picture is always much the same. Apart from cutaneous and ungual lesions, which have their specific features, hand tumours are usually evidenced by a local or diffuse swelling with little or no pain, that develops progressively without signs of compression. Radiography reveals the osteocartilaginous tumours but has little to show in tumours of the soft parts, apart from possible calcification or outlines. Arteriography is particularly useful in defining the nature of a tumour of vascular origin, while scintigraphy is necessary in the diagnosis of certain bony tumours, such as osteoid osteoma. The outlines of an ill-defined tumour may be better visualised with the aid of tomography. For many tumours examination by specialists in different disciplines is advisable, making use of thorough clinical experience. Collaboration is also often advisable in interpreting the histology, which may be extremely difficult (fibrous and bony tumours, melanomata).
Biopsy excision or partial biopsy of the lesion under regional anaesthesia is always possible without hospitalising the patient. Histological or cytological examination by aspiration is too imprecise and should be rejected. If the lesion is small, circumscribed and clinically benign, it is removed as a whole and an exact diagnosis is made by postoperative histological examination. If any doubt exists as to the benign nature of the tumour, it is important that the preoperative diagnosis should be as precise as possible. The radical treatment of a malignant tumour is always mutilating, especially in the case of a melanoma or malignant tumour of the soft parts, and it should not be undertaken without a warning to the patient.

Bibliography

Alter AH, Haber MH (1967) Hand tumors: A review. Hawaii Med 26: 403-408

Beasley RW, Ristow BB (1976) Malignant tumors of the hand. In: Andrade R et al (eds) Cancer of the skin. Saunders, Philadelphia, pp 1435-1448

Bogumill GP, Sullivan DJ, Baker GI (1975) Tumors of the hand. Clin Orthop 108: 214-222

Boyes JH (1970) Tumors. In: Bunnell S (ed) Surgery of the hand, 5th edn. Lippincott, Philadelphia Toronto, pp 666-710

Burton DW, Littler JW (1975) Non trauma soft tissues afflictions. Curr Probl Surg 1: 56

Butler ED, Hamill JP, Seipel RS, De Lorimier AA (1960) Tumors of the hand. A ten-year survey and report of 437 cases. Am J Surg 100: 293-302

Byrne JJ (1954) Hand surgery. Am J Surg 88: 470-472

Clifford R, Kelly A (1955) Primary malignant tumors of the hand. Plast Reconstr Surg 15: 227-232

Cooley SGE (1964) Tumors of the hand and forearm. In: Converse JM (ed) Reconstructive plastic surgery. Saunders, Philadelphia London, pp 1740-1786

Cornet L, Loubiere R (1974) Les tumeurs des parties molles de la main et des doigts. J Chir (Paris) 108: 143-154

Gaisford JC (1966) Tumors of the hand. Surg Clin North Am 40: 549-566

Glicenstein J (1977) Tumeurs des parties molles de la main. Ann Chir 31/12: 1087-1101

Greene RG (1964) Soft tissue tumors of the hand and wrist: A 10 years survey. J Med Soc NJ 61: 495

Haber MH, Alter AH, Wheelock MC (1965) Tumors of the hand. Surg Gynecol Obstet 121: 1073-1080

Iselin M, Iselin F (eds) (1967) Tumeurs de la main. In: Traité de chirurgie de la main. Flammarion, Paris, pp 739-769

Kendall TE, Robinson DW, Masters FW (1969) Primary malignant tumors of the hand. Plast Reconstr Surg 44: 37-40

Leung PC (1981) Tumors of hand. Hand 13: 169

Mason ML (1937) Tumors of the hand. Surg Gynecol Obstet 64: 129-148

Matton G, Defloor E (1974) Benign tumors of the hand. Acta Orthop Belg 40: 172-198

Pack GT (1939) Symposium: Tumors of the hands and feet. Surgery 1: 5

Posch JL (1956) Tumors of the hand. J Bone Joint Surg [Am] 38: 517-539

Posch JL (1966) Soft tissue tumors of the hand. In: Flynn JE (ed) Hand surgery. Williams & Wilkins, Baltimore, pp 999-1026

Stack G (1960) Tumors of the hand. Br Med J I: 517, 919-922

Stout AP (1950) Report on 1346 soft tumors (of the hand). (News Notes). J Bone Joint Surg [Am] 32: 457

Stout AP (1953) Tumors of the soft tissues. In: Atlas of tumor pathology. Armed Forces Institute of Pathology, section III, fascicule 5. Washington, pp 29-35

Woods JE, Murray JE, Vawter GF (1970) Hand tumors in children. Plast Reconstr Surg 46: 130-139

I. Skin Tumours

Warts (Verrucas)

Warts are benign skin lesions of viral origin which mainly affect the extremities. The very wide range of therapies suggested indicates that treatment may not be a simple matter.

Clinical Features

Warts tend more often to affect the young, but not exclusively. On the hands, there are two types: the common wart (verruca vulgaris) and the plane wart.

The common wart may be single or multiple; in the latter case there may be several dozen lesions, sometimes confluent in sheets, with one larger, older or "mother" wart (Fig. 1). It is usually situated on the back of the hand, appearing as a small, keratinised, greyish tumour with an irregular surface traversed by projections (Fig. 2). These projections are less obvious when the lesion is young, while older lesions have a creviced, blackish appearance. It is often painless, and the adjacent skin is healthy. On the front of the hand the wart is much more deeply embedded in the palm, often blackish and surrounded by a keratotic rim (Fig. 3). Warts at the fingertips are identical in appearance but are often sensitive to pressure.
One special type is the periungual wart, embedded in the nail fold and forming a more or less continuous rim at the periphery of the nail. Nail growth may be affected, with longitudinal or transverse striae (Leclercq and Wibault 1972). The site of implantation of the base of the wart is often inflamed and painful (see p. 213).

Plane warts may or may not be associated with common warts. They are macules of very small, even punctiform dimensions, buff-coloured, showing a finely granular appearance under the lens that is identical in miniature with that of the common wart. They are most often multiple, even profuse.

Differential Diagnosis

The common form is quite characteristic, but a single lesion on the back of the hand may resemble a senile keratosis. Warts that have already been treated are sometimes difficult to diagnose, suggesting a keratosis, a radiodermatitis, or even a squamous-cell epithelioma. A periungually situated wart may be confused with other tumours of the nail bed (see p. 115). Plane warts may simulate Darier's disease or Hopf's acrokeratosis verruciformis (in which verrucous lesions are associated with keratodermia and changes in the nails). Warts are common in states of immunodepression (as in renal transplant cases), as are squamous-cell epitheliomata, and it is sometimes difficult to distinguish them. Biopsy is indicated in every case in which the diagnosis is doubtful.

Histology

Warts have a characteristic histological appearance, showing bands of hyperkeratosis and parakeratosis overlying an irregular granular layer. In this layer and in the malpighian layer are found cells arranged in groups so as to produce a cavitary appearance, with a small, peripheral, condensed nucleus surrounded by a clear halo. Plane warts contain less parakeratosis and more vacuolated cells.
Intracellular inclusions are regarded as evidence of viral infection and have long been known (Dubreuilh 1895, cited by Leclercq and Wibault 1972). They are not constantly present and are more often found in the deeply embedded warts of the palm and fingertips. They may be intranuclear or intracytoplasmic.

Aetiology

The viral aetiology of warts has long been known and much research of an immunochemical and histochemical nature is in progress. It has been possible to demonstrate antigens and antibodies and to assess the lymphocytic response accompanying cure. There are current studies on the synthesis of an "anti-verrucal" vaccine (Stoughton 1977). However, many other factors play a part in the appearance and development of warts, particularly transpiration, maceration and psychological factors.

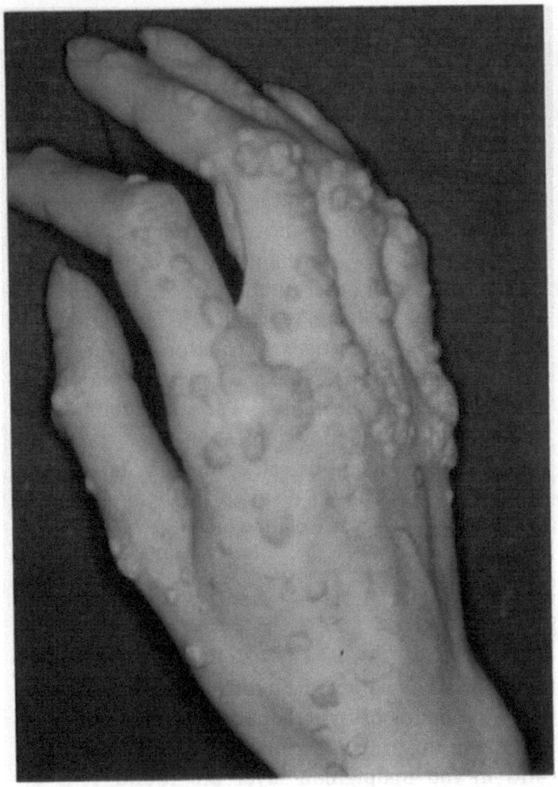

Fig. 1. Multiple common warts: note the confluent patches on the dorsum of the metacarpophalangeal joints

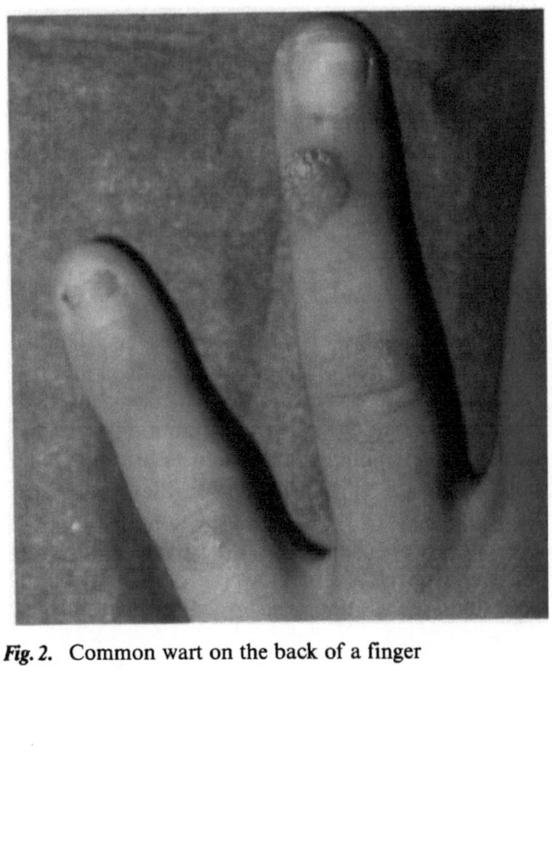

Fig. 2. Common wart on the back of a finger

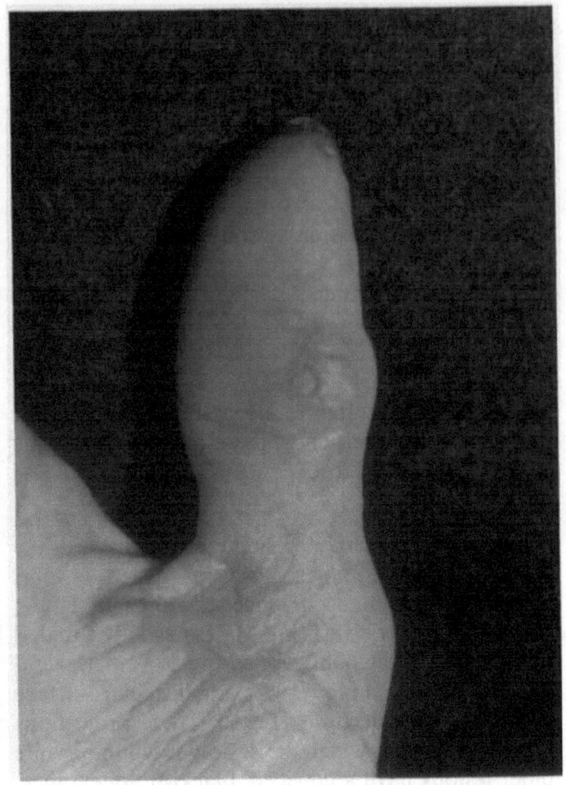

Fig. 3. Common wart on the palmar aspect of the thumb

Psychology can be exploited in certain forms of treatment.

Development

Incubation of a wart lasts 1-8 months. Spontaneous cure may occur in a few months, but it is usually accompanied by the appearance of other warts in the same area.

Treatment

A great many methods of treatment have been suggested:

1. Psychological methods are employed more frequently for children aged 6-12 years and are as various as the imagination of the doctor (persuasion, rituals, "magic", etc.). Thus Litt (1978) treats all periungual warts by application of an adhesive dressing left in place for 15 days.
2. Medical treatments include local applications, the most used being a preparation of salicylic ac-

id with lactic acid and collodion, and cryotherapy, either with liquid nitrogen or carbon dioxide snow. Bunney et al. (1976) in a comparative study of eleven types of treatment in 1800 patients found liquid nitrogen (three applications at 3-week intervals) most effective; another large-scale British study reported similar results with both liquid nitrogen and the salicylic-lactic acid-collodion mixture.

3. Surgical treatment may be by electrocoagulation, with circular ablation of the wart by a diathermy knife. This can be done under local anaesthesia or general anaesthesia for multiple lesions. It may be combined with curettage, provided care is taken to avoid bleeding with its risk of contamination. Simple surgical excision carries a risk of recurrence.
4. Radiotherapy is contraindicated for this benign lesion.

Indications

Treatment is not always justified for this benign lesion, which may regress spontaneously. It is indicated for lesions that are a nuisance, painful, cosmetically embarrassing or multiple. Most authors advocate medical treatment, such as cryotherapy, and reserve surgery for resistant or multiple lesions.

Bibliography

Baran R (1978) Tumeurs unguéales. In: Pierre M (ed) L'ongle. Monographie du Groupe d'Etude de la Main. L'Expansion Scientifique, Paris, pp 76-84

Bunney MH, Nolan MW et al. (1976) An assessment of methods of treating viral warts by comparative treatment trials based on a standard design. Br J Dermatol 94: 667

Leclercq R, Wibault C (1972) Tumeurs bénignes infectieuses. In: Encyclopédie médico-chirurgicale-Dermatologie. Paris, pp 12725 - A10: 1-5

Litt JZ (1978) Don't excise: Exorcise. Treatment for subungual and peri-ungueal warts. Cutis 22: 673-676

Morison WL (1974) In vitro assay of cell-mediated immunity to human wart antigen. Br J Dermatol 90: 531

Rees RB (1979) Warts. Cutis 23: 588-593

Stoughton RB (1977) Advances in dermatology: 1976. Bull Am Acad Dermatol, p 1

Naevi

Pigmented naevi are very common skin lesions, present from birth (congenital) or developing in the child or adult (acquired). They are relatively rare in the hand. They consist of accumulations of naevus cells or melanocytes grouped in the dermis or at the dermo-epidermal junction. Their appearance, extent, shape and colour are extremely variable. Lentigos, or beauty spots, may be very small - only 1-2 mm in diameter - while giant naevi, which are often hairy, may cover the whole of one surface of the hand or even a large part of the upper limb or of the body surface.

Naevi are extremely common and may affect any white-skinned individual and also, though more rarely, Asiatics, black people and albinos. The incidence seems to be the same in boys and girls. The palm is rarely involved, lesions of the hands being nearly always lentigos. The larger naevi mainly affect the back of the hand, to which they are restricted. The distal part of the finger (outside the nail bed) is rarely involved.

Clinical Features

Naevi differ in appearance. Murray (1983) classifies naevi and melanotic lesions as follows:

A. Lesions with increased activity of epidermal melanocytes:
   Freckles (ephelides)

B. Lesions with an increased number of epidermal melanocytes:
   Lentigo simplex
   Lentigo maligna (Dubreuilh's melanosis)

C. Lesions with naevus cells
   Common:
   1. Junctional pigmented naevus
   2. Compound pigmented (mixed) naevus
   3. Intradermal pigmented naevus
   Rare:
   1. Congenital pigmented naevus
   2. Halo naevus (Sutton's naevus)
   3. Juvenile melanoma
   4. Balloon-cell naevus (naevus with distended cells)

D. Lesions with intradermal melanocytes:
   Mongolian spot
   Ito's naevus
   Ota's naevus
   Blue naevus
   Cellular blue naevus

Not all these types are found in the hand. The most common clinical types are described below.

Lentigo is a small, brown or black macule, flat or slightly elevated, not suppressing the epidermal furrows in the palm. Lentigos are habitually acquired, smooth and glabrous (Fig. 4).

Congenital naevi, smooth or hairy, flat or elevated, rarely verrucous in appearance, are patches extending over several square millimeters or centimeters. They usually have a uniform light or dark-brown colour. They are common on the wrist and the back of the hand (Fig. 5).

Giant naevi, also congenital, affect the whole of the back of the hand and wrist, often extending to the proximal phalanges of the fingers and the base of the thenar and hypothenar eminences (Fig. 6). The skin is often thickened, wrinkled and hairy. Rarely, this naevus may involve the whole of the upper limb, extending to the middle of the arm or even as far as the shoulder, but not usually affecting the volar aspect of the arm and elbow. Sometimes there is a "seed bed" of smaller naevi, which are pigmented or just hairy. Exceptionally, naevi of the hand and upper limb are only components of a disseminated naevus formation extending over the entire body surface and sometimes to the leptomeninges.

The *blue naevus* is quite common on the back of the hand or wrist, one of its preferred sites. It presents as a small, bluish nodule, deeply embedded in the dermis (Fig. 7).

The *juvenile melanoma* of Spitz (1948) is seen more often in the child than in the adult as a pink, grey or ochre papule; diagnosis is by histological examination.

Naevi of the nail matrix constitute a particular form, often very small but associated with a band of pigmentation of the nail bed. Common in black-skinned subjects, they are also acquired in whites (Fig. 8; p. 215).

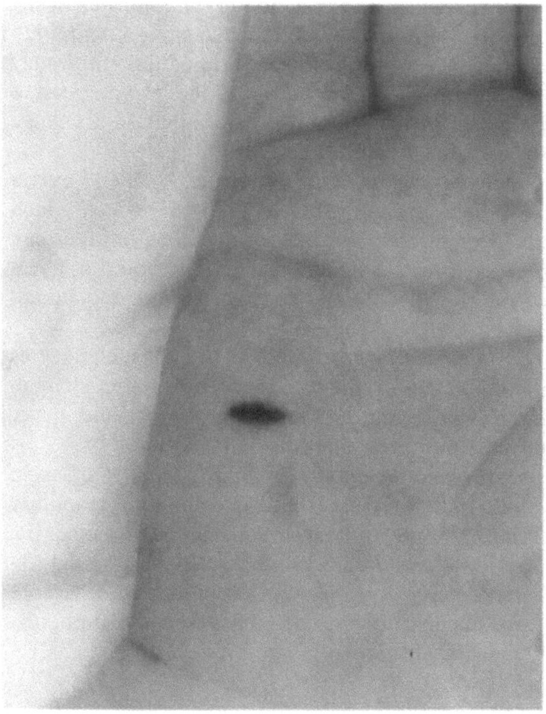

Fig. 4. Lentigo of the ulnar border of the hand

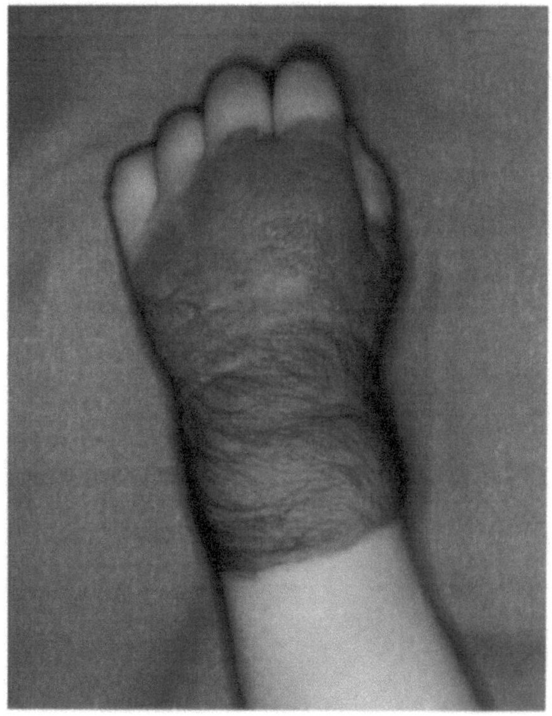

Fig. 6. Giant naevus

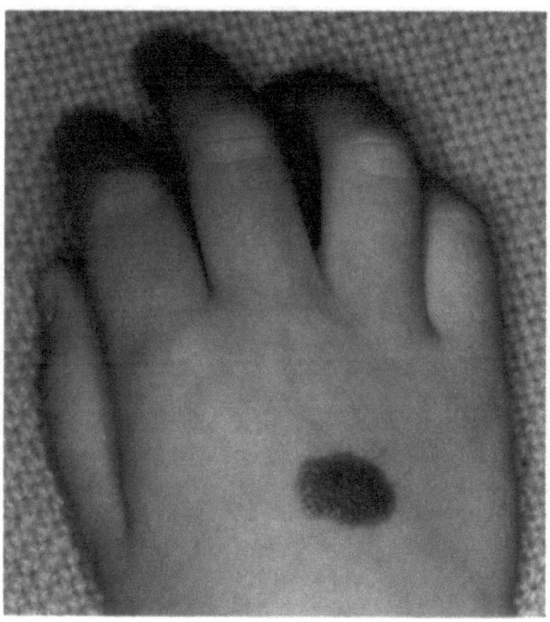

Fig. 5. Congenital naevus

Fig. 7. Blue naevus *(right)*

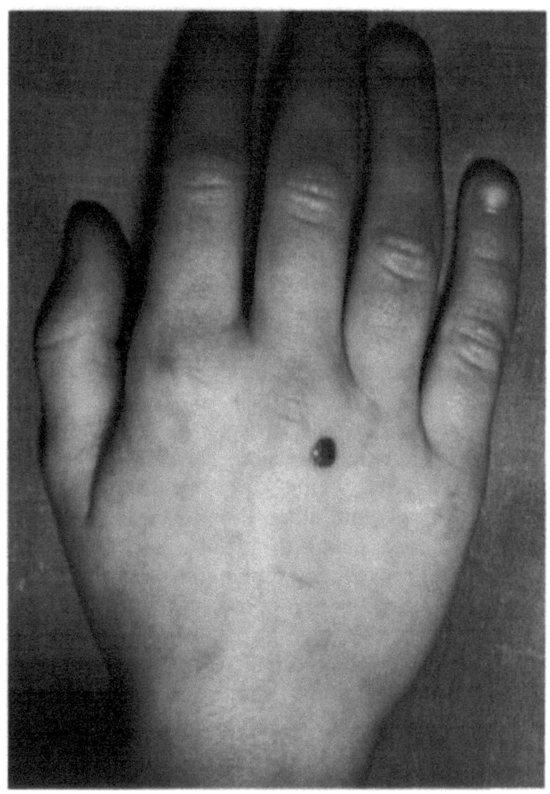

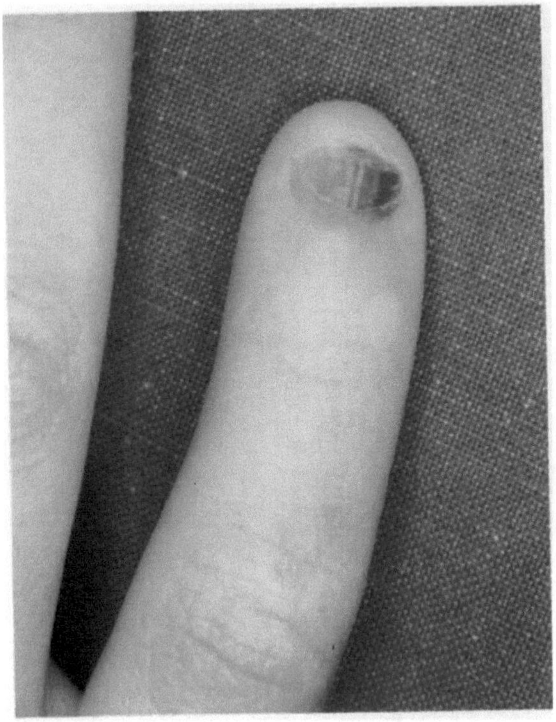

Fig. 8. Naevus of the nail matrix

Pathology

In normal skin the melanocytes or pigmented cells are found in the basal layer of the epidermis, where they are 5-10 times as rare as the keratinocytes. Ephelides (freckles) are formed by a greater concentration of pigment in the basal layer of the epidermis, but without any increase in the number of melanocytes. Such an increase is characteristic of lentigo simplex, in which the melanocytes replace the keratinocytes almost entirely. True pigmented naevi are associated with clumps or nests of melanocytes.

In *junctional naevi* these cell groups are situated at the dermo-epidermal junction. The cells of the naeval nests have an epithelioid appearance. Naeval nests are also found along the pilosebaceous follicles or the sweat glands. Clinically, junctional naevi are flat and pigmented, resembling simple lentigos. They may extend and give rise to nests in the dermis.

The *compound naevi* so formed thus have a dermal as well as an epidermal component. Owing to the extension of the naeval focus and the connective tissue proliferation, the naevus is often elevated. The cells are frequently epithelioid in the papillary dermis, smaller and resembling lymphocytes in the reticular dermis. The deepest cells have a neural appearance. With the passage of time these naevi may become purely *intradermal,* and the lesion ceases to progress. In many mature naevi there is a change towards fatty infiltration and fibrosis.

Congenital melanotic naevi frequently extend over a wide area and are often hairy. These naevi, which are present from birth, are mixed or intradermal, their cells always being present in the inferior layers of the reticular dermis and even in the subcutaneous fat.

The *blue naevus* is formed of dermal melanocytes dissimilar to ordinary naevus cells (star shaped, larger, occupying the reticular dermis) and never extends into the epidermis.

The *juvenile melanoma* of Spitz actually arises in adults in 20% of cases and is commonly a solitary lesion. The site of the lesion is habitually mixed, i.e. junctional and dermal, and the cells may be suggestive of malignant melanoma (epithelioid cells with large nuclei and nucleoli, mitoses and lymphohistiocytic infiltration).

Differential Diagnosis

Diagnosis is normally simple when the lesion is extensive and congenital. It is more difficult when the lesion is small, acquired, not much pigmented or, on the contrary, very black. In such cases one would consider it to be a seborrhoeic wart or an actinic keratosis in the case of a poorly pigmented lesion, a pigmented basal-cell epithelioma or an angioma if very pigmented. Diagnosis is confirmed by biopsy excision.

Development

The very great majority of congenital and acquired naevi are benign and remain so; 80% of malignant melanomata appear in healthy skin. However, any naevus may undergo malignant transformation. Two varieties appear more liable to such degeneration: the very large congenital ones associated with a seed bed of naevi, in which malignant change occurs in 10%-30% cases, and the subungual type, seen as bands or sheets which are difficult to distinguish from Dubreuilh's melanosis.

Treatment

Excision of naevi of the hand may be desirable for two reasons, prophylactically due to the risk of malignant degeneration, and aesthetically. Prophylactic excision is preferred for naevi over 1 cm^2 in size, for congenital naevi, for those with changes in size, colour and thickness and for those which are bleeding. The routine excision of palmar naevi does not seem essential. On the other hand, naevi of the nail matrix may undergo malignant change and should always be removed.

Technique

Surgical excision followed by suture, skin grafting or, more rarely, by autoplasty is indicated in the great majority of cases. However, Bruce Williams (1983) advised dermabrasion of large congenital naevi before the age of 1 year; one of his cases involved the hand, and there was marked improvement in the palm. The technique of excision is simple; the lesion is removed with a skin margin of 2-3 mm, including the entire thickness of the skin. However, the technique of repair may be much more difficult. Only naevi of a very small size can be excised with simple suture. Loss of substance on the back of the hand not exceeding 1-2 cm^2 can be dealt with by a double Z-plasty. In other cases a graft is necessary; full-thickness skin grafts from the arm or groin give the best functional results in terms of supple, mobile skin and the best aesthetic appearance in terms of colour. Unfortunately, the grafts often undergo pigmentation, and the grafted area should be protected from the sun for several years.

Bibliography

Arens MS, Hurwitz S (1983) Congenital nevocellular naevus. A review of the treatment controversy and a report of 46 cases. Plast Reconstr Surg 72: 355-365

Bruce WH (ed) (1983) The use of dermabrasion in giant pigmented nevi. Symposium on vascular malformations and melanotic lesions. Mosby, St Louis, pp 321-326

Jackson R (1983) Treatment of nevi: General principles. In: Bruce WH (ed) Symposium on vascular malformations and melanotic lesions. Mosby, St Louis, pp 294-303

Murray D (1983) Classification and pathology of pigmented nevi. In: Bruce WH (ed) Symposium on vascular malformations and melanotic lesions. Mosby, St Louis, pp 253-267

Solemen LM (1980) The management of congenital melanocytics nevi. Arch Dermatol 116: 1017

Spitz S (1948) Melanomas of childhood. J Pathol 24: 591

Epidermal Cysts

Epidermal cysts are relatively common benign tumours, essentially localised to the hand; though known for a long time, very little has been written about them. Wernher reported the first case in 1855, followed by Rizet in 1866. Further reports came from Muron (1868), Gillette (1875), Masse (cited in Reverdin 1887), Troquart (1881) and Gibier (1881). In 1887, Reverdin produced the first clinical description and published a comprehensive study of the literature hitherto. Other authors reported additional cases, and different terms were successively proposed to designate this tumour: dermoid cyst (Rizet 1866), epidermal cyst (Reverdin 1887; Troquart 1881), epidermoid cyst (Francke 1894), traumatic epithelial cyst (Garré 1894), traumatic epidermoid cyst (Blond 1922), implantation cyst (Sutton 1895), implantation dermoid (Burrows 1926).
Epidermal cysts constitute less than 10% of tumours of the soft parts of the hand: 8.4% (Boyes 1970), 4.1% (Posch 1956), 10% (Stack 1960), 5.6% (Haber et al. 1965). Virtually all these cysts are situated in the dermis and subcutaneous tissue. Cases of intraosseous epidermal cysts are rare and form a related entity to which we shall return after considering cysts of the soft parts.

Cysts of the Soft Parts

These have been known since the times of Wernher (1855) and Rizet (1866) and are situated almost exclusively on the palmar aspect of the hand and fingers, occurring particularly in manual workers (metalworkers, carpenters, watch-makers and surgeons). Owing to the nature of these occupations there has long been a marked predominance in males, though the entry of women into manual labour has considerably reduced the difference. The age of patients affected by epidermal cysts also coincides with vocational activity, the majority of cases being seen in adult life with a peak around 43 years (Stack 1960). Most authors are agreed in ascribing the origin of the condition to microtrauma and on the deferred appearance of the tumour. From this aspect it is essential to discover whether or not there has been any injury, sometimes long ago. Such a history is found in half the cases but is of varying severity. Sometimes it is so slight as to have passed unnoticed; it may have been a prick, a crush, a contusion or a surgical operation. Behrens (1931) and Wien and Caro (1934) report a greater incidence of contusional lesions than perforating wounds. There is a variable interval before the appearance of the epidermal cyst; the lesion may develop in a few weeks or, on the contrary, not appear until 30 years later. It is in these latter cases, especially if the trauma was minor, that the history will not uncover any special antecedents.

Clinical Features

Once formed, the tumour is spherical or ovoid, situated subcutaneously on the palmar aspect of the hand or fingers (Fig. 9). It presents as a nodule of varying size, from that of a small pea to an almond, hard, tense, elastic, adherent to the superficial layers, sometimes with a small skin cap but mobile in relation to the deep structures (Fig. 10). It is usually painless but may be sensitive to pressure. There may be, in relation to the cyst, a scar recalling the initial trauma. Sometimes the cyst ruptures and opens superficially to discharge its yellowish-white contents which have a greasy appearance; the cyst subsides and then reforms and reappears.
Although the typical site of epidermal cysts of the soft parts is in the dermis and subcutaneous cellular tissue, a very rare alternative localisation has been reported by Berkheiser (1955) and Saïto et al. (1979). This is an intratendinous development of the cystic lesion. It appears following a closed injury and is evidenced by the development of a subcutaneous swelling 1-2 cm in diameter, following the movements of the corresponding tendon and limiting its excursion to a varying degree. The tumour presents as a yellowish-brown cyst within the tendon (superficial flexor) and may occupy its entire thickness. Diagnosis in this case, as in the numerous cases without history of preceding trauma, is made by histological examination.

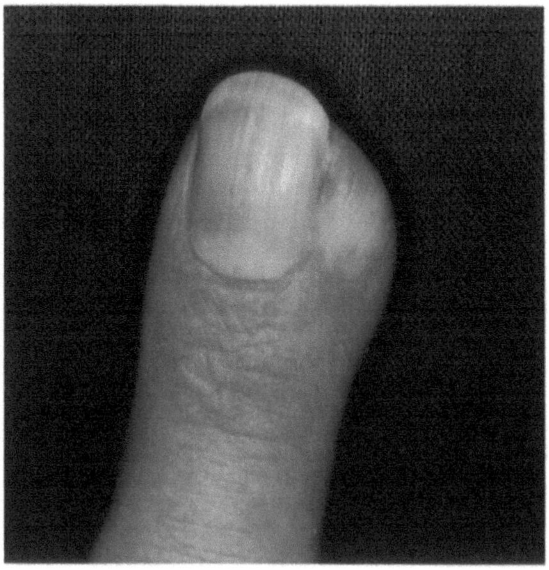

Fig. 9. Epidermal cyst of fingertip

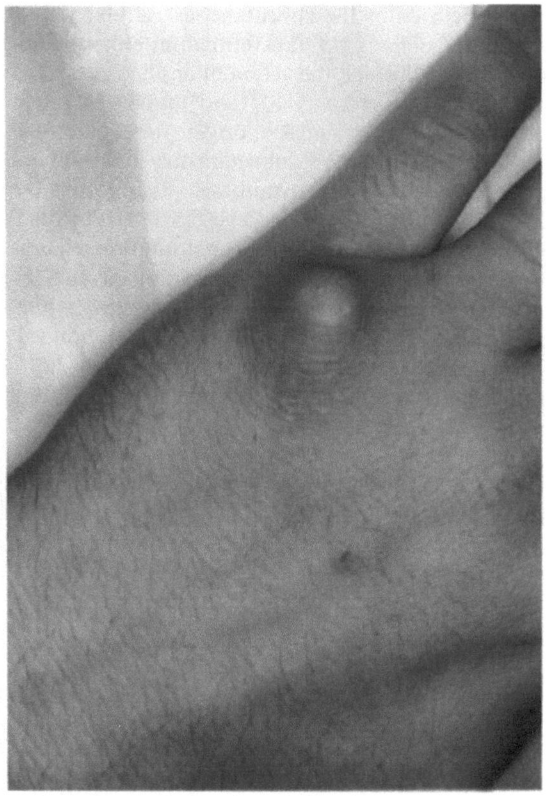

Fig. 10. Epidermal cyst: note adherence to the skin *(right)*

Histology

Histological examination of the tumour provides essential information. Grossly, the cyst is whitish in colour and has a wall 1-2 mm thick with contents similar to those of a sebaceous cyst (Fig. 11). Microscopically, the cystic cavity is surrounded by fibrous tissue compressed outwardly, but there is no true connective tissue capsule. The cyst wall is composed of malpighian epithelium of an epidermoid keratinising nature, covered by a stratum granulosum indicative of a cutaneous type of keratinisation; this in turn is covered by layered orthokeratotic strata. The basal layer faces outwards, and the horny layer faces the cystic cavity. There are no ancillary structures arising from sweat or sebaceous glands or hair follicles. The cavity contains a semi fluid substance staining pink with eosin and containing much keratotic and keratohyaline debris.

Aetiopathogenesis

Opinions as to the aetiopathogenesis have long been divided. The dysembryoplastic theory advanced by Francke (1894) argues in favour of embryonic heterotypy. An embryonic epidermal remnant, originally destined for the formation of cutaneous glands and appendages, is imagined as

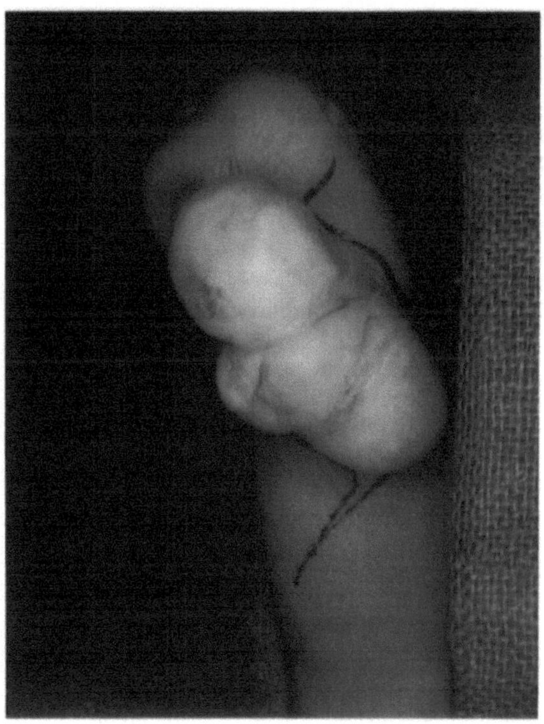

Fig. 11. Epidermal cyst, gross appearance

being enclosed in the subcutaneous cellular tissue and undergoing cystic transformation, either spontaneously or through the action of applied pressure. In contrast, Reverdin (1887) and numerous other authors avow a traumatic origin for epidermal cysts, with dermal and subcutaneous inclusion of epidermal debris. This hypothesis is supported by the work of Gross (cited by Reverdin 1887). In 1885 Masse (cited by Reverdin 1887) reported on experimental studies leading to the formation of epidermal cysts at the French Surgical Congress while Kauffmann (1884) and Schweninger (1884) induced the formation of epidermal cysts by inserting little oval strips of skin into the subcutaneous areolar tissue.

Finally, there is the theory of King (1933, cited by Zadek and Cohen 1953), based on the frequent absence of a perforating wound, which postulates that cyst formation is due to the epithelialisation of a haematoma or inflammatory focus by cells derived from sweat glands involved in blunt injury.

Treatment

This is purely surgical. The course of the incision is determined by the site and size of the lesion. Excision of the tumour is generally simple and should include the peripheral fibrous shell in order to avoid any risk of recurrence.

Intraosseous Cysts

Sonntag's initial report of 1923 gives a complete and detailed description of an intraosseous epidermal cyst situated in the middle phalanx of the left middle finger arising in a young woman 24 years after the original injury. Since then, numerous other cases have been reported in the literature, and this condition has become a true pathological entity (Burrows 1926; Friedlander 1930; Harris 1930; Hamman 1930; Behrens 1931, etc.). While the intraosseous location of the lesion endows it with certain clinical and therapeutic peculiarities, the condition remains entirely analogous to epidermal cysts of the soft parts, especially as regards its histology and aetiopathogenesis.

Designated by Bissel and Brunschwig (1937) as a post-traumatic epithelial cyst of the phalanx, it develops mainly in adults (age range 13–83 years) without sexual preference. The initial injury is a prick, cut, contusion or crush, and the latent period before the tumour appears varies from a few months to several years. The occupational risks concerned are the same as those already cited.

Clinical Features

The lesion is evidenced by progressive swelling and sensitivity of the fingertip. Spontaneous pain is more or less acute and is markedly increased by handling. The finger acquires an inflammatory appearance, and there is a functional handicap of varying severity. Ungual lesions are often associated, coincident with the initial trauma. Sometimes a pathological fracture caused by a minor injury leads to discovery of the lesion, which is almost always situated in the distal phalanx of the finger. The middle finger is most likely to be affected, followed by the index finger and the thumb. Right and left hands are equally involved.

Radiography

The punched-out appearance of the phalanx is very typical (Fig. 12). There is a rounded or oval cavity within the bone, perfectly circumscribed and with regular margins, that may rarely involve the entire phalanx. If this is so, the peritumoral bony layer is sometimes so thin that mechanical infraction may appear at one or more points. There is no destruction of the cortex suggestive of a malignant tumoral process.

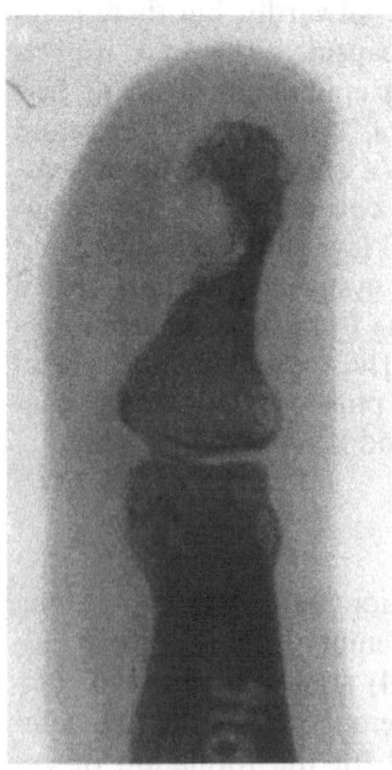

Fig. 12. Epidermal cyst, intraosseous location

Histology

At histological examination the characteristics of the wall of the cyst and its contents are identical with those already described. The bony shell around the tumour shows erosive lesions due to compression, and the soft tissues around the phalanx are intact.

Aetiopathogenesis

The invariable absence of adnexal structures tells against the theory of dysembryoplasia. The traumatic hypothesis based on the intraosseous inclusion of a fragment of epidermal tissue is currently accepted by all workers.

Differential Diagnosis

This may be tricky. Clinically, an osteoid osteoma may be suspected, but a radiograph would then show sclerosis around the cavity. A central phalangeal chondroma may be considered, but the clear, rounded, uniform image revealing, in most cases, fine septa which gives it a mottled appearance is very characteristic. A central myxoma is usually less well defined and less geometric than an epidermal cyst, and this is also true for giant-cell tumours. Finally, an idiopathic cyst or an osteitis cystica may have to be considered.

Treatment

This is surgical. The thin layer of bone around the tumour is easily opened and discloses the firm, whitish cyst wall. For smaller lesions, curettage suffices; with larger cysts it may be useful to make up for the loss of substance by inserting a bone graft.

Bibliography

Behrens A (1931) Über traumatische Epithelzysten. Virchows Arch Pathol Anat 280: 144

Berkheiser SW (1955) Epithelial inclusion cysts of tendon. Clin Bull 24: 147-149

Bissel AD, Brunschwig A (1937) Squamous epithelial bone cysts of the terminal phalanx. JAMA 108: 1702-1704

Blond K (1922) Zur Kenntnis der traumatischen Epithelzyste. Arch Klin Chir 120: 695

Boyes JH (1970) Tumors. In: Bunnell S (ed) Surgery of the hand, 5th edn. Lippincott, Philadelphia Toronto, pp 666-710

Burrows H (1926) Implantation dermoid of the terminal phalanx of the thumb. Br J Surg 13: 761

Byers P, Mantle J, Salm R (1966) Epidermal cysts of phlanges. J Bone Joint Surg [Br] 48/3: 577-581

Caparosa RJ (1955) Epidermoid cyst of the frontal bone. Arch Otolaryngol 62: 218-220

Carroll RE (1953) Epidermoid (epithelial) cyst of the hand skeleton. Am J Surg 85: 327-334

Chakravorty RC, Coley BL (1954) Epidermoid (epithelial) cyst of the phalanx. Am J Surg 87: 917-920

Curtis FE, Owens CI (1933) Sebaceous cyst of the distal phalanx. J Bone Joint Surg 15: 998-1000

Delarue J, Chomette G (1954) Un kyste épidermoïde de la phalangette. Société Anatomique, Paris

Duperrat, Gandrille (1948) Les kystes épithéliaux intra-osseux de la phalangette. Sem Hôp Paris 24/3: 78-79

Feulner RC, Marks JL (1958) Epidermoid (epithelial) cyst of the hand. AJR 79/4: 645-647

Fisher ER, Gruhn J, Skerrett P (1958) Epidermal cyst in bone. Cancer 11: 643-648

Francke F (1894) Über Epidermoide. Dtsch Z Chir 40: 197

Friedlander C (1930) Traumatische Epithelzyste im Fingerendglied. Zentralbl Chir 57: 203-212

Garré C (1894) Über traumatische Epithelzysten der Finger. Beitr Klin Chir 11: 524

Gibier P (1881) Extraits des procès verbaux. Kystes de la face palmaire de la deuxième phalange de l'index droit. Bull Soc Anat (Paris) 6: 661-662

Gillette (1875) Des difficultés du diagnostic dans certaines tumeurs enkystées des régions palmaire et plantaire. Société de Médecine, Paris

Haber MH, Alter AH, Wheelock MC (1965) Tumors of the hand. Surg Gynecol Obstet 121: 1073-1080

Haig PV (1956) Primary epidermoids of the skull, including a case with malignant change. AJR 76: 1076-1080

Hamman (1930) Traumatische Epithelzysten an den Fingerknochen. Dtsch Z Chir 223: 308

Harris RI (1930) Sebaceous cyst of terminal phalanx. An unusual form of bone tumor. J Bone Joint Surg 12: 647-648

Heidt RS, Hammond G (1955) Epidermoid cyst of a phalanx: A case report. Clin Bull 9: 151-154

Hinrichs RA (1965) Epidermoid cyst of the terminal phalanx of the hand. Case report and brief review. JAMA 194/11: 195

Iselin M, Iselin F (eds) (1967) Kystes épidermoïdes de la phalangette. In: Traité de chirurgie de la main. Flammarion, Paris, pp 765-766

Jouanneau P, Thazard MR (1958) Kyste épidermoïde intra-osseux de la 2e phalange du pouce. Acad Chir 84: 566-571

Kauffmann E (1884) Über Enkatarrhaphie von Epithel. Virchows Arch Pathol 97: 236

Kelly AP, Clifford RH (1956) Epidermoid cysts of the bony phalanges. Plast Reconstr Surg 17/4: 309-313

Knicker Bocker TW, Reilly R (1954) Traumatic epidermoid cyst of the terminal phalanx. Radiology 63: 550

Kotcamp WW, Cesarano FL (1962) Epidermoid cyst of

the terminal phalanx of the finger. J Bone Joint Surg [Am] 44/2: 377-379

Lazorthes G, Anduze Acher H (1957) Les signes radiologiques des kystes épidermoïdes cranio-encéphaliques d'après 9 cas opérés. Neurochirurgie 3: 336-338

Martin P (1959) Kyste épidermoïde intra-osseux. Thèse de Paris

Masseboeuf A, Acquaviva R, Thevenot C, Benjelloun M (1959) La forme calvarienne du kyste épidermoïde de la voute. Remarques à propos de trois observations. Maroc Med 38: 1612-1614

Muron M (1868) Extraits des procès verbaux. Kystes dermoïdes. Bull Soc Anat (Paris) 13: 539

Pohlmann HF, Wachstein M (1944) Epidermoid (squamous epithelial) bone cyst of phalanx. Ann Surg 119: 148-154

Posch JL (1956) Tumors of the hand. J Bone Joint Surg [Am] 38: 517-539

Reverdin JL (1887) Des kystes épidermiques des doigts. Rev Med Suisse Romande 7: 121

Rizet F (1866) Kystes dermoïdes. Arch Gén Méd (Paris) 2: 615

Roth SI (1964) Squamous cysts involving the skull and distal phalanges. J Bone Joint Surg [Am] 46/7: 1442

Saïto H, Tajima T, Watanabe H, Yamamoto H (1979) An epidermoid cyst of the tendon. J Hand Surg 4/5: 448-450

Schweninger E (1884) Beitrag zur experimentellen Erzeugung von Hautgeschwülsten. Charité Ann 11: 643

Siebner M (1931) Traumatic epithelial cysts of finger bones. Chirurg 3: 810-814

Sieracki JC, Kelly AP (1959) Traumatic epidermoid cysts involving digital bones. Epidermoid cysts of the distal phalanx. Arch Surg 78: 957

Skandalakis JE, Godwin JT, Mabon RF (1958) Epidermoid cyst of the skull. Report of four cases and review of the literature. Surgery 43: 990-1001

Sonntag (1923) Traumatische Epithelzyste im Knochenende an einem Fingerstumpf als Unfallfolge. Münch Med Wochenschr 70: 1055-1056

Stack G (1960) Tumors of the hand. Br Med J I: 517, 919-922

Sutton JB (1895) Implantation cysts. Br Med J 1: 461

Troquart R (1881) Kyste épidermique de la pulpe du doigt. J Méd Bordeaux 11: 52-53

Wernher A (1855) Das Atherom, ein eingebalgtes Epithelium. Virchows Arch Pathol Anat 8: 221

Wien MS, Caro MR (1934) Traumatic epithelial cysts of the skin. JAMA 102/3: 197-199

Yachnin SC, Summerill F (1941) Traumatic implantation of epithelial cyst in a phalanx. JAMA 116: 1215-1218

Zadek I, Cohen HG (1953) Epidermoid cyst of the terminal phalanx of a finger, with a review of the literature. Am J Surg 85: 771-774

Interdigital Sinus

This rare lesion, characterised by the development of interdigital cysts or sinuses, is seen almost exclusively in the hands of men's barbers. Hodges, in 1880 (cited by Currie et al. 1953 a) when describing sacrococcygeal cysts, was the first to note that short hairs prick the skin and penetrate into the epidermis of the fingers and hands of barbers. However, it was not until 1942 that the first case of interdigital sinus was reported by Allington and Templeton (cited by Currie et al. 1953), Patey and Scarff (cited by Currie et al. 1953) described the disorder in 1948, with subsequent case reports by Ewing (1947), Jenkins (1951) and Warren (cited by Currie et al. 1953). In 1953 Currie et al. reviewed the cases reported to that date and tackled the problems of aetiology, nosology and treatment. It emerged from their study that all the cases occur in men; moreover, of 77 barbers examined 10 were suffering from the lesion in some stage of its development.

Interdigital sinuses approximate to the pilonidal cysts found at other sites: the sacrococcygeal region in particular, the perineum and, much more rarely, the sole of the foot, the umbilicus, the axillary fossa and the suprapubic region.

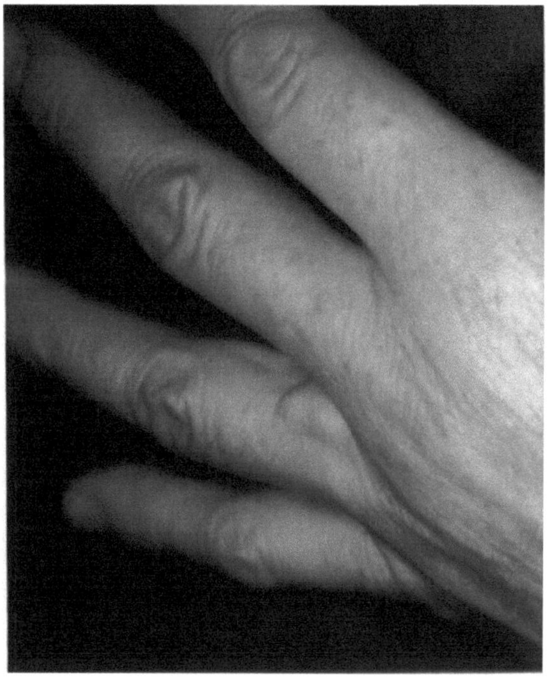

Fig. 13. Interdigital sinus of the third cleft

Clinical Features

The development of an interdigital sinus is promoted by permanent and repeated contact with freshly cut short hairs that penetrate the epidermis of the interdigital spaces. There then appear, on the dorsal aspect of the interdigital clefts, lesions marked by excoriation and depression, with small openings maintained and aggravated by the persistence of the irritant and by secondary inflammatory manifestations (Fig. 13). New and deeper sinus tracts are thus formed into which new hairs make their way, perpetuating the pathological cycle and sometimes giving rise to secondary infection and abscess formation.

Histology

The lumen of the sinus track is epithelialised. Deep inflammatory reactions of a nonspecific nature exist, indicative of chronic foreign body irritation. Also, hairs are found both within the lumen at the opening and also deep inside, embedded in fibrous connective tissue and surrounded by macrophages and lymphocytes.

Aetiology

This is not precisely understood. According to Currie et al. (1953 a) it is a matter of the direct action of short-cut hairs that stick into the skin, excoriating the epidermis and creating small skin depressions that may give rise to abscesses or sinuses. Sames (cited by Currie et al. 1953 a) stresses the role of the comb, and Ewing (1947) that of shampooing, in explaining the action of the hair on the skin.

Treatment

This is essentially preventive. In the early stages prophylactic measures suffice: scrupulous cleansing of the hands and daily removal of hair fragments. Such measures may stabilise a lesion that is not yet advanced, but for well-developed lesions with an epithelialised sinus track, especially with suppuration and recurrent infection, surgical excision is the only treatment.

Bibliography

Currie AR, Gibson T, Goodall AL (1953a) Interdigital sinuses of barbers' hands. Br J Surg 156: 278–286

Currie AR, Gibson T, Goodall AL (1953b) Pilonidal sinus of the suprapubic region. Br Med J, April 25: 936–937

Ewing MR (1947) Hair-bearing sinus. Lancet, March 29: 427

Ian Aird CM (1952) Pilonidal sinus of the axilla. Br Med J, April 26: 902–903

Jenkins DH (1951) Pilonidal sinus in a barber. Lancet, March 31: 746–747

Mac Leod RG (1953) Pilonidal sinus of the supra-pubic region. Br Med J, March 28: 710–711

Raw SC (1951) Pilonidal sinus in a barber. Lancet, April 21: 912

Keratoacanthoma

Keratoacanthoma, described by Hueston (1956) as "a sheep in wolf's clothing", is a benign skin tumour very characteristic in its tendency to regress but capable of posing tricky problems of differentiation from squamous-cell epithelioma, with all the prognostic and therapeutic problems attached to this. Described by Hutchinson in 1888 as "a crateriform ulcer of the face", keratoacanthoma was recognised as a lesion resembling squamous-cell skin cancer but spontaneously regressive by Dunn and Smith (1934) and Levy et al. (1954). Baer and Kopf (1963) made a global review of 883 keratoacanthomata derived from 18 series and defined the characteristics. They found that the hand was affected in 14.7% of cases, which is not surprising since keratoacanthomata develop mainly on exposed areas bearing hair follicles. Carroll and Bowers (1976) reported 15 lesions on the back of the hand, while Shapiro and Baraf (1970) added 3 cases of subungual lesions to 3 other cases from the literature (Lamp et al. 1964). Localisation of keratoacanthoma in the hand is therefore far from rare, amounting to 10%–15% (Cohen et al. 1972; Carroll and Bowers 1976). However, other workers consider that keratoacanthoma constitutes only a small proportion of tumours of the hand: 5 of 437 for Butler et al. (1960) and 9 of 2161 for Schultz and Kearns (1983). No cases were reported by Mason (1937), Posch (1956) or Stack (1960).

Clinical Features

Two clinical types may be distinguished in the hand, dorsal and subungual.

Lesions of the dorsum have the same features as other cutaneous keratoacanthomata (Fig. 14). A red papule appears in healthy skin and rapidly enlarges over a period of 6–8 weeks to attain a diameter of 1–2 cm or even more. Umbilication of its central portion leads to a crater filled with keratin that is difficult to remove, while the adjacent skin often has an inflammatory appearance. After remaining unchanged for 2–8 weeks the lesion begins to regress spontaneously, with flattening and formation of a retracted, depigmented scar. This phase of cicatrisation lasts 1–3 months, but the speed of the process varies, and some lesions may not regress for a year or more. Keratoacanthomata may affect the entire hairy region of the hand and the dorsum of the metacarpals and proximal phalanges (Fig. 15).

Subungual lesions have an uncharacteristic symptomatology (see p. 218), with localised pain, swelling and inflammation suggestive of whitlow or paronychia, while the periungual and subungual hyperkeratosis is suggestive of verruca. However, removal of the nail and opening of the nail bed reveal a keratotic lesion or a nodule. Radiography often shows a well-localised erosion of the tuft of the terminal phalanx; the bone reforms after excision of the lesion.

Pathology

Morbid anatomy examination is the essential step in diagnosis, but the lesion must be examined in toto and throughout its thickness since its resemblance to squamous-cell epithelioma often makes distinction by partial biopsy impossible. Macroscopically, there is a soft tumour with a central crater filled with keratin which is difficult to extract, whereas squamous-cell carcinoma is firmer with a central crater due to minor ulceration. The histological examination looks for features in favour of keratoacanthoma: an epithelial covering at the crater margin, acanthosis, epithelial pseudohyperplasia with epithelial ridges penetrating deeply into the adjacent tissues but rarely involving the entire thickness of the skin, and papillary hyperplasia beneath the thick epidermal layer which respects the basal membrane.

The cytology shows atypical nuclei with abnormal and few mitoses, chromosomal fragmentation, loss of polarisation and variability of cell size. The inflammatory reaction is very intense, forming inflammatory plaques. The acanthotic zone is invaded by inflammatory cells, unlike squamous-cell carcinoma in which the inflammatory infiltrate is repelled by the tumour. Involvement of the reticular and papillary dermis brings about the formation of a retractile scar.

When the nail bed is affected, its epithelium and

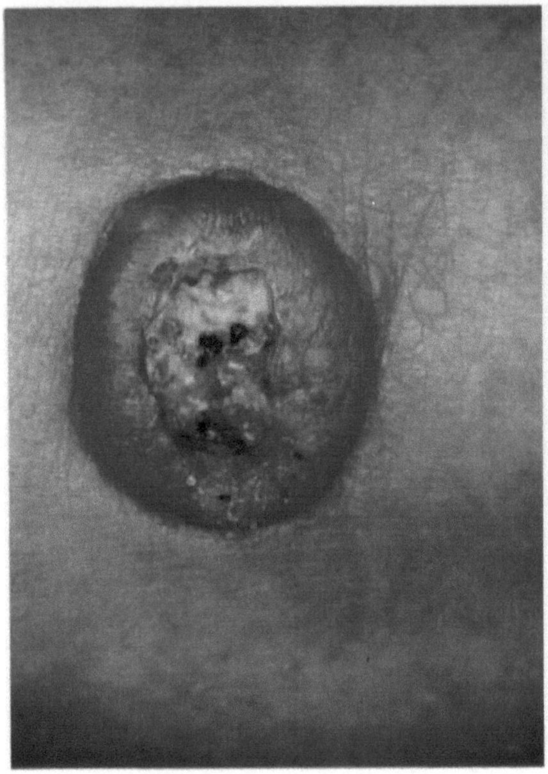

Fig. 14. Keratoacanthona, clinical appearance

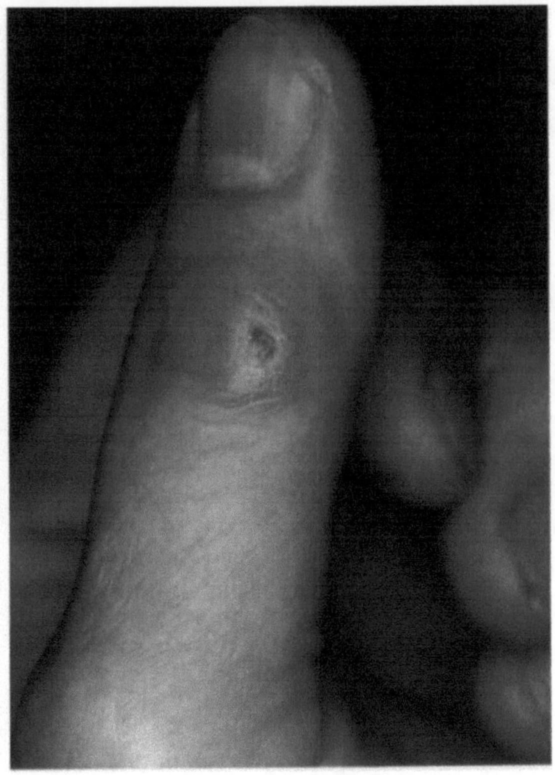

Fig. 15. Keratoacanthoma of the thumb

the tumour itself are raised, accompanied by hyperkeratosis, parakeratosis, pseudoepithelial hyperplasia, central depression and a crater or crypts with keratin residues.

Differential Diagnosis

Diagnosis is based on the clinical appearance, course and histology, all of which may be needed to exclude a squamous-cell carcinoma. In addition to this, with subungual lesions one must exclude a whitlow or chronic paronychia, a verruca or an epidermal cyst of the distal phalanx. Radiological examination and operation are only adjuncts to the histological examination. At this site the course of the lesion is not helpful.

Aetiology

Keratoacanthoma affects both sexes, but in the hand males are affected much more often than females (13 of 15 on the dorsum according to Carroll and Bowers 1976, 6 of 7 in the subungual region according to Shapiro and Baraf 1970). Most of these cases arise after the age of 60 years, but sometimes before 40 with subungual keratoacanthomata. Even very exhaustive research (Ghadially et al. 1963) has not revealed any precipitating factors, though recurrent tumours have been described in workers in contact with tar, and the incidence of the disorder is higher in smokers. Ghadially et al. (1963) also found that 25 keratoacanthomata of a series of 250 were subsequent to injury such as a burn, blow or grease injection; 7 of these lesions were in the hand, 1 following injection of oil under pressure. The development of the lesion in exposed areas implies that the sun may play a part (Belisario 1965), while the course suggests a possible viral aetiology.

The most debated problem is that of the relation between keratoacanthoma and squamous-cell epithelioma. Three hypotheses have been advanced. (1) The two lesions have clinical and histological similarities, but one is entirely benign while the other is malignant and may metastasise. (2) Keratoacanthoma is a squamous-cell cancer that heals spontaneously. (3) Keratoacanthoma is a benign lesion that may, in certain cases, undergo malignant change.

Treatment

This is based on one's ideas as to pathogenesis. When the clinical appearance is typical, and there is a risk that excision may result in mutilation, some simply advise abstention (Stranc and Roberston 1978). Conversely, most of those who fear a malignant course or an erroneous diagnosis (Schnurr and Bozzo 1978; Van den Bussche et al. 1972) prefer total excision with either simple closure or a split skin graft (Carroll and Bowers 1976). On the dorsum of the hand, the use of a thin graft leaves little cosmetic impairment. In the ungual region a precise histological diagnosis is essential before deciding on treatment, which in this case must be conservative with simple ablation of the tumour and the nail bed and local curettage. Digital amputation, lymphnode eradication and radiotherapy have no place in the treatment of keratoacanthoma of the hand, as the prognosis is excellent.

Bibliography

Baer RL, Kopf AW (1963) Keratoacanthoma. Year Book of dermatology. Year Book Medical Publishers, Chicago, p 741

Belisario JC (1965) Keratoacanthoma and its incidence as affected by sunlight in the tropical and subtropical areas of Australia. Derm Int 4: 53-59

Butler ED, Hamill JP, Seipel RS, de Lorimier AA (1960) Tumors of the hand. A ten-year survey and report of 437 cases. Am J Surg 100: 293-302

Carroll RE, Bowers WH (1976) Keratoacanthoma - An unusual hand tumor. Clin Orthop 118: 173-179

Cohen N, Plaschkes Y, Pevzner S, Loewenthal M (1972) Review of 57 cases of keratoacanthoma. Plast Reconstr Surg 49: 138-142

Dunn JS, Smith JF (1934) Self healing primary squamous carcinoma of the skin. Br J Dermatol 46: 519

Ghadially FN, Barton BW, Kerridge DF (1963) The etiology of keratoacanthoma. Cancer 16: 603-611

Hueston JT (1956) Molluscum Sebaceum. Aust NZ J Surg 26: 25

Hutchinson J (1888) The crateriform ulcer of the face: A form of epithelial cancer. Tr Pathol Soc (London) 40: 275-281

Lamp JC, Graham JH, Burgoon CF Jr (1964) Keratoacanthoma of the subungueal region. J Bone Joint Surg [Am] 46: 1721-1732

Levy EJ, Kahn MM, Schaffer B, Beerman H (1954) Keratoacanthoma. JAMA 155: 562

Mason ML (1937) Tumors of the hand. Surg Gynecol Obstet 64: 129-148

Posch JL (1956) Tumors of the hand. J Bone Joint Surg [Am] 38: 517-539

Schnurr PL, Bozzo P (1978) Metastazing keratoacanthomas. Plast Reconstr Surg 62: 258-262

Schultz RJ, Kearns RJ (1983) Tumors in the hand. J Hand Surg 8: 803-806

Shapiro L, Baraf CS (1970) Subungueal epidermodi carcinoma and keratoacanthoma. Cancer 25: 141-152

Stack G (1960) Tumors of the hand. Br Med J I: 517, 919-922

Stranc MF, Roberston GA (1978) Conservative treatment of keratoacanthoma. Ann Plast Surg 2: 525-530

Van den Bussche F, Meresse B, Lagache G (1972) Kéra-toacanthomes: Piege diagnostique en pathologie tumorale cutanée - Attitude thérapeutique rationnelle. Ann Chir Plast 17: 289-292

Actinic Keratosis

Better known under the name of solar keratosis or senile keratosis, actinic keratosis owes its name to Becker (1959). It affects exposed sites, particularly the dorsum of the hand, and carries a risk of degeneration into squamous-cell epithelioma.

Clinical Features

Actinic keratosis preferentially affects subjects of mature age with white skin, exposure to sunlight, and working in certain occupations associated with ionising radiation or the treatment of coal. It is evidenced at the outset by the appearance of telangiectases, which are progressively replaced by a dry, adherent, crusted zone of a yellowish-brown colour. The lesion is initially a few millimeters across but gradually extends and becomes progressively thickened and keratinised. It has very well-defined margins, and removal of the crust produces immediate bleeding. In the hand, this dermatosis typically affects the dorsum (Fig. 16). It is habitually associated with a loss of skin elasticity.

Histology

Histologically the lesions are sharply demarcated from the adjacent healthy tissue. There is a modification of the epidermal cells, which exhibit a pale cytoplasm and progress to the formation of a parakeratotic crust. There is hyperacanthosis and hyperkeratosis. The basal membrane remains intact.

Development

The course of the lesion is very slow, with progressive increase in the extent of the keratosis; conversely, it may regress if the patient avoids the sun's rays. Malignant degeneration is suggested by the appearance of induration at the base of the keratosis, which may be complicated by secondary infection. This change, which is generally delayed, results in a squamous-cell epithelioma that is not very invasive, progresses slowly and rarely metastasises. On average it is 10 years before this develops.

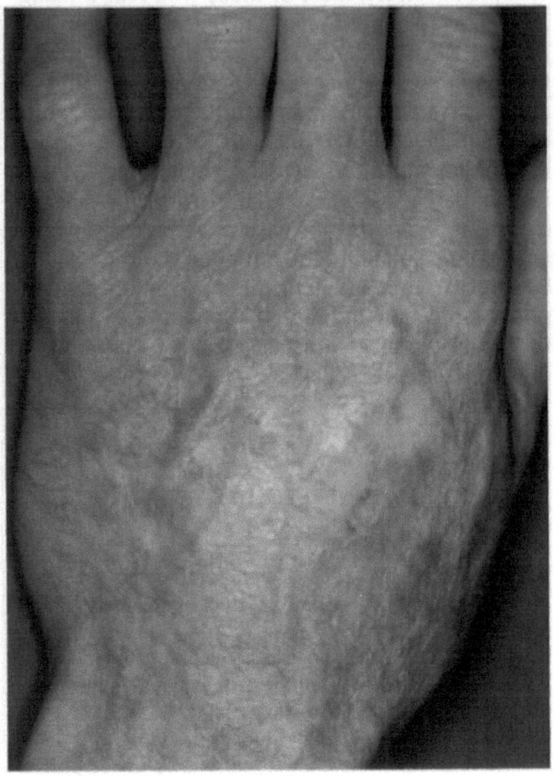

Fig. 16. Actinic keratosis

Differential Diagnosis

The clinical appearance of the lesion may resemble that of other keratoses (Fig. 17) or of Bowen's disease. The histology may be confused with that of epithelioma, lichen or even lupus.

Treatment

The superficial lesions respond to brief cryotherapy. Diathermy coagulation, with or without curettage, is equally effective but leaves more scarring. Thicker, indurated lesions may be surgically excised. For disseminated lesions it is better to use locally applied cytotoxic agents, such as 5-fluorouracil. However, in certain cases, the often advanced age of the

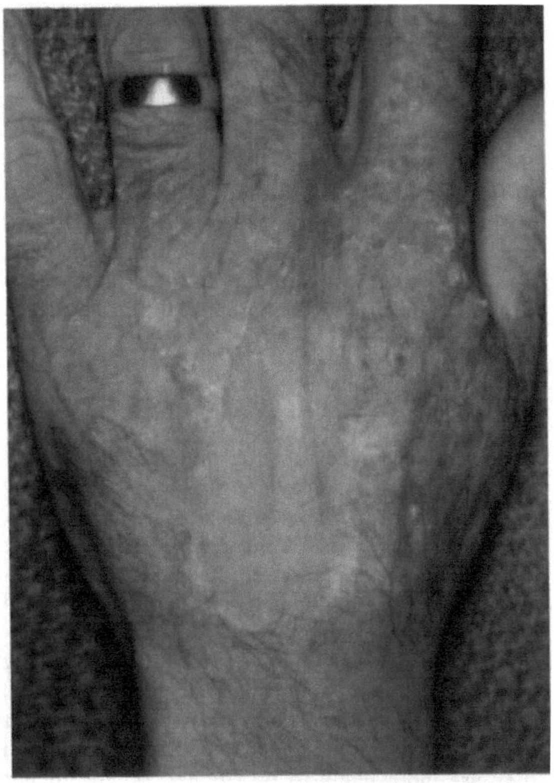

Fig. 17. A rare form of keratosis: immunosuppression in renal transplantation. Note the full-thickness skin graft centrally ▷

patient justifies simple surveillance associated with prophylactic measures (avoidance of exposure to sunlight), with treatment reserved for lesions suspected of malignant degeneration.

Bibliography

Becker SW (1959) Dermatological nomenclature. Arch Dermatol 80: 778

Dubreuilh W (1907) Epithéliomatose d'origine solaire. Ann Dermatol Syphiligr (Paris) 8: 386

Kostel J Jr (1970) Treatment of actinic keratosis. Arch Dermatol 102: 351

Pinkus H (1976) Actinic keratosis - actinic skin. In: Andrade et al. (eds) Cancer of the skin. Saunders, Philadelphia, pp 437-457

Schnitzler L, Degos R, Civatte J (1968) Taches pigmentées séniles du dos des mains; étude histologique de 50 cas. Bull Soc Fr Dermatol Syphiligr 75: 550

Arsenical Keratosis

In the skin arsenical poisoning can be manifested by three types of lesions: keratosis, Bowen's disease and squamous- or basal-cell epithelioma. Of these, keratosis is the most common, but it is hardly seen any longer because of the greatly decreased use of arsenical derivatives.

Clinical Features

The palms and soles are the preferential sites for arsenical keratoses. They are initially evidenced by the development of a punctate hyperkeratosis, sometimes barely visible but readily palpable. The lesions increase progressively in size to reach a diameter of several millimeters (10 at most) and then remain static for several years. Their site of implantation may be erythematous, and the surrounding skin is sometimes abnormally dry and wrinkled, with fissuring in the flexion creases. These lesions are generally multiple and symmetrical. In the hand they are more likely to be situated on the thenar and hypothenar eminences and are often associated with lesions on the dorsum. They may be accompanied by other keratotic localisations (extensor aspects of the joints), lesions of a hyperpigmented type (especially in dark-skinned subjects), telangiectases and, sometimes, peripheral vascular insufficiency.

Development

The course of arsenical keratoses is very gradual, and they generally remain static for many years. The lesions may regress if poisoning ceases at an early stage. Malignant degeneration is possible in the shape of squamous-cell carcinoma; basal-cell epitheliomata are also seen in arsenical poisoning, but these are primary.

Aetiopathogenesis

The skin toxicity of arsenic has long been familiar (Paris 1820). Poisoning is most often due to drugs (Fowler's solution) but may be occupational, as in farm labourers and vine growers using weed killers or defoliants, or occasionally it is associated with the presence of arsenic in drinking water, as reported by Tseng et al. (1968) in Taiwan. Studies of large patient series (Tseng et al. 1968; Yeh 1973) show that individual sensitivity to the agent is clearly variable, some patients being much more vulnerable than others.

The carcinogenic action of arsenic is not universally accepted (Frost 1967). In any case, it has been demonstrated in only one animal experiment (Knoth 1966). The keratoses may certainly undergo degeneration in the shape of squamous-cell epithelioma, as we ourselves have noted, but it is questionable whether the arsenic is responsible at this stage. Sanderson (1976) reports from his own experience a relatively frequent association with deeper cancers.

Histology

Arsenical keratosis appears histologically as a minor and relatively orderly cellular dysplasia associated with a delay in cell maturation. These abnormalities are situated mainly in the horny layer, with the malpighian layer usually intact. They are seen in more marked degree in Bowen's disease, in which the dysplasia is irregular and severe. Transitional forms between the two exist, known as "bowenoid keratoses", and this may make diagnosis difficult.

Differential Diagnosis

It is often difficult to prove the fact of arsenical poisoning, and close enquiry into drug use and occupation may be required. If this diagnosis does not come to mind, psoriasis, Darier's disease or keratosis of some other origin may suggest themselves. Once the aetiological diagnosis has been made, the distinction between a keratotic lesion and Bowen's disease is not always simple and cannot always be settled by histological examination.

Treatment

The arsenical origin of the keratosis does not contribute any special feature to the treatment. Klein et al. (1970) report an appreciable effect of 5-fluorouracil on the keratotic lesions. A suspicion of malignant degeneration, as may be aroused by persistent fissuring or ulceration, calls for biopsy. If the lesions are multiple, surgical excision may be difficult, and destruction by diathermy or topical application of cytotoxic agents may be preferable.

Bibliography

Frost DV (1967) Arsenical in biology: Retrospect and prospect. Fed Proc 26: 194

Klein E, Stoll H et al. (1970) The effects of 5 fluoro-uracile ointment in the treatment of neoplasic dermatoses. Dermatologica [Suppl 1] 140: 21

Knoth W (1966) Psoriasis vulgaris. Arsenbehandlung. Arch Klin Exp Dermatol 227: 228

Sanderson KV (1976) Arsenic and skin cancer. In: Andrade R et al. (eds) Cancer of the skin. Saunders, Philadelphia, pp 473-491

Thiers H, Colomb D, Moulin G, Cohn L (1967) Le cancer cutané arsenical des viticulteurs du Beaujolais. Ann Dermatol Syphiligr 94: 133

Tseng WP, Chu HM et al. (1968) Prevalence of skin cancer in an endemic area of chronic arsenicism in Taïwan. J Natl Cancer Inst 40: 453

Yeh S (1973) Skin cancer in chronic arsenicism. Hum Pathol 4: 469

Radiodermatitis

Although very common during the first three-quarters of this century, radiodermatitis of the hands is now happily on the decrease. Its main cause was occupational irradiation, which has greatly diminished thanks to progress in radiological techniques, legislation and, above all, the awareness by doctors of its dangers.

The first case of radiodermatitis of the fingers was reported in 1896, only 1 year after Roentgen's discovery. The first case of malignant degeneration was published 6 years later, in 1902. Subsequently, impressive series were published confirming its essentially occupational nature - Porter (1909), 47 cases; Leddy and Rigos (1941), 80 cases; Saunders and Montgomery (1938), 295 cases - and the frequency of malignant degeneration - Ledoux-Lebard (1922) with 77 cases. A pioneer in his time, Porter from 1900 onwards advised excision of the lesions followed by skin grafting, at a time when many watched the spread of the lesions helplessly until amputation, sometimes repeated, became necessary. It took nearly half a century to convince surgeons (Lagrot et al. 1951) of the importance of Porter's technique, which is now generally accepted (Lagrot 1974).

Radiodermatitides of the hands account for 5% of all cases of radiodermatitis seen.

Aetiology

There are two types of radiodermatitis: acute, which is related either to irradiation for therapeutic purposes (as for verruca) or, more rarely, to accidental and often massive irradiation; and chronic, which is much commoner and was originally almost exclusively occupational, affecting primarily radiologists but also physicians (tuberculosis specialists) and (orthopaedic) surgeons in their practice of screening. Other affected occupations include watch-makers, who handle luminous paint (Poulenas and Verdan 1975). Lagrot, in his exhaustive study published in 1974, reported 109 cases of radiodermatitis of the hands, of which 95 were professional, 11 post-therapeutic and only 3 accidental.

Clinical Features

Here it is useful to distinguish the acute from the chronic forms as neither the clinical picture nor the course correspond.

Acute Radiodermatitis

The first signs appear 15-30 days after irradiation. In the superficial forms there is merely erythema (radiodermal "sunburn") which regresses spontaneously in a few days. With the moderate type the onset is identical, but the erythema reappears 1 week later, leading to superficial erosions (Fig. 18) which heal more or less rapidly, leaving the skin smooth, atrophic and very fragile, since the slightest injury may bring about chronic ulceration. Severe cases have a sudden onset, with congestive erythema and vesicles that rapidly develop into painful ulceration of the dermis. Spontaneous scarring is very slow and unstable. In some types onset is delayed, and the progress is towards necrosis that is irreversible and extensive in depth.

Chronic Radiodermatitis

The first clinical features appear several years after beginning regular irradiation, usually occupational. They are always insidious, develop very slowly and progress in three stages. The first stage is indicated by the appearance, usually on the dorsal aspect of the middle and terminal phalanges, of minor lesions marked by anhidrosis, pallor, discoloration and thinning of the skin, with desquamation, loss of hair and minor changes in the nails such as longitudinal striation, fragility and cracking of the tips. This stage constitutes an alarm signal calling for absolute and immediate cessation of exposure to X-rays. In that case the lesions regress spontaneously in a few months; if not, by the end of a year the condition will have progressed to the second stage.

This second stage is marked by the appearance of disseminated brownish patches which progressively thicken and keratinise to form actual warts (hyperdyskeratosis; Fig. 19). The adjacent skin becomes sclerosed, and telangiectases appear as a fine, pink, capillary network. Changes in the nails now predominate (Fig. 20). The nail assumes a brownish colour and either thickens or becomes very thin and dry, with breakage and splitting. It is marked by black dots or "coal spots" and sometimes by transverse furrows (Beau) if the onset of the irradiation was massive. These features, prominent mainly over the middle and distal phalanges of the middle fingers, give rise to very little pain, but they may be complicated by superadded eczema.

The third stage is marked by exacerbation of the lesions and episodes of secondary infection. The warts become confluent and are furrowed by very painful fissures at the flexion creases. Ulcers develop which are at first regressive and then become established and painful and may extend deeply to expose tendons and bone, which undergo necrosis. These ulcers may form the starting point for malignant degeneration.

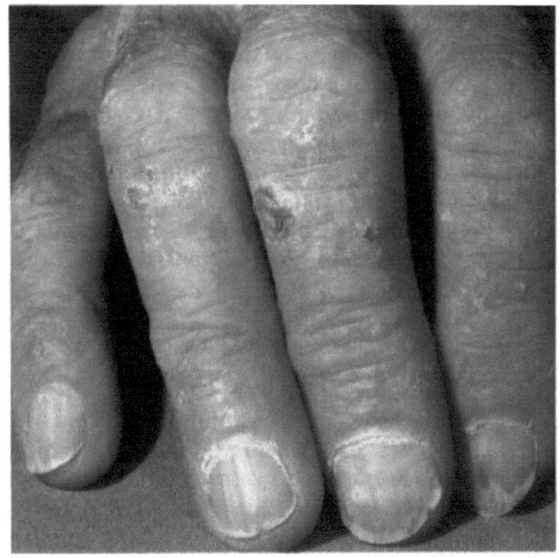

Fig. 19. Chronic radiodermatitis, stage II

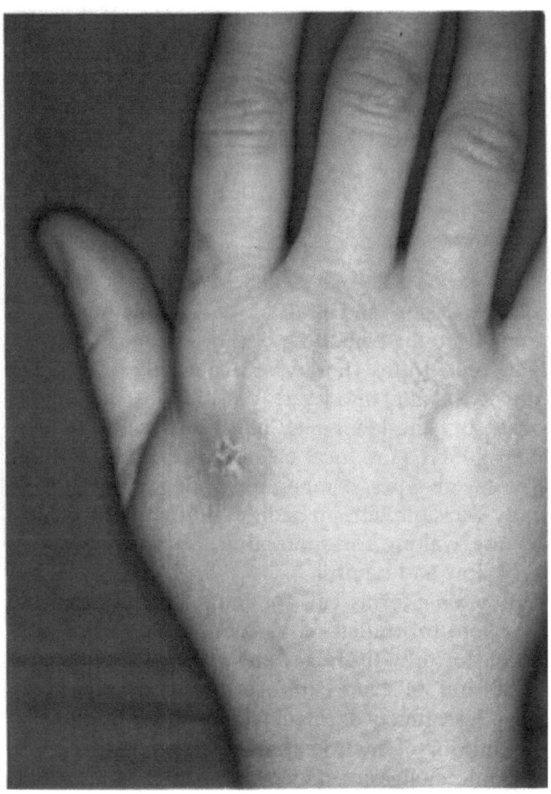

Fig. 18. Acute radiodermatitis after irradiation of a wart

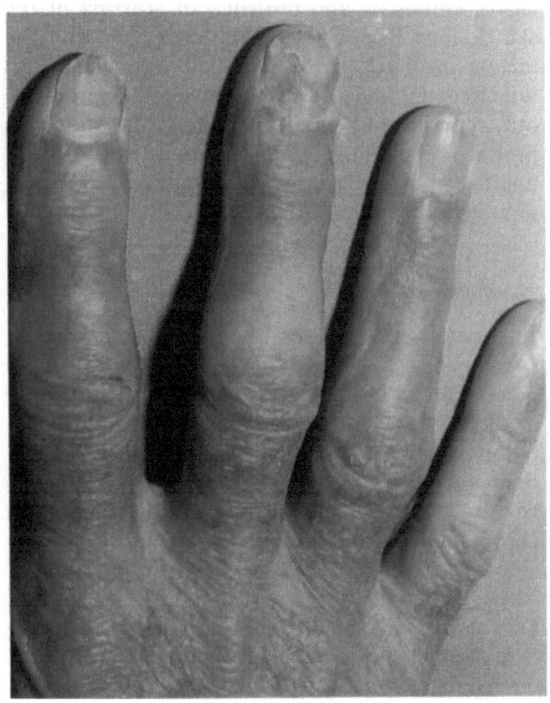

Fig. 20. Radiodermatitis, stage II, subungual changes; note the fissures and the transverse groove (index finger). Same patient as in Fig. 22. Dorsal aspects of the phalanges have been grafted

Clinical Types

Other sites with lesions may be seen that are related to the aetiological circumstances. Thus, the handling of tubes of radium by gynaecologists produces lesions of the thumb and fingertips, and the latter are the main site of disease in dentists who themselves hold films in their patients' mouths. Lagrot (1974) distinguishes two types of lesions: a verrucous or "pachydermatous" type and a sclero-atrophic type.

Pathology

The tissue changes found in the different stages of radiodermatitis are associated with modifications of cell permeability, altered synthesis of various enzymes and nucleic acids, changes in cell division and modifications of blood supply at the cellular level. The blood supply is initially increased in the inflammatory stage and then decreased by vascular thrombosis, which is responsible for the tissue necrosis, chronicity and difficulty of scarring of the skin lesions. The lesions themselves vary with the factors involved: if the radiation responsible has little penetration (soft rays, ultraviolet), it produces only superficial lesions, while more penetrating rays produce deep-seated lesions and may leave the skin unaffected. Again, tissues with rapid cell reproduction are more affected than others.

Development

In general terms lesions of the first stage regress spontaneously. Second stage lesions can do so, but much more slowly and leave the skin fragile. Those of the third stage never regress. Malignant degeneration is only exceptionally seen in acute radiodermatitis (fibrosarcoma after massive high-intensity irradiation). On the other hand it is common in chronic cases, the incidence varying from 20% (Mouly and Bureau 1970) to 75% (Porter 1909) of cases. Lagrot (1974) found it in 55% of his surgical patients. The incidence is much higher in the hands than at other sites. Malignant change is usually late, the average onset being 20-25 years after the start of irradiation and 10-12 years after the onset of clinical features of radiodermatitis. It may be provoked by a precipitating factor, notably by a marked additional dose of radiation. It appears as a supervention on two types of lesions, keratotic warts and ulcers. The diagnosis of such change in a verrucous lesion is mostly histological, as the clinical changes are minor or absent (fissuring, episodic bleeding). On the other hand, it is more readily suggested when an ulcer becomes painful or bleeds, or when there is resistance to local treatment. Any progressive change should be an alarm signal and considered suspicious of malignant degeneration until proved otherwise.

In the very great majority of cases the tumour is a squamous-cell epithelioma; much more rarely, and then generally in the context of therapeutic irradiation, it is a basal-cell lesion. Progress is slow and essentially local, and the mortality is minor and delayed (Lagrot 1974: 2 deaths in 59 post-radiodermatitis epitheliomata). Histological examination distinguishes three stages of evolution: epithelioma in situ (intraepithelial), invasive epithelioma with disruption of the basal membrane, and the infiltration stage, in which the dermis is invaded in depth, and mutilating, excisional surgery is called for.

Differential Diagnosis

Given the context, the diagnosis of radiodermatitis of the hands usually poses no problem. If the context is not recognised, the condition may rather suggest an eczema, a keratotic type of mycosis or a dyshidrosis. The lesions of the nails, while they remain isolated, may suggest eczema or Darier's disease.

Treatment

Here we owe a great deal to Porter, who had advised excision and grafting of radiodermatitis lesions from the beginning of the century.

The *acute* forms should be treated like burns. The superficial and moderate varieties heal spontaneously, but the latter may ulcerate secondarily, and then they require local cleansing and skin grafting. The severe types, comparable to deep burns, justify early excision with immediate skin-graft cover; it is no use waiting for spontaneous healing, which is very slow and painful.

The *chronic* forms call for immediate cessation of exposure to irradiation as soon as they are recognised, for only this gives any hope of spontaneous regression of the lesions in the early stages. In stage I treatment is medical: local applications of vaseline or of hydrocortisone in some cases, with systemic analgesics if the lesions are painful. It is sometimes necessary to excise and graft the residual area of atrophic skin.

In stages II and III, treatment is surgical and consists of excision and grafting of the entire macroscopically affected area. The excision must be performed through healthy tissue, generally removing the cutaneous tissues of the dorsal aspect of the 2nd and 3rd phalanges en bloc with the matrix and nail bed. Coverage is by immediate grafting or with a flap if necessary, avoiding locally raised flaps whose vascularisation is often precarious (Beasley 1975). Lagrot (1974) has devised a system of holding the grafts in place with a simple tubular sheath (tube gauze). The nail may be left in place if it is simply striated and not dystrophic; the coal spots should be diathermised. Histological examination should address the whole of the excised specimen in a search for evidence of malignancy, the discovery of a squamous-cell epithelioma in the invasive stage calling for mutilating excision. Though surgical intervention may cure the lesions in the grafted zones, it does not exclude the later appearance of other local lesions; long-term surveillance is therefore necessary.

Bibliography

Antkowiack JG (1980) Scatter radiation to the hands from insertion of transvenous pacemakers. Ann Thorac Surg 30: 405-406

Beasley RW (1976) Malignant tumors of the hand. In: Andrade R (ed) Cancers of the skin. Saunders, Philadelphia, pp 1435-1448

Ducourtioux M, Civatte J (1955) Étude anatomo-clinique des radiodermites. Ann Chir 31: 123-132

Dufourmentel C, Tergueu J (1957) Le traitement chirurgical des radiodermites. Presse Méd 61: 1239

Fryer MP, Brown JB (1962) Repair of atomic, cathoderay, cyclotron and X-ray burns of the hand. Am J Surg 103: 688-691

Hansen FC, Edgerton MT (1966) Radiation burns. In: Flynn JE (ed) Hand surgery. Williams & Wilkins, Baltimore, pp 597-614

Lagrot F (1974) Radiodermite des mains. Doin, Paris, p 227

Lagrot F, Greco J, Py N (1951) Traitement chirurgical des radiodermites. Greffe cutanée. Lyon Chir 46: 469

Leddy ET, Rigos FJ (1941) Radiodermatitis among physicians. AJR 45: 696

Ledoux-Lebard R (1922) Le cancer des radiologistes. Paris Méd 12: 299-303

Mouly R, Bureau H (1970) Les radiodermites chirurgicales. Rapport au Congrès de Chirurgie Plastique, Paris

Porter CA (1909) The surgical treatment of X-ray carcinoma and other severe X-ray lesions, based upon an analysis of 47 cases. J Med Res 21: 357

Poulenas I, Verdan C (1975) Radiodermite et cancérisation secondaire par usage de produits radio-actifs dans l'horlogerie. Ann Chir 29: 509-512

Saunders TS, Montgomery H (1938) Chronic roentgen and radium dermatitis: An analysis of 259 cases. JAMA 110: 23

Dubreuilh's Melanosis

This condition is also known by the name of senile lentigo, circumscribed precancerous melanosis, melanotic freckle (Hutchinson), lentigo maligna, etc. Dubreuilh's circumscribed, precancerous melanosis is a noncongenital naevoid patch appearing after middle life and particularly affecting the face, more rarely the hands and legs. The importance of its study relates to its frequency and the major risk of transformation into malignant melanoma.

Hutchinson (1892) and Dubreuilh (1894) seem to have been the first to describe this disorder in detail. Since then, the frequency of the lesion has been demonstrated in numerous studies (Clark and Mihm 1969). In contrast, localisation of the melanosis in the hand has only rarely been mentioned (Andrade 1972, 1976; Lupulescu et al. 1973). Most of the reports deal with cases affecting the face, and only exceptionally the legs, feet, oral and genital mucosae. The aetiology is unknown; the predominance of the lesions on the face suggests the influence of sunlight. Development based on preexisting naevoid cells has also been suggested. Pinkus and Mishima (1963) believe that the lesion develops from epidermal dendritic melanocytes that are distinct from naevus cells, which would explain its biologically different behaviour. The disorder appears to affect females twice as often as males, especially after the age of 50 years (Andrade 1976).

Clinical Features

The isolation by Clark and Mihm (1969) of melanomata with superficial extension has made it possible to distinguish these from Dubreuilh's melanosis. Initially, there is a patch 8-10 mm in diameter which is brown, sepia or blackish, oval or rounded, and becomes progressively clearer. As it evolves it becomes polymorphous and appears as an expansive patch with brown, black, grey or bluish areas and ingrowths of normally coloured skin. Phases of advance alternate with phases of regression, and the pigmented areas may lighten or even fade to leave a picture of atrophic skin. After several years thickened areas appear which are verrucous, encrusted or nodular, even ulcerated, and it is then that malignant degeneration is to be feared.

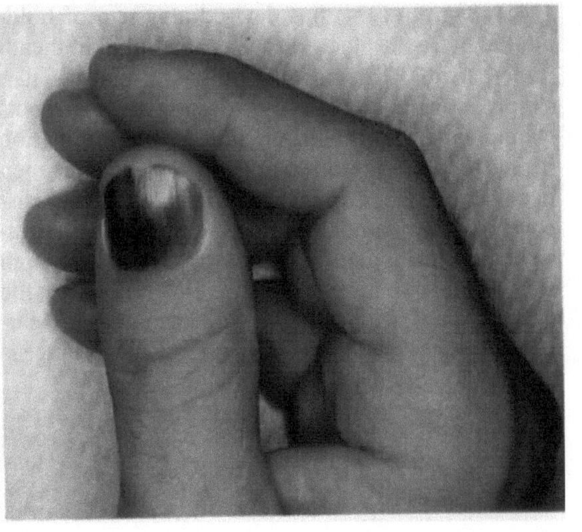

Fig. 21. Dubreuilh's melanosis: exceptional location in the nail matrix

Accounts of Dubreuilh's circumscribed melanosis in the hand are rare. Andrade (1972) reported two cases. One involved a woman 75 years old who had had a pigmented patch on the dorsum of the interphalangeal joint of her right 4th finger for 3 years; as this spread progressively, its appearance became irregular and crumbling. The other was that of a man 58 years old, who for 8 years had had a pigmented lesion on the dorsum of his hand, unaccompanied by any functional disturbance; the irregular, unevenly brown or black patch was slightly elevated on palpation. Two cases of lesions of the fingertip (Lupulescu et al. 1973) come into the category of melanotic whitlow (see p. 44). Our sole personal case of Dubreuilh's melanosis in the hand concerned the nail matrix, the nail being brownish, a site that we have not found previously reported in the literature (Fig. 21).

Histology

The diagnosis of Dubreuilh's melanosis is based on both clinical and histological features. The histological evidence was established in 1912 by Dubreuilh.

There is a functional proliferation, the extent of which varies with the age of the lesion and its clinical appearance, the older and darker lesions combining more marked cell proliferation and pigment accumulation. The cells contain an often bulky and hyperchromatic nucleus and may be grouped in honeycomb fashion, possibly also in nests. The signs of malignant transformation include poor delimitation of the functional proliferation with invasion of the superficial dermis, mitoses, cell irregularities and an intense inflammatory reaction in the dermis.

Differential Diagnosis

This is from non-naevoid lesions, naevi and certain types of malignant melanoma. The non-naevoid lesions are seborrhoeic warts and solar keratoses or certain superficial basal-cell epitheliomata. Punctate naevi are difficult to distinguish from Dubreuilh's melanoma at the outset. The former affect younger subjects, in areas both exposed and unexposed to the sun and change very little. Melanomata with superficial spread affect mainly the trunk and lower limbs, have a relatively short course (1-5 years) and present as polycyclic lesions with pink or depigmented zones. Biopsy of the lesion is essential in every case.

Development

Evolution of the precancerous melanoses is very slow, and the lesions may attain a considerable size of several centimeters in diameter. The incidence of cancerous degeneration is difficult to determine, as the patients involved are often over 70 years of age. Andrade (1976) believes that it develops in less than 30% of cases, after an average course of 10-14 years. Total excision of the lesion is justified, even though the melanomata that develop in circumscribed melanosis advance slowly and only give rise to metastases late (Clark and Mihm 1969).

Treatment

Although superficial destruction of the lesion by diathermy curettage has been advised, it is much better to perform complete excision of the affected skin with 0.5-cm margin if benign or 2-cm if not. Loss of substance may require skin grafting for closure. Prophylactic eradication of the lymph nodes, chemotherapy and pre- or postoperative radiotherapy are not indicated.

Bibliography

Andrade R (1972) Lentigo maligna (Hutchinson's freckle). In: Demis JD, Crounse RG, Dobson RL, Mc Guires (eds) Clinical dermatology, vol 2, unit 11-54. Harper & Row, New York

Andrade R (1976) Circumscribed precancerous melanosis (Dubreuilh). In: Andrade R, Gumport SL, Popkin GL, Rees TD (eds) Cancer of the skin. Saunders, Philadelphia, pp 679-702

Clark WH Jr, Mihm MC Jr (1969) Lentigo maligna and lentigo maligna melanoma. Am J Pathol 55: 39

Dubreuilh MW (1894) Lentigo malin des vieillards. Bull Soc Fr Dermatol Syphiligr, pp 460-467

Dubreuilh MW (1912) De la mélanose circonscrite précancéreuse. Ann Dermatol Syphiligr (Paris) 3: 129

Hutchinson J (1892) On senile moles and senile freckles and on their relationship to cancerous processes. Arch Surg 2: 218

Lupulescu A, Pinkus H, Birmingham DJ, Windekhe H, Posch JL (1973) Lentigo maligna of the fingertip - Clinical histologic and ultrastructural studies of two cases. Arch Dermatol 107: 717

Pinkus H, Mishima Y (1963) Benign and precancerous nonnevoid melanocytic tumors. Ann NY Acad Sci 100: 256

Bowen's Disease

Bowen's disease, or intra-epidermal squamous-cell epithelioma, is a potentially malignant intra-epithelial carcinoma, presenting as a dyskeratosis. It may develop at any site in the skin and genital or buccal mucosae; localisation in the hand is not unusual. Involvement of the ungual region is not considered here, but is dealt with in the section on other ungual tumours.

The condition was initially described in 1912 by Bowen, who established its precancerous nature; it was definitively categorised by Darier in 1914 (cited by Pelisse 1981) under the name of "discoid lenticular dyskeratosis". To our knowledge, the first report of involvement of the hand was that of Duperrat and Meyer in 1947. The incidence at this site is given as 14.5% by Graham and Helwig (1959) and 13.5% by Letessier (1972) in a thesis on 243 cases of Bowen's disease. Some 15 separate reports of lesions in the hand have been published (Table 2), including Hyman (1963), Duperrat et al. (1966), Harmel et al. (1968), Achten and van Oost (1969), Wagers et al. (1973), Larrègue et al. (1974) and Kern and Schiff (1977), to which must be added some twenty ungual lesions (see p. 220: tumours of the ungual region).

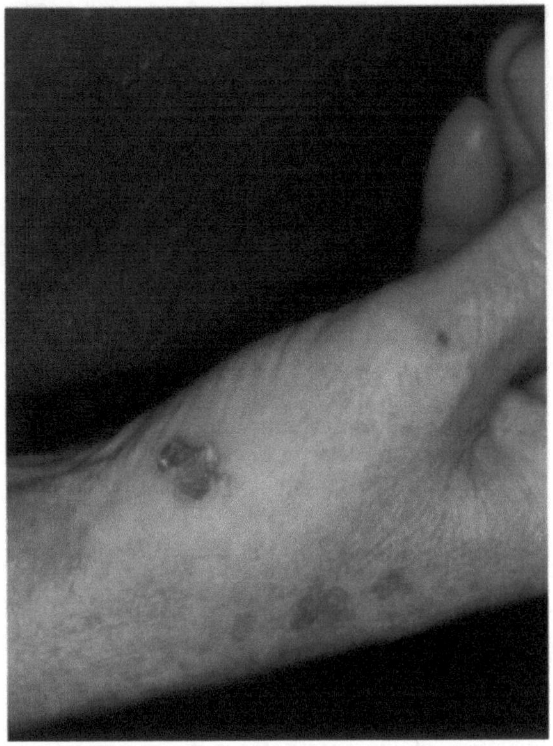

Fig. 22. Bowen's disease: note the rim of desquamation

Clinical Features

Involvement of the dorsum of the hand is classical (Knox and Joseph 1976; Pelisse 1981) and accounts for 88% of hand lesions in the thesis of Letessier (1972). However, the published cases tend to be palmar (Duperrat et al. 1966; Hyman 1963; Achten and van Oost 1969; Wagers et al. 1973) or digital (Harmel et al. 1968; Wagers et al. 1973; Larrègue et al. 1974; Kern and Schiff 1977). The onset of clinical signs is insidious, and several years may elapse before the first consultation. The disease is evidenced by the appearance of one or more pink lenticular spots which progressively extend and merge into an erythematosquamous plaque with sharp margins, surrounded by a little collar of desquamation (Fig. 22). Its size varies with the duration of the lesion and may reach several centimeters in diameter (or even involve the entire palm, as in the case studied by Achten and van Oost 1969, which developed over 20 years). The lesion gradually thickens and assumes a psoriasiform or keratinised appearance, developing a crust beneath which is a pink, oozing but not haemorrhagic patch (Fig. 23). Elsewhere, the lesion may become fissured, erosive (Fig. 24) or pseudoverrucous, and its outline may be less sharp (Larrègue et al. 1974).

Bowen's disease occurs in adults (aged 30-60 years). A review of the published cases shows a striking and very definite female predominance (71%) for lesions of the hand (excluding the nail), with a contrasting male predominance (89%) if the nail is involved. This may be only a coincidence related to the small number of cases reported. The literature also indicates a slight left-sided predominance already noted by Harmel et al. in 1968. Multiple lesions are very rare and, when they occur

Table 2. Reported sites of Bowen's disease in the hand (nail bed excluded)

Reference	Site	Sex	Age	Lesion	Treatment	Course
Duperrat and Meyer (1947)	Palm L	M	56	Unknown	Unknown	Unknown
Schonberg and Litt (1961)	Palm	Unknown		Unknown	Irradiation	Cured
Hyman (1963)	Thenar L	M	66	Unknown	Unknown	Unknown
Duperrat et al. (1966)	Palm L	F	79	Unknown	Unknown	Unknown
Harmel et al. (1968)	Palm, thumb L	M	37	Patch of hyperkeratosis	Diathermy	Cured at 1 year
Achten and van Oost (1969)	Entire palm	F	80	Erythematosquamous lesion	5-fluorouracil	Cured at 1 year
Wagers et al. (1973)	Palm L	F	20	Verrucous plaque	5-fluorouracil	
	5th finger L	F	46	Hyperpigmented plaque	5-fluorouracil	
Larregue et al. (1974)	4th finger R	F	65	Verrucous lesion	Diathermy and curettage	Cured at 18 months
	Thumb L	F	70	Crusted, erosive	Excision and graft	Cured at 1 year
	Index L	F	41	Erythemato-verrucous	Excision and graft	Cured at 6 months
Kern and Schiff (1977)	Thumb R	F	45	Erythematokeratotic lesion	Excision	
	4th finger L	F	48	Erythematous plaque	Excision and graft	
	R Index	F	54	Erythematous plaque	5-fluorouracil	Recurrence excision graft

in the hand, are usually ungual (Coskey et al. 1972; Strong 1983).

Histology

The histological features of Bowen's disease are those of an intra-epithelial, squamous-cell epithelioma. The epidermis is hyperacanthotic, with lesions of hyperkeratosis and parakeratosis. In the malpighian layer the cellular arrangement is anarchic, with epithelial giant cells and large round cells with a nucleus that is either pyknotic or vice versa very large and hyperchromatic, numerous mitoses and cellular keratinisation. These abnormal cells, which correspond to Darier's poikilocarynosis, are randomly distributed among the normal epidermal cells. The hair follicles are often affected. The basal membrane is intact in all cases. It is essential to examine the entire specimen to look for any breach in continuity; this would considerably affect the prognosis, as it is indicative of an invasive stage progressing like a squamous-cell epithelioma.

Aetiopathogenesis

Although the aetiology cannot be stated with any certainty, several factors appear to favour the development of Bowen's disease:

1. Arsenic is found proportionally more often in the cutaneous tissues of patients affected by Bowen's disease than in the general population (Graham et al. 1961), but it is only in exceptional instances that enquiry reveals exposure to arsenic.
2. The viral aetiology suggested by Mierowski et al. in 1954 is supported by more recent workers (Olson et al. 1968; Nordquist et al. 1970) who have found suggestive intracytoplasmic particles.
3. The influence of solar radiation must be discounted, since Bowen's disease affects exposed and unexposed regions equally; however, such radiation is certainly responsible for actinic keratoses resembling Bowen's disease.
4. Several cases have been reported in which Bowen's disease has developed after trauma (Graham et al. 1961).

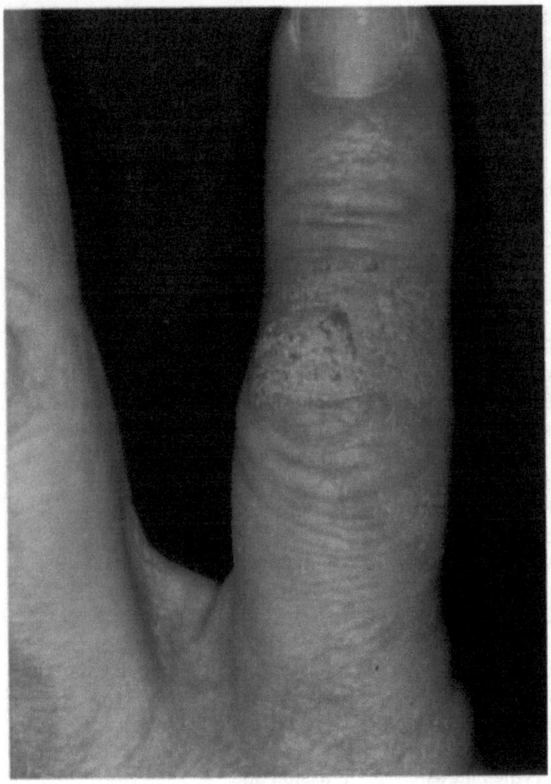

Fig. 23. Bowen's disease: psoriasiform appearance

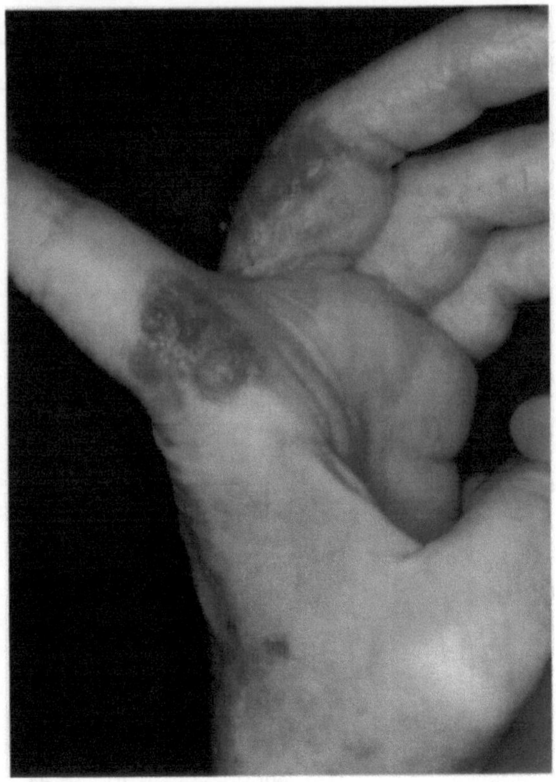

Fig. 24. Bowen's disease: erosive form

5. Finally, one should stress the frequent association of Bowen's disease with another skin tumour (15% of cases in Letessier's 1972 series) of the nature of an epithelioma or pre-epitheliomatous keratosis, suggesting the concept of an "epitheliomatous predisposition".

Differential Diagnosis

A dyskeratotic lesion of this nature may be confused with many other conditions. Psoriasis is often difficult to distinguish, but removal of the psoriatic crust produces immediate bleeding. The superficial type of basal-cell epithelioma is very much like Bowen's disease, but in the former the epidermis is atrophic, and there is a characteristic fine peripheral border. As for the other keratoses, particularly actinic keratosis, only exposed areas are affected, producing drier and more adherent lesions which are less sharply outlined and rest on an erythematous base. There are also the encrusted form of eczema, lichen and even the early stage of a seborrhoeic wart. In every case this established, slowly progressive lesion requires biopsy, and this will allow diagnosis. Nevertheless, the histological picture is sometimes deceptive, suggesting Paget's disease of the skin (insofar as the localisation allows), senile keratosis or the intra-epidermal epithelioma of Bordst-Jadassohn.

Development

The spontaneous course of Bowen's disease is very slow, extending over several decades as the tumour progressively increases in size. Sometimes, however, a much more rapid course is seen. The crust erodes, granulates, ulcerates or infiltrates, and the lesion grows very rapidly, suggesting the invasive form of Bowen's disease. Such a course is rare (5% in Graham's series; Graham and Helwig 1959) and usually late, after 15-20 years' progress, possibly elicited by an incomplete initial excision. Invasive Bowen's disease progresses like a squamous-cell epithelioma, with a risk of metastases in the lymph nodes and viscera. Moreover, the disease may be associated with another cancer, either cutaneous, as already mentioned, or visceral. The incidence of such association has been assessed rather variously: Pelisse (1981) found it in 12%-24% of his series, whereas Graham and Helwig (1959) found a deeper-lying cancer at autopsy in 80% of 35 patients with Bowen's disease. These figures justify the further search for another malignant condition once Bowen's disease has been diagnosed. The association gravely worsens the prognosis, with a mean survival of 4.5 years in Graham's series (Graham and Helwig 1959).

Treatment

Surgical excision, where possible, remains the treatment of choice for Bowen's disease (78% of patients cured in Letessier's series, 80% in Graham's). The excision must be well outside the visibly affected tissues by a 3-mm margin, and include all the subcutaneous areolar tissue in depth. In the hand, skin cover is usually obtained by free grafting, sometimes by a local flap. Other methods of treatment tend to be reserved for cases in which surgery is contraindicated, as when lesions are multiple or skin replacement is difficult. Cytotoxic agents (5-fluorouracil) must be applied twice daily for 4-6 weeks and histological examination repeated after treatment. As for radiotherapy, Bowen's disease is sensitive to several types of radiation: electron, Bucky therapy and contact radiotherapy, but with a risk of subsequent radiodermatitis (Letessier 1972). The results of diathermy combined with curettage appear unsatisfactory (only 27% of cures in Graham's series). The invasive cases call for the same treatment as a squamous-cell epithelioma that is malignant from the outset.

Bibliography

Achten G, van Oost A (1969) Bowen étendu de la paume de la main. Bull Soc Fr Dermatol Syphiligr 76: 322-323

Beasley B (1976) (Bowen's disease) Malignant tumors of the hand. In: Andrade R, Gumport SL, Popkin GL, Rees TD (eds) Cancer of the skin, vol 2. Saunders, Philadelphia, pp 1438-1439

Bowen JT (1912) Precancerous dermatoses: A study of two cases of chronic atypical epithelial proliferation. J Cut Dis 30: 241

Coskey RJ, Mehregan A et al. (1972) Bowen's disease of the nail bed. Arch Dermatol 106: 79-80

Duperrat B, Meyer J (1947) Maladie de Bowen de la paume avec épithélioma sudoripare de la cuisse. Bull Soc Fr Dermatol Syphiligr 54: 29-30

Duperrat B et al. (1966) A propos des localisations peu fréquentes de l'epithélioma de Bowen. Arch Belg Dermatol Syphiligr 22: 217-220

Graham JH, Helwig EB (1959) Bowen's disease and its relationship to systemic cancer. Arch Dermatol 80: 133

Graham JH, Mezzanti G, Helwig EB (1961) Chemistry of Bowen's disease: Relationship to arsenic. J Invest Dermatol 37: 317

Harmel L, Kalis B, Mallet P (1968) Maladie de Bowen localisée à la face palmaire d'un pouce. Bull Soc Fr Dermatol Syphiligr 75: 651-653

Hutchinson J (1892) On senile moles and senile freckles and on their relationship to cancerous processes. Arch Surg 2: 218

Hyman AB (1963) Bowen's disease of finger tips. Arch Dermatol 88: 86-88

Kern AB, Schiff BL (1977) Bowen's disease of the finger. Cutis 19: 229-232

Knox JM, Joseph LM (1976) Bowen's disease and erythroplasia. In: Andrade R, Gumport SL, Popkin GL, Rees TD (eds) Cancer of the skin, vol 2. Saunders, Philadelphia, pp 646-661

Larrègue M, Civatte J et al. (1974) Maladie de Bowen digitale. Bull Soc Fr Dermatol Syphiligr 81: 24-26

Lengyel J (1958) Cancer intra-épithélial de type Bowenien. Ann Dermatol 85: 299-301

Letessier S (1972) La maladie de Bowen. A propos de 243 cas. Thèse, Paris, p 76

Michaelides P, Hyman AB (1964) Bowen's disease of the palm: Report of a case. Dermatologica 128: 239-244

Mierowski E, Freeman LW, Woodward J (1954) The response of embryonic chick membrane. Bowen's intraepithelial cancer. J Invest Dermatol 22: 417

Nordquist RE, Olson RL, Everett MA, Condit PT (1970) Virus-like particles in Bowen's disease. Cancer Res 30: 288

Olson RL, Nordquist RE, Everett MA (1968) An ultrastructural study of Bowen's disease. Cancer Res 30: 288

Pelisse M (1981) Maladie de Bowen. In: Encyclopédie medico-chirurgicale: Dermatologie. Paris, 12755-C 10

Schonberg IL, Litt JZ (1961) Bowen's disease of the palm treated with Grenz rays. Society Transactions. Arch Dermatol 84: 144

Strong ML (1983) Bowen's disease in multiple nails beds. Case report. J Hand Surg 8: 329-330

Wagers LT, Shapiro L et al. (1973) Bowen's disease of the hand. Arch Dermatol 107: 745-746

Basal-Cell Epithelioma

Basal-cell epitheliomata - also known as basal-cell carcinoma, rodent ulcer, basaloma and atypical malpighian or non-malpighian epithelioma - are the most common malignant tumours of the skin and yet are very rare in the hand. They are characterised by their slow course and entirely local malignancy.

Known from antiquity, basal-cell epitheliomata are much commoner on the face than the limbs (1.2% for the limbs taken together as against over 80% for the face: Civatte and Blanchet 1975). Basal-cell epitheliomata are quite rare in the dorsum of the hand and exceptional in the palm (Golitz et al. 1980) and subungual region. Males appear to be affected more often than females (Gellin et al. 1965).

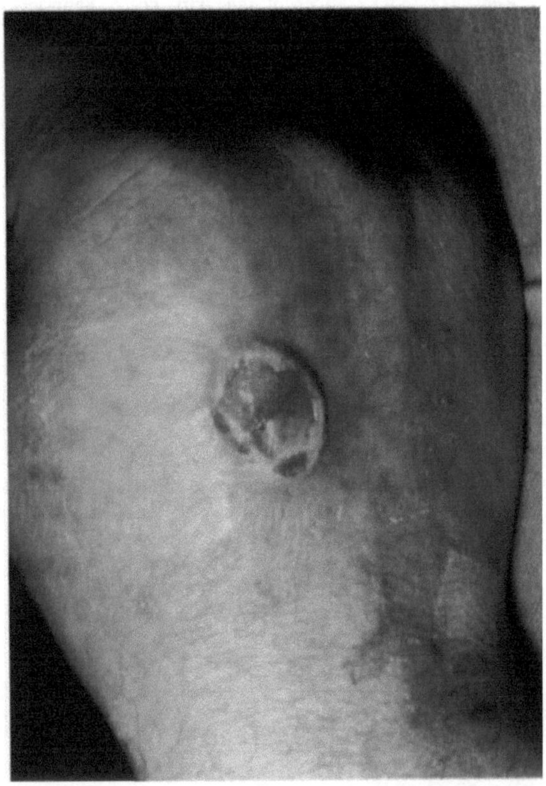

Fig. 25. Basal-cell epithelioma: note the peripheral cushion

Clinical Features

Basal-cell epitheliomata are characterised by their clinical polymorphism. They may develop on a lesion of senile keratosis which thickens and ulcerates, the ulceration becoming surrounded by small nodules. Often, the lesion appears in healthy skin in the form of little nodular prominences a few millimeters in size that are pink or red and shiny, resembling "pearls". After the lapse of some weeks or months, various appearances may develop: a slightly infiltrated, squamous, hyperkeratotic lesion; a superficial erosion; an ulcerated lesion surrounded by a peripheral pearly rim (Fig. 25); a rounded, fungating, papillomatous and sometimes ulcerated tumour (Fig. 26); a multinodular lesion; a pigmented tumour with black or grey patches; or a sclerosed, infiltrated plaque. Another characteristic of basal-cell epitheliomata is their slow progress; there may be an interval of months or even years between the appearance of the first signs and medical consultation. As with other cutaneous epitheliomata, fair-haired males with light complexion and blue eyes are those most often affected (Gellin et al. 1965). The majority of cases is seen after the age of 40 years.

Aetiopathogenesis

As with other skin cancers, light and ionising radiation seem likely to play a part. Schreiber et al. (1971) note that at Tucson, Arizona, there are 422 cases of skin cancer per 100000 inhabitants as against 350 in Queensland, Australia, and that this very high incidence is associated with sunlight, low levels of humidity, high temperature and altitude. Similarly, the role of ionising radiation has long been familiar, and malignant degeneration in radiodermatitis may be basal or squamous celled. Basal-cell epitheliomata of the palm are sometimes noted in the syndrome of basal-cell naevus.

Histology

The tumour cells contain a reduced amount of clear and poorly defined cytoplasm and an oval or rounded, very basophilic nucleus with one or two rounded nucleoli. The architecture of the tumour is conditioned by grouping of the cells into clumps or strands of variable size that are very basophilic, sharply circumscribed, rounded or irregular and often with lateral outgrowths. The stroma is often quite characteristic, with a senile elastosis; the generally rather scanty inflammatory infiltrate is compressed peripherally and contains elastic fibres and collagen. There is cylindromatous degeneration of the stroma, which is variably invaded by epitheliomatous nests. However, there is no very precise correlation between the clinical appearance and the microscopic findings (Civatte and Blanchet 1975).

Differential Diagnosis

Diagnosis is easy if the tumour is solitary, develops steadily and is surrounded by a pearly border. Otherwise, it may be suggestive of solar keratosis, squamous-cell epithelioma or keratoacanthoma. If there is any doubt, partial or excision biopsy of the lesion establishes the diagnosis.

Development

The course of basal-cell epithelioma is very slow and progressive. Involvement of the subcutaneous layer is very late, while lymph node invasion and metastases, if they occur at all, are exceptional. Lattes and Kessler (1951) found only 0.09% of cases had metastases in a series of 9050 basal-cell epitheliomata at all sites.

Treatment

Complete excision of the tumour is essential and adequate. On the dorsum of the hand, if the lesion is not very large (less than a centimeter in diameter), excision followed by suture is practicable. In other cases a skin graft will fill the gap. Radiotherapy is effective for basal-cell epithelioma but, unlike surgery, does not provide the certainty of histologically verified complete excision.

Basal-Cell Naevus Syndrome or Naevomatosis

This is a dysembryoplastic syndrome with dominant autosomal transmission, combining five main features:

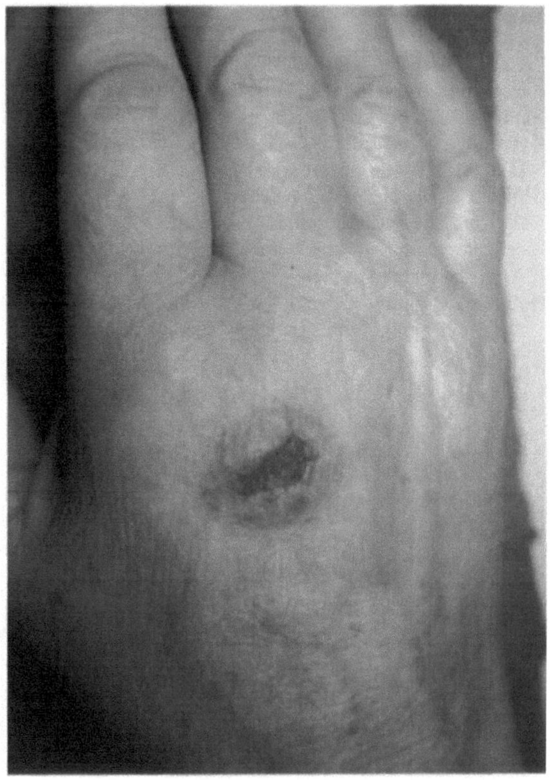

Fig. 26. Basal-cell epithelioma

1. Basal-cell epitheliomata at an early stage
2. Epithelial cysts of the mandible
3. Pitting of the palms and soles
4. Skeletal manifestations (rib anomalies such as synostoses and bifid ribs, vertebral malformations, short 4th metacarpal)
5. Anomalies of the neuroendocrine system (calcification of the falx cerebri, meningiomata)

The other lesions are mainly cutaneous: hemispheric translucent skin tumours and molluscoid epidermal cysts. These skin lesions often appear very early, from the age of 2 years, and especially in the second decade of life.

The basal-cell naevus syndrome affects the skin of the hand as two kinds of lesion: the very characteristic palmar "pits", 1-2 mm in diameter and only 1-mm deep, punched out in the palmar skin (Nicolai 1979) and basal-cell epitheliomata of the palm, which are relatively frequent. Golitz et al. (1980) collected four cases from the literature of eight published cases; in one of his oun patients there were 28 basal-cell lesions of the palms and 11 of the back of the hands, but she had been treated several times with radiotherapy in infancy.

Bibliography

Civatte J, Blanchet P (1975) Epithéliomas cutanés. Encyclop Med Chir 12750 - A10-A30
Gellin GA, Kepf AW, Garfinkel L (1965) Basal cell epithelioma, a controlled study of associated factors. Arch Dermatol 91: 38
Golitz LE, Nerris DA, Lueckens I Jr, David MD (1980) Nevoid basal cell carcinoma syndrome. Multiple basal cell carcinomas of the palms after radiation therapy. Arch Dermatol 116: 1159-1163
Holubar K, Matras H, Smalik AV (1970) Multiple palmar basal cell épithélioma in basal cell nevus syndrome. Arch Dermatol 102: 586-597
Johnson DE (1960) Basal cell epithelioma. A report of a case. Arch Dermatol 82: 253-256
Lattes R, Kessler RW (1951) Metastazing basal cell epithelioma of the skin: Report of 2 cases. Cancer 4: 866
Nicolai JP (1979) The basal cell naevi syndrome and palmar cysts. Hand 11: 98-101
Popkin GE, Defeo CP (1976) Basal cell epithelioma. In: Andrade R, Gumfort SL, Popkinaud GL, Rees TD (eds) Cancer of the skin. Saunders, Philadelphia
Schreiber MM, Shapiro SI, Berry C, Dahlim R, Szurl (1971) The incidence of skin cancer in South Arizona. Arch Dermatol 104: 124

Squamous-Cell Epithelioma

This is by far the most common malignant tumour of the hand, mainly affecting the dorsum and nail bed. It is a malpighian epithelioma, also known as an epidermoid carcinoma, containing squamous cells. Normally, they arise on a solar keratosis, sometimes on a chronic lesion (an old burn scar or osteomyelitis sinus), in other cases in healthy skin.

Known from antiquity (Celsius, see Grier 1976) as malignant tumours arising on burn scars, there have been numerous descriptions of secondary squamous-cell epitheliomata since the times of Pott (1775) and Marjolin (1828) see Grier 1976). Although, as a whole, these lesions are rarer than basal-cell epitheliomata, they predominate in the hand. Forsythe et al. (1978) found 41 patients treated at the Oklahoma Health Sciences Center in 15 years for squamous-cell cancer of the hand; Rayner (1981) treated 273 cancers of the hand in 10 years at Aberdeen, Scotland, while Clifford and Kelly (1955) found 47 squamous-cell cancers in 63 malignant tumours of the hand. All the authors insist on the part played by exposure to sunlight and the fact that the incidence is twice as great in males as in females; 80% of patients are over the age of 60 years (Rayner 1981).

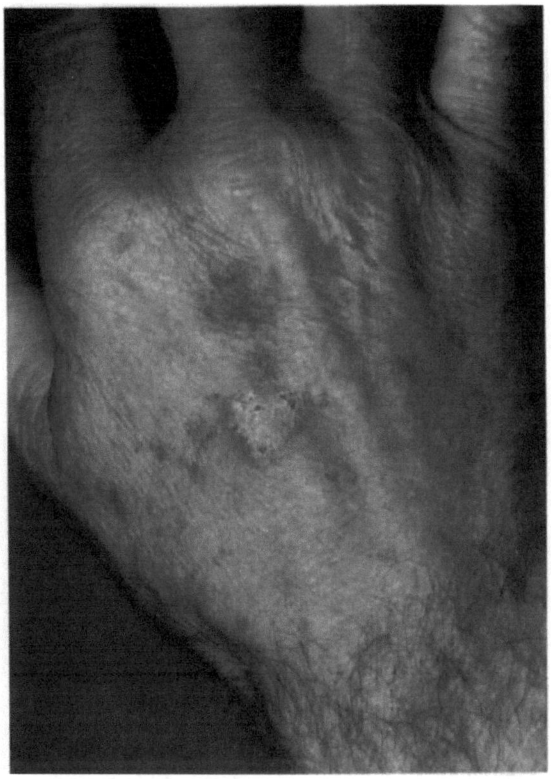

Fig. 27. Squamous-cell epithelioma

Clinical Features

Rayner (1981), while studying 273 cases, noted that only 7 were situated in the palm of the hand; the others affected the dorsum, mostly at the metacarpal level (64%), 16% at the thumb cleft, 15% in the finger clefts and 1.5% on the dorsum of the fingers. While squamous-cell epitheliomata of the dorsum of the hand may arise on a solar keratosis, an old scar or healthy skin, the appearance of a tumour on a solar keratosis is the most common eventuality. The affected individuals are usually males over 60 working out-of-doors (sailors, farmworkers) whose skin has been modified by the action of the sun over many years. Clinical suspicion is aroused by a zone of induration on an old keratosis (Fig. 27); reactive hyperaemia, crusting or even ulceration are often present. The typical ulcer has an indurated zone with a heaped-up rim and a central crater containing purulent exudate. More rarely, there is an exophytic tumour, a reddish-pink nodule. When the tumour arises on an old lesion such as a burn scar, it presents as a chronic ulceration within a zone of induration (Fig. 28). The burn, frequently situated in a flexion crease, has healed spontaneously with a long history of chronic ulceration, often extending over decades. Exceptionally, in the hand squamous-cell epithelioma may develop in a sinus of chronic osteomyelitis or on an epidermolysis bullosa (Gipson 1975). Squamous-cell epitheliomata differ from basal-cell lesions in their tendency to local spread, but the frequency and speed of invasion vary with the site of the cancer, its aetiology and its degree of cellular differentiation. Secondary infection is the rule in the ulcerated

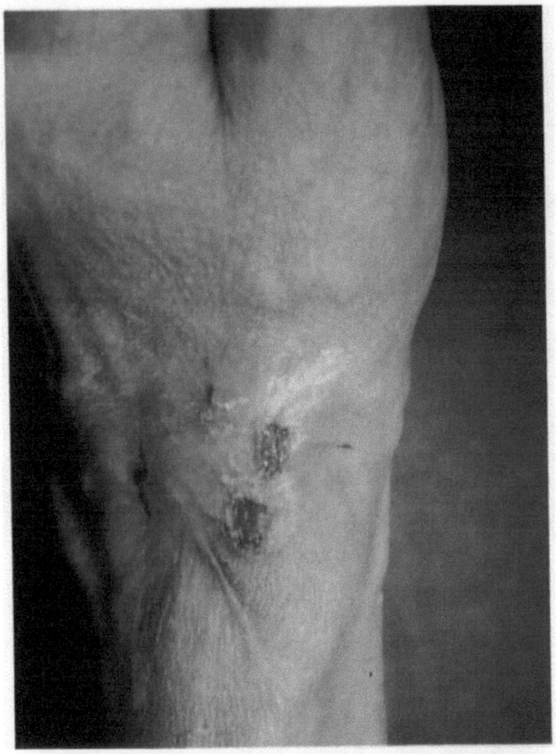

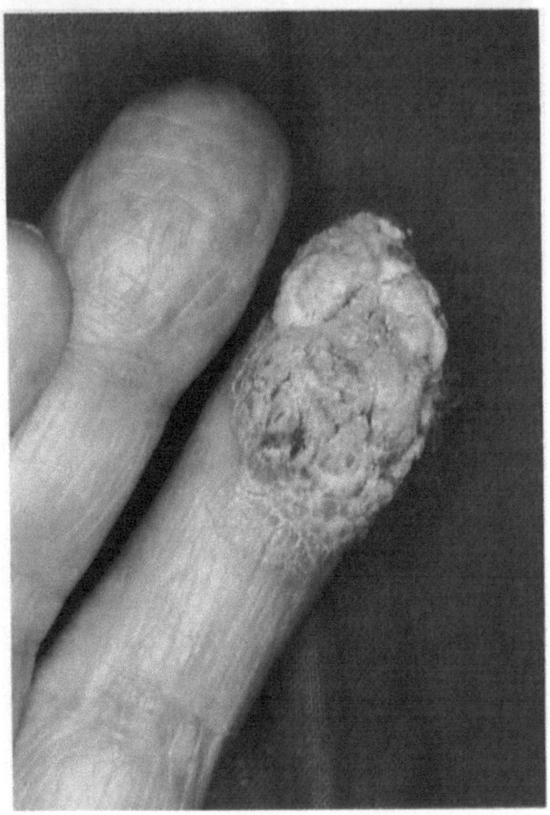

forms. Quite often there is an inflammatory lymphadenopathy, but nodal metastasis appears only after a varying interval. The frequency of nodal involvement is difficult to establish. Lund (1965) examined 3700 squamous-cell epitheliomata and found 700 (19%) to be invasive, but only 4 caused metastases. In the 6900 cases of Epstein et al. (1968), there were 142 (2%) with metastases, of which 115 were nodal. Of 72 cancers of the fingers, hands and forearms, Johnson and Ackermann (1950) found ten metastases in the axillary nodes and one in the epitrochlear nodes, but no metastases in cases in which the lesion had not invaded the dermis more deeply than the sweat glands. Palmar lesions are exceptional (Fig. 29). Rayner (1981), reporting on seven cases, considers this site to have a poor prognosis. Binet et al. (1980) reported a case associated with Bowen's disease. Periungual lesions are of a very specific nature (see p. 223).

Aetiopathogenesis

Squamous-cell epithelioma appears mainly on skin areas exposed to sunlight, the risk being greater in individuals with greater sensitivity (xeroderma pigmentosa, albinism). Arsenical poisoning has long been recognised as a promoting factor: Shu Yeh et al. (1968) studied a group of people in Taiwan drinking arsenic-rich water and found 238 squamous-cell epitheliomata, 10 of which affected the hands. Numerous other aetiologies have been suggested: old scars of burns or osteitis, psoriasis, chronic dermatoses, exposure to tar or ionising radiation (radiodermatitis, see p. 24), immunodepression (Fig. 30), etc.

Pathology

Grossly, the tumour presents either as an exophytic outgrowth or, vice versa, as an ulceration with infiltration of the subjacent tissue. On the dorsum of the hand the latter is marked by tethering in the plane of the extensor tendons, sometimes associated with simple inflammation.

Fig. 28. Squamous-cell epithelioma on burn scar; note ulceration *(above)*

Fig. 29. Squamous-cell epithelioma of the tip of the index finger: a very rare case *(left)*

Histologically, the characteristic cellular feature is the malpighian type of cell with abundant cytoplasm rich in mitochondria and a well-rounded nucleus rich in chromatin. The tumour cells are arranged in superimposed layers to give a paved appearance. The tumour is formed of more or less bulky neoplastic masses, with the epithelial clumps separated by a connective tissue stroma of varying extent. Superficially, the epidermis progressively disappears, is replaced by neoplastic epithelium and ulcerates. More deeply, the dermis, hypodermis and underlying planes are progressively invaded. There exists considerable cytological pleomorphism: mature forms with very advanced epidermoid differentiation, the cells arranged in layers resembling the basal layer of the normal epidermis, or, on the contrary, unregulated layers replaced by irregular squamous cells. Often a central zone exists marked by keratinisation, the keratinised spheres. The stroma undergoes changes marked by inflammatory infiltration with dissociation and destruction of the elastic fibres and collagen. On the basis of the cytological and architectural appearances, the lesions may be classified as lobulated, trabecular or dyskeratotic epitheliomata. Broders (1921) uses the degree of cell differentiation to classify squamous cell epitheliomata into four stages, considering that the tumour is less malignant the more differentiated it is.

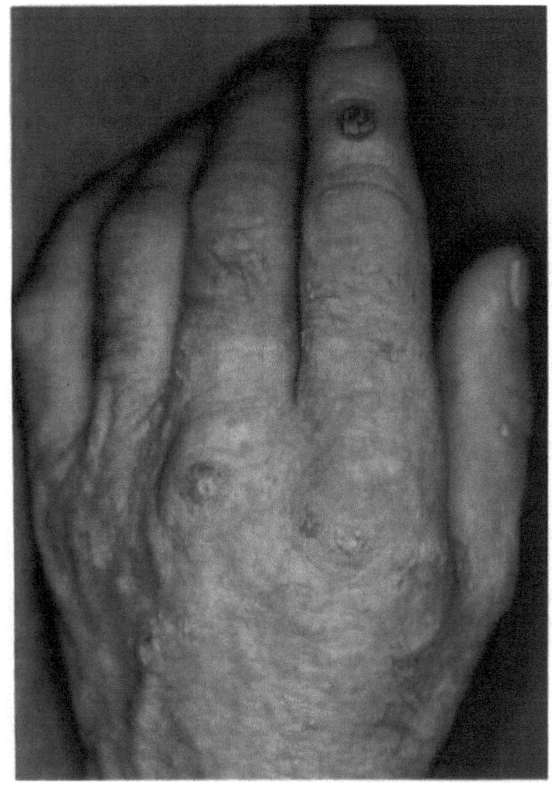

Fig. 30. Multiple squamous-cell epitheliomata: a renal transplant patient on immunosuppressants

Differential Diagnosis

This is difficult only at the outset, at the stage of a horny papillary epithelioma which may be confused with a wart, a seborrhoeic wart, a horny lichen or, most likely, an actinic or senile keratosis. When the epithelioma is ulcerated or fungating, diagnosis is easier, but there may be confusion with an ulcerated, fungating, basal-cell epithelioma, a mycosis or tuberculous lesion, fungating pyoderma, keratoacanthoma or pseudoepithelial hyperplasia. The two last diagnoses may be difficult to substantiate and require biopsy of the entire thickness of the lesion.

Development

After surgical excision in the absence of nodal involvement, cure is to be expected, but recurrences are possible in elderly or immunodepressed subjects. The appearance of metastatic lymphadenopathy is a bad prognostic sign, especially after cancer in a scar (Arons et al. 1965). According to Epstein et al. (1968), 5% of patients exhibiting adenopathy die within 5 years. Visceral metastases are rare; they affect the lungs, liver and bones.

Treatment

The treatment of squamous-cell epithelioma may consist of surgery or physical agents. Surgical methods include: amputation in lesions of the fingers, particularly at the distal phalanx; simple excision with primary closure for very small lesions; wide excision followed by split-skin grafts, especially for dorsal lesions; and clearance of axillary or, more rarely, epitrochlear nodes. The physical agents comprise simple radiotherapy (Forsythe et al. 1978) and radioactive gold or radon (Rayner 1981) using a moulded applicator. The type of treatment chosen depends on the site and extent of the lesion. Rayner emphasises the gravity and difficulty of treatment of lesions of the interdigital clefts; of 33 patients with such a lesion who had surgical excision, 5 died, and 14 of his 16 cases of radionecrosis occurred at this site. Forsythe et al. stress the gravity of local recurrences: one of his patients with a local recurrence died despite disarticulation of the finger, eradication of the lymph nodes and radiotherapy. Of 44 patients treated by Forsythe and co-workers 13 involved an amputation; local recurrences were responsible in 4 cases for amputation of the forearm, in 2 cases for amputation of the arm and in 2 for disarticulation of the shoulder.

In sum, when the lesion is solitary and limited and not adherent to the deeper layers, wide excision with a 1-cm margin followed by split-skin grafting is all that is needed. If the lesion is extensive, adherent to the deep layers or recurrent, amputation is preferable, and in these cases prophylactic clearance of the lymph nodes may be considered (Forsythe et al. 1978) 2-3 weeks after excision of the primary lesion. If palpable adenopathy already exists, wide excision followed by grafting and nodal eradication is usually advised (Butler et al. 1960; Kendall et al. 1969). Radiotherapy or ionising radiation may also be used, either for the primary lesion (Rayner 1981) or for recurrences.

Bibliography

Arons MS, Lynch JB, Lewis SR, Blocker TG (1965) Scar tissue carcinoma. Ann Surg 161: 170

Binet O, Beltzer-Garelli E, Elbaz JS, Aron-Brunetiere R (1980) Bowen's disease and squamous cell carcinoma of the palm. Dermatologica 161: 136-139

Broders AC (1921) Squamous cell epithelioma of the skin. 73: 141

Butler ED, Hamill JP, Seipel RS, de Lorimier AA (1960) Tumors of the hand. A ten year survey of 437 cases. Ann J Surg 100: 293-302

Clifford RH, Kelly AP (1955) Primary malignant tumors of the hand. Plast Reconstr Surg 15: 227-232

Epstein E, Epstein NN, Brag K, Linden G (1968) Metastases from squamous cell carcinoma of the skin. Arch Dermatol 97: 245

Forsythe RL, Bajav P, Engeron ON, Shadid EA (1978) The treatment of squamous cell carcinoma of the hand. Hand 10: 104-108

Gipson M (1975) Squamous cell carcinoma in épidermolysis bullosa dystrophica. Hand 7: 179-182

Grier WRN (1976) Squamous cell carcinoma of the body and extremities cancer of the skin. In: Andrade R, Gumport SL, Popkin GL, Rees TD (eds) Cancer of the skin. Saunders, Philadelphia, pp 916-932

Johnson, Ackermann LV (1950) Epidermoid carcinoma of the hand. Cancer 3: 657

Kendall TE, Robinson DW, Masters FW (1969) Primary malignant tumors of the hand. Plast Reconstr Surg 44: 37-40

Lund HZ (1965) How often does squamous cell carcinoma of the skin metastasize. Arch Dermatol 92: 635

Rayner CRW (1981) The results of treatment of two hundred and seventy three carcinomas of the hand. Hand 13: 183-186

Shu Yeh, How SW, Lin CS (1968) Arsenical cancer of the skin. Cancer 25: 141

Wilhel M, Konz B (1977) Squamous cell carcinoma of the hand. Handchir Mikrochir Plast Chir 9: 225

Sweat-Gland Epithelioma

Carcinomata or epitheliomata of the sweat glands are very rare, and their precise nature is still a matter of debate, no more than a dozen cases having been observed in the hand. Simard (1938), Stout and Cooley (1951) and Levitt et al. (1972) have reported solitary cases. Lucas and Nordby (1974) reported a case of a recurrent tumour of the palm, not invading the deep layers, classified histologically as a mixed tumour of the sweat glands. Clinically, the tumour has no features suggestive of a lipoma of the palm (Lucas and Nordby 1974) and may be located at the wrist, the hypothenar eminence or on the dorsum. There seems to be a capacity for metastasis (Teloh and Grier 1957). All authors are agreed in advising very wide excision, with a possible need for amputation, lymph-node eradication and postoperative radiotherapy.

Bibliography

Levitt RL, Russin DJ, Teal FF (1972) Sweat gland carcinoma of the hand. A case report. Clin Orthop 83: 95-98

Lucas GL, Nordby EJ (1974) Sweat gland carcinoma of the hand. Hand 6: 98-102

Simard LC (1938) Tumor of palm having the structure of a mixed tumour of the salivary gland. Ann J Cancer 33: 182-195

Stout AP, Cooley SGE (1951) Carcinoma of sweat gland. Cancer 4: 521-536

Teloh HA, Grier JP (1957) Metastasizing sweat gland carcinoma. Arch Dermatol 76: 80-86

Kaposi's Sarcoma

This is a tumoral disease that is always fatal but develops at a very variable pace; it is considered to be very rare outside certain ethnic groups. The disease has recently acquired a high incidence in the course of the acquired immune deficiency syndrome (AIDS). Involvement of the hands is very often the first sign of the condition.

First described in 1872 by Moricz Kaposi, a Hungarian dermatologist who subsequently amplified and clarified the description (1887-1897), Kaposi's syndrome has suddenly aroused acute interest in the light of its mode of origin. It was originally thought that the condition affected mainly Eastern European Jews after middle age. However, the discovery of cases in East Africa (Hallenberger 1914) revived interest in the matter to the extent that actual foci of the disorder affecting both children and adults were located in Uganda and Central Africa. By 1976 over 4000 cases had been discovered, the majority in Africa, the disease remaining rare in the United States of America and Western Europe. In 1955, Dorn and Cutler found only 0.02% in 33000 cases of cancer in America, and these principally in whites; from 1980 onwards, an increasing number of cases appeared in young and often homosexual subjects, sometimes of Haitian origin, or transfused haemophiliacs, during the course of AIDS. However this may be, the disease retains a largely male predominance (3:1 in children and 15:1 in adults).

Clinical Features

There are three modes of presentation: nodular, ulcerated and generalised.

1. At the outset *the nodular form* is a simple change of colour in the skin in the form of a purplish or brownish macule which may thicken and form a nodule; sometimes there are several nodules adherent to the skin but not to the underlying layers. The lesions are first unilateral, then bilateral. The hands and forearms are initially affected in 15% of cases, more rarely than the lower limbs. The course is inexorable and progressive, but slow and with remissions; plaques are formed by coalescence of the nodules. A hard oedema appears, followed by lymphadenopathy and skeletal involvement (osteoporosis with cortical erosions), and death occurs after 10-15 years.
2. *The ulcerated form* begins with invasion of the deep layers and tumoral necrosis and granulation. Extension is slow but inexorable. This course is rare in the upper limb.
3. *The generalised form* affects mainly young Africans.

Aetiopathogenesis

Certain factors are very specific: the racial and geographical links, the enormous male predominance, the histological finding of fusiform cells and vascular tissue, the frequent association with malignant lymphoma, the initial involvement of the limbs and the appearance in immunodepressed subjects (AIDS, renal transplantation). Several factors are certainly involved in a subject rendered receptive by medical treatment, viral infection, etc. Some authors (Camain and Quenum 1962) think more in terms of an angioblastic reticulosis than a true sarcoma. The origin of the tumour cells is controversial: vascular, lymphatic or perivascular. Electron microscopy reveals at least two types of cells: endothelial cells and phagocytic fibroblasts, derived from the reticuloendothelial system.

Histology

Two kinds of cells are seen at histological examination: fusiform cells very much like fibroblasts, of moderate size with rare mitoses, grouped in nodules or sheets; and a vascular component, often very marked, with well-formed vessels at the periphery of the nodule. In some cases a major part of the tumour is occupied by large cavernous spaces. Extravascular red corpuscles are also found around the fusiform cells and their macrophages. Inflammatory tissue is present at the periphery and within the tumour, especially at the outset.

Differential Diagnosis

This is simple in light of the particular context of Kaposi's sarcoma (Africans, AIDS, etc.). In Africans differentiation is from pyogenic granuloma, leprosy and neurofibromatosis. In Europeans the diagnosis is much more difficult, and distinction must be made from mycosis fungoides, sarcoidosis, fibrosarcoma, pyogenic granuloma and malignant lymphoma.

Treatment

This is very limited. In the localised forms surgical excision is rarely possible. Otherwise, radiotherapy and chemotherapy may be used, but with inconstant success. The long-term prognosis is always fatal, but the course in elderly subjects often exceeds 10 years.

Bibliography

Camain R, Quenum A (1962) Histopathologie et histogénèse de la maladie de Kaposi. Acta Un Int Cancer 18: 453

Dorn HF, Cutler C (1955) Morbidity from cancer in the United States. Part I: Variations in incidence by age, sex, marital status and geographic region. Monograph n° 29. Public Health, Washington, p 121

Hallenberger O (1914) Multiple Angiosarkome der Haut bei einem Kamerunneger. Arch Schiffs Trop Hyg 18: 647

Kaposi M (1872) Idiopathisches multiples Pigmentsarkom der Haut. Arch Dermatol Syphiligr 4: 265

Kaposi M (1887) Pathologie und Therapie der Hautkrankheiten. Urban & Schwarzenberg, Wien Leipzig

Kaposi M (1894) Zur Nomenclatur der idiopathischen Pigmentsarkome Kaposi. Arch Dermatol Syphiligr 29: 164

Templeton AC (1976) Kaposi's Sarcoma. In: Cancer of the skin. In: Andrade R, Gumport SL, Porkin GL, Rees TD (eds) Saunders, Philadelphia, pp 183-225

Melanomata

Malignant melanomata are tumours arising from the pigmented cells (melanocytes) of the basal layer of the epidermis and may affect the hand as often as any other part of the skin. Hutchinson, in 1886, gave the name of "melanotic whitlow" to a bulky tumour of the base of the nail with a characteristic pigmentation of the eponychium and insisted on the urgency of rapid radical treatment by amputation of the affected digit. In certain anatomical sites melanomata of the hand have a specific appearance. Periungual and subungual tumours (melanotic whitlow) often deceptively resemble an inflammatory lesion. Some melanomata of the palmar aspect of the hand and fingers have been specifically labelled as acral. The complexity of the diagnosis explains why melanotic whitlow is so often diagnosed late, as it is difficult to biopsy. The problems of including this tumour in a prognostically based classification (Clark et al. 1969; Breslow 1970) and the fear of recurrence after treatment that is too conservative justify singling out melanotic whitlow among the melanomata of the hand.

Melanotic Whitlow

Localisation of a malignant melanoma at the end of a digit is rather rare: 11% of melanomata of the upper limb and 2% for the entire body in one study (Das Gupta and McNeer 1965) or 20% and 3%, respectively, in another (Glicenstein 1966). Melanotic whitlow is certainly rare among malignant tumours of the hand in general (Kendall et al. 1969; Aulicino and Hunter 1982). In his original description, a simple clinical note of a few lines, Hutchinson (1886) defined the chief characteristics of the tumour: the frequent role of trauma, the late diagnosis, the malignancy from the outset and the pigmented margin around the nail. He advised amputation. However, before him, Boyer (in Duperrat 1962) had reported a case. Hertzler (1922) and Pack and Adair (1939) reported the first series of melanotic whitlow. Gibson et al. (1957) collected 2 cases of submucous melanomata of the toes and fingers and added 38 cases from the Mayo Clinic, while Das Gupta and Brasfield (1964) listed 34 subungual melanomata treated over 25 years at the Memorial Hospital of New York. Pack and Oropeza (1967) found 72 of a total of 2384 melanomata situated in the fingers and toes. Dufourmentel et al. (1968) collected 22 cases. Most of the articles combine reports of melanotic whitlow of the fingers and toes since the features of the lesions are identical, but some authors separate the digital lesions (Das Gupta and Brasfield 1964; Dufourmentel et al. 1968; Banzet et al. 1975).

Clinical Features

Although the symptoms of melanotic whitlow rapidly become apparent, the diagnosis is often delayed because the lesion seems commonplace, infective or due to injury. Patients usually seek advice for deformity of the nail and a patch around the nail, rarely for bleeding, pain or discomfort. As with melanomata at other sites, the patients are usually fair or red-haired, with a clear complexion and blue or brown eyes. Although melanomata are very rare in black- or dark-skinned people, the site of melanotic whitlow is in fact one of those most frequently affected (Pack and Adair 1939). These lesions are the most common of all the melanomata of the hand; Dufourmentel et al. (1968) found 17 of 22 cases. The thumb is affected in 50%-60% of cases (Das Gupta and Brasfield 1964; Pack and Oropeza 1967). Three modes of clinical presentation are distinguished, depending on whether the lesion affects the nail bed, the nail matrix or the skin around the nail.

Involvement of the Nail Bed

At the outset subungual melanoma resembles a simple haematoma, and the diagnosis is difficult because there is often actually a history of previous injury (Duperrat et al. 1968). However, certain clinical features should arouse suspicion of this subungual patch (Fig. 31): the persistence of the brown spot without passing through the successive colour changes of a bruise and the progressive extension of the lesion, which spreads to reach the nail tip while persisting at the matrix and possibly overlap-

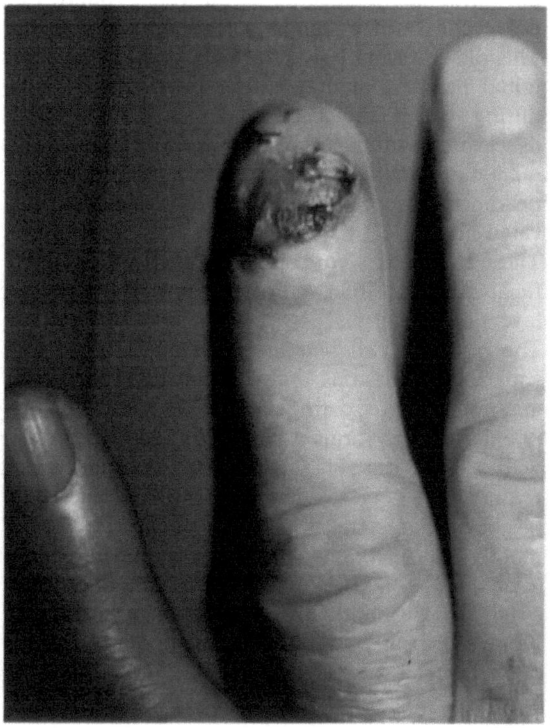

Fig. 31. Advanced subungual melanoma

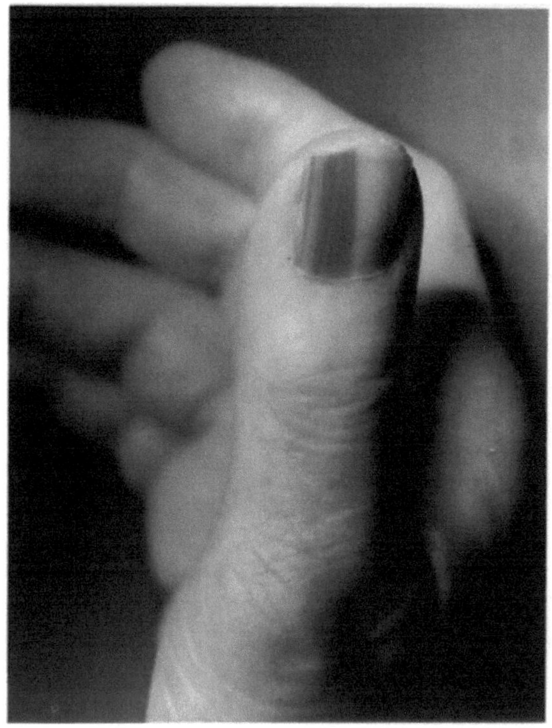

Fig. 32. Subungual melanoma presenting with black band

ping onto the nail fold. A subungual melanoma may also present in two other fashions (Duperrat 1962): a long, pigmented, longitudinal streak in a single nail (Fig. 32) or a black or achromatic longitudinal fissure. One must not await progressive changes in the subungual spot but remove the melanotic nail in whole or part without delay. This makes it possible to settle whether the lesion is a blood residue in process of spontaneous resolution or, on the contrary, a black spot in the matrix or an outgrowth very suggestive of melanoma. It also eliminates certain rarer affections: the very painful glomus tumour, malignant haemangioendothelioma (diagnosed only at histology) or the coffee bean of radiodermatitis accompanied by other evidence of radiation lesions.

Periungual Melanoma

This is easy to recognise. It begins as a black or achromatic outgrowth or melanotic whitlow. This rapidly develops a fleshy rim which extends towards the nail, which it strips up, and the base of the phalanx. When the lesion is unpigmented, it may suggest a whitlow, an oozing chronic paronychia or a botryomycoma. The fully developed lesion elevates the nail, which is split or destroyed, forming a bulky, purplish-red outgrowth.

Sub- and Periungual Melanoma

As well as these two types, described long ago by Hutchinson (1886), there have been recent descriptions (Lupulescu et al. 1973; Braitman 1979) of a macular periungual affection extending to both the pulp and the nail bed. This appears as a black patch whose initial progress is very slow, comparable to that of Dubreuilh's precancerous melanosis (lentigo maligna) or a particular variety of the Peutz-Jeghers syndrome combined with lentiginosis of the oral mucosa and intestinal polyposis.

Other Melanomata

Apart from the fingertip, localisation of a melanoma in the hand itself is rare (6 of 23 cases, Dufourmentel et al. 1968). The lesion then assumes the usual features of superficial spreading melanoma or of nodular melanoma (Clark et al. 1969). It is the superficial spreading melanoma that seems most common in the hand, forming a slowly extending,

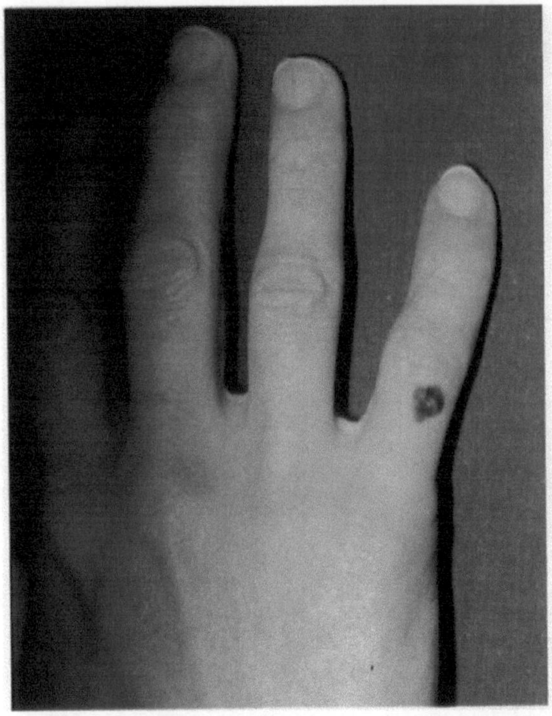

polychromatic pigmented lesion whose rounded or polycyclic border stands out somewhat (Fig. 33). This old or recent lesion develops in two stages: an initial period that may extend over several years, during which the patch enlarges and becomes dyschromic with zones of decoloration or hyperpigmentation, and a second period marked by the appearance of specks or a darker or achromatic elevated nodule.

These two phases correspond to a period of superficial extension followed by invasion of the deeper layers of the dermis.

The *malignant nodular melanomata* usually correspond to the lesions that develop in healthy skin and not on a preexisting naevus. The tumour rapidly enlarges and forms a black or uncolored nodule with a black rim (Fig. 34) that may resemble a botryomycoma (Fig. 35). More recently, another type has been characterised, *acral melanoma,* developing mainly in the palms and soles of black people (Coleman 1980). The lesion appears as a pigmented patch on a smooth area on which, after several months or years, a rapidly enlarging nodule develops.

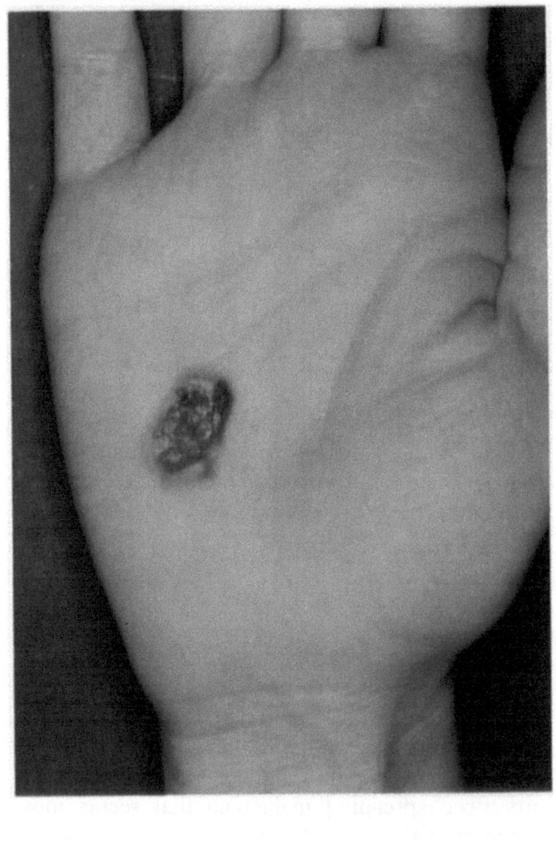

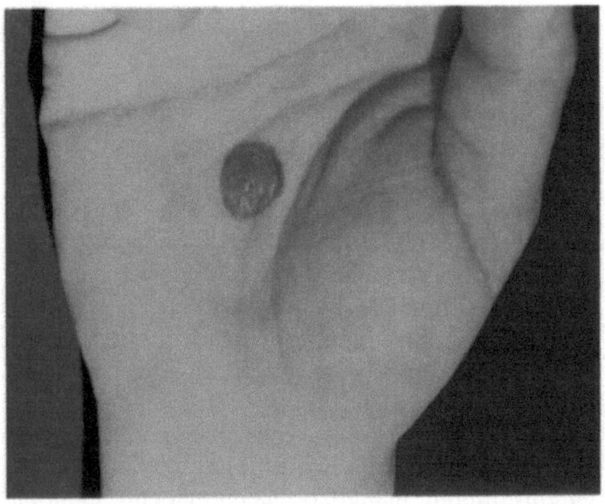

Fig. 35. Malignant nodular melanoma: resemblance to botryomycoma

Fig. 33. Melanoma with superficial spread *(above left)*

Fig. 34. Malignant nodular melanoma: note the peripheral rim *(left)*

Three factors must be defined in deciding diagnosis and treatment: the type of melanoma, the degree of dermal invasion and the thickness of the lesion.

Histological Type

Malignant melanomata arise from intra-epidermal melanocytes or preexisting naeval nests and seem to develop in two stages. First, there is a phase of horizontal spread that may be very brief or, alternatively, last several years, during which the tumour cells migrate into the adjacent epidermis; in this stage the purely intra-epidermal lesion is incapable of metastasising. A phase of extension in depth follows more or less rapidly during which the tumour progressively invades the dermis and acquires the ability to metastasise. Clark and Mihm (1969) postulate two kinds of tumours. There are those with a horizontal component: Dubreuilh's melanosis, superficial spreading melanoma and acral melanoma. There are also those extending vertically from the outset, with a very brief or non-existent horizontal phase, the malignant nodular melanomata. However, certain melanotic whitlows are impossible to classify.

Malignant Melanoma Arising on Dubreuilh's Melanosis (Lentigo Malignant Melanoma)

This is found mainly on the face, but also on the dorsum of the hand and fingers (Andrade 1976). The non-invasive or precancerous stage of Dubreuilh's melanosis may extend over a number of years. Initially, there is a brown macule which continuously increases in size and forms a multicolored patch. Malignant transformation, which occurs in 25% of cases, is signalled by the appearance of a nodule, often black. Microscopic examination shows that the intra-epidermal component of the tumour consists of a proliferation of atypical melanocytes in the basal layer of the epidermis plus a stromal reaction in the papillary dermis with an excess of pigment.

Superficial Spreading Melanoma (S.S.M.)

This is an irregular patch that develops on the back of the hand or fingers. This macule has polycyclic borders and is variously coloured brown, pink, black or even bluish. It is flat or slightly thickened and has a course of several months or years, sometimes reaching a diameter of several centimeters. In some cases there exist whitish zones of cicatricial appearance indicative of a local regressive tendency. Eventually, the development of one or more ulcerated nodules betrays invasion in depth. Microscopic examination shows an intra-epidermal component with abnormal melanocytes, monomorphs of 9 type. The vertical component consists of fusiform cells or a combination of the two types of cells.

Acral Lentiginous Melanoma

This is found only in the palms and soles, at the fingertips or sometimes in the nail bed. There is an increase in the melanocytes in the basal epidermis, with large nucleolated nuclei and dendritic extensions. Epidermal hyperplasia is present. During the stage of vertical proliferation, the cells often become fusiform.

Nodular Melanoma

This may affect any part of the hand, including the periungual region. It has a nodular, black, grey, bluish or uncoloured appearance, often with a pigmented rim. The main histological feature is vertical proliferation; there is no lateral intra-epidermal component.

Degree of Dermal Invasion

Clark et al. (1969) produced a classification based on the level of dermal invasion and accorded a prognostic value:
- Level 1: the tumour is localised to the epidermis (melanoma in situ)
- Level 2: solitary and partial lesion of the papillary dermis
- Level 3: involvement of the entire papillary dermis
- Level 4: infiltration of the reticular dermis
- Level 5: involvement of the hypodermis

Thickness of the Lesion

Breslow (1970) was the first to measure with the aid of a micrometer the thickness of the lesion between the most superficial cell of the epidermis and the deepest cell of the dermis, a procedure calling for sections strictly perpendicular to the epidermis. Interpretation is always difficult and may be falsified by regressive phenomena.
To conclude, the tumoural prognosis can be evaluated only by a joint study of the various clinical and histological factors: histological type, degree of dermal invasion, tumour thickness and regressive phenomena which introduce a pejorative element.

Clinical Features

Definitive treatment of melanotic whitlow and melanomata of the hand cannot be undertaken unless the diagnosis is clearly established by an experienced histologist. The lesion is excised in toto with a 0.5-cm margin under local anaesthesia or regional plexus block. If the loss of substance is too great for simple suture, the wound should be just dressed. Immediate histological examination is too imprecise to permit definition of the type of tumour, the degree of dermal invasion or the thickness of the lesion. It is very difficult to perform excision biopsy of a melanotic whitlow, so that partial biopsy becomes essential, although its use in cases of melanoma is debatable in view of the hypothetical risk of dissemination. An adequate fragment of tissue should be removed from where the tumour seems thickest.

Checking for any spread of the neoplastic process requires clinical examination for epitrochlear, axillary or cervical lymphadenopathy, hepatomegaly and cutaneous or subcutaneous tumours. A chest X-ray is essential. In doubtful cases, echography of the liver or testing for melanuria may be required.

The malignant melanoma can then be defined according to its stage: stage I or T, purely local tumour; stage II or N, presence of regional adenopathy; stage III or M, metastatic dissemination or remote lymphadenopathy (Das Gupta and McNeer 1965). The course of the melanoma depends on the type of tumour, its thickness and extent of nodal involvement. Keyhani (1977) studied 2268 cases of melanoma collected at the New York Memorial Cancer Center over a 28-year period, of which 55 affected the hands and 43 were followed-up. Seven patients had adenopathy at the first examination, and nine had secondary nodal involvement after an average interval of 23 months. The mean survival rate for these sixteen patients at 5 years was 31.2% as opposed to 60.5% for the group as a whole, indicating the gravity of nodal involvement.

After treatment the course of the tumour may be towards cure, local recurrences, lymphatic recurrences in the upper limb (in transit metastases), nodal involvement or cutaneous or visceral metastases.

Treatment

Melanotic Whitlows, Stage I. The treatment advised is transmetacarpal amputation or carpometacarpal disarticulation, since more conservative treatment is often followed by local recurrence (Pack and Oropeza 1967), the tumour often having invaded the periosteum and subjacent bone. However, a distal amputation may be considered in cases of acral melanoma or lesions of the nail bed which are less than a millimeter in thickness. The indications for prophylactic nodal eradication are debatable. The discovery of microscopic nodal invasion in 25% of clinically normal axillary nodes argues in favour of elective nodal eradication (Pack and Oropeza 1967). However, the possible development of skin tumours or lymphatic nodules in the upper limb following operation may be linked with the lymphatic blockage resulting from nodal eradication - delayed for 4-6 weeks with the theoretical aim of "trapping" the neoplastic cells. Finally, many workers reserve eradication for secondary involvement of the nodes.

Malignant Melanomata, Stage I. The therapeutic programme for the hand is identical with that for lesions at other sites. If the thickness of the lesion is less than 0.76 mm, simple excision with a margin of 1 cm suffices (Breslow 1970). If the thickness lies between 0.76 and 2 mm, a margin of 3 cm is necessary (Roses et al. 1983). The excision should include the skin, the subcutaneous tissue and the underlying fascia except in the region of the extensor tendons (Keyhani 1977). The loss of substance is covered by a thin graft taken from another limb (for the dorsal aspect of the hand and fingers) or a full-thickness skin graft (for the palm of the hand). The question of prophylactic nodal eradication follows the same lines as for melanotic whitlow.

Stage II Tumours. Pack and Oropeza (1967) treated 19 melanotic whitlows at stage II of tumoral evolution (an as yet untreated tumour with nodal metastases, or a local recurrence with adenopathy, or a solitary tumoral adenopathy without local recurrence). The treatment consisted of amputation or disarticulation limited to the affected finger and axillary nodal eradication. Disarticulation of the shoulder, while seeming logical to the authors, was not suggested in view of the reluctance of the patients. The addition of local perfusion by the arterial route (Banzet et al. 1975) or via the lymphatics seems useful for preventing in transit metastases. Of these 19 patients, only 1 was alive without recurrence at the end of 5 years, and 8 had died from melanoma.

Stage III Tumours. These are those with subcutaneous, cutaneous and visceral metastases, in the affected limb or more remotely. The presence of in transit metastases is a bad prognostic factor and is an indication for local perfusion with antimitotic agents. In cases of remote metastases, a solitary tumour may still be operated on (such as a pulmonary secondary). The prognosis is very grave, even though the patient may live "in equilibrium" with his tumour for several years.

Bibliography

Andrade R (1976) Circumscribed precancerous melanosis (Dubreuilh). In: Andrade R, Gumport SL, Popkin GL, Rees TP (eds) Cancer of the skin. Saunders, Philadelphia, pp 679-702

Aulicino PI, Hunter JM (1982) Subungueal melanoma. Case report and literature review. J Hand Surg 6: 167-169

Banzet P, Glicenstein J, Dufourmentel C (1975) Melanotic tumors of the hand. Hand 7: 183-184

Braitman M (1979) Subungueal malignant melanoma. Cutis 23: 617-623

Breslow A (1970) Thickness, cross sectional area and depth of invasion in the prognosis of cutaneous melanoma. Ann Surg 171: 902

Clark WH, Mihm MC (1969) Lentigo maligna and lentigo maligna melanoma. Am J Pathol 55: 39-67

Clark WH Jr, Frem L, Bernardino EA, Mihm MC (1969) The histogenesis and biologic behavior of primary human malignant melanoma of the skin. Cancer Res 29: 705-726

Coleman WP, Loria PR, Kremetz ET (1980) Acral lentigous melanoma. Arch Dermatol 116: 773

Das Gupta T, Brasfield R (1964) Subungueal melanoma. 25 years review of cases. Ann Surg 161: 545-552

Das Gupta T, McNeer G (1965) The incidence of metastasis to accessible lymphnodes from melanoma of the trunk and extremities. Its therapeutic significance. Cancer 17: 197-911

Dufourmentel C, Mouly R, Banzet P, Preaux J, Glicenstein J (1968) Le panaris mélanique, mélanome malin sous- et péri-unguéal. A propos de 22 cas. Mém Acad Chir 94: 309-314

Duperrat B (1962) Panaris mélanique. Etude basée sur quatorze cas personnels. Sem Hôp Paris 38: 304-313

Duperrat B, Puissant A, Glicenstein J, Cherif Cheikh JL (1968) Le panaris mélanique post-traumatique (A propos de six cas). Bull Soc Fr Dermatol Syphiligr 75: 582-584

Gibson SH, Hamilton M, Woolner LB, Bursting CA (1957) Melanotic whit low (Subungueal melanoma). J Invest Dermatol 29: 119-125

Glicenstein J (1966) Les narvocarcinomes cutanés. A propos de 341 cas. Thèse, Paris

Hertzler AE (1922) Melanoblastoma of the nail bed (melanotic whit low). Arch Dermatol 6: 701

Hutchinson J (1886) Melanotic whit low. Br Med J I: 491

Kendall TE, Robinson DW, Masters FW (1969) Primary malignant tumors of the hand. Plast Reconstr Surg 44: 37-40

Keyhani A (1977) Comparison of clinical behavior of melanoma of the hands and feet. A study of 283 patients. Cancer 40: 3168-3173

Lupulescu A, Pinkus H, Birmingham D, Usndek H, Posch J (1973) Lentigo maligna of the finger tip. Arch Dermatol 107: 717-722

Oropeza R (1976) Melanomas of special sites. In: Andrade R, Gumport SL, Popkin GL, Rees TD (eds) Cancer of the skin. Saunders, Philadelphia, pp 974-986

Pack G, Adair FR (1939) Subungueal melanoma. Surgery 46: 447

Pack G, Oropeza R (1967) Subungueal melanoma. Surg Gynecol Obstet 124: 571

Roses DF, Harris MN, Ackerman AP (1983) Diagnosis and management of cutaneous malignant melanoma. Major problems in clinical surgery, vol 27. Saunders, Philadelphia

II. Tumours of the Soft Parts

Synovial Cysts (Ganglia)

Synovial cyst is the most common of the tumours of the soft parts of the hand, a benign and rarely troublesome condition situated adjacent to a joint or tendon. Its precise aetiology, and in particular its relations with adjacent structures, have long been and are still debated. Arthrographic studies and intraoperative microscopic examination have afforded a better approach to this lesion. These have shown that radical surgical excision under the best conditions may be expected to prove satisfactory, whereas this is not the case when - as so often - the operation is hurried or performed under local anaesthesia without a tourniquet. That this condition is so commonplace should not let us underestimate the relative difficulty of the surgical treatment: the high proportion of recurrences after certain types of treatment is proof, if some were needed.

Ganglia have been known since the time of Hippocrates and were the object of ancient descriptions (Celsus, Cyprianus) and studied by Ambroise Paré. The problem of their aetiology and pathogenesis aroused early interest, and different hypotheses were proposed. Eller (1746) considered that there was a rupture of the synovial sheaths. Begin (1832) was the first to mention synovial hernia, an idea taken up and developed by Denonvilliers. Gosselin in 1850 thought that the cysts developed from synoviparous follicles.

Subsequent hypotheses derived more exact support from the examination of histological sections. In 1894, Ledderhose confirmed the connective tissue origin of these tumours. Lecène (1927) postulated a fibrotic process developing around a zone of degeneration in connective tissue. This terminological imprecision is the result of the plurality of hypotheses, the terms used being so numerous: synovial cast, pseudocyst, gelatinous cyst, colloid cyst, parasynovial cyst or ganglion. In all cases these are benign new formations containing a viscous fluidlike apple jelly that appear and develop in contact with or adjacent to a joint or tendon. They have a cystic structure and their wall more or less resembles synovial tissue devoid of any specific pathological process.

Their incidence is certainly underestimated, as many patients do not seek medical advice. Stack (1960) and Posch (1966) state that they constitute 60% of tumours of the hand. The topographical distribution of synovial cysts, as studied by McEvedy (1962) in 513 cases (84% were in the hand), is as follows: 61% on the dorsum of the wrist, 13% on the palmar aspect of the wrist, 9% on the flexor tendons and 1% on the dorsum of the fingers. The distribution in terms of age and sex is equally striking. This tumour is found mainly between the ages of 20 and 60 years, with maximum occurrence arising between 20 and 30 years (Nelson et al. 1972. Of 569 reported cases these authors noted the following incidences:

0– 9 years: 1.2%
10–19 years: 8.5%
20–29 years: 24.5%
30–39 years: 19.7%
40–49 years: 16.5%
50–59 years: 19.7%
60–69 years: 8.4%
70–79 years: 1.2%
80–89 years: 0.35%

Also in this study Nelson et al. (1972) confirmed the marked predominance of cases observed in women: 363 females to 180 males, a proportion of 2 to 1. In a series of 4000 tumours of the hand reported by Alter and Haber (1967), including 1000 synovial cysts, 20 cases (only 2%) were observed between the ages of 10 and 20 years.

Clinical Features

Many people affected by a synovial cyst do not seek advice as the lesion produces little functional disturbance. This moderately sized swelling, often appearing spontaneously and suddenly or after an injury, is usually no more than a cosmetic blemish. However, it may sometimes be the site of moderate and persistent pain, more marked during exertion and repetitive movements, causing a sensation of weakness in the hand or fingers. The tumour may be hard, tense and bulky; conversely, it may subside partially or even completely. Sometimes the site and size of the cyst may produce particular symptoms, due to compression of vessels or nerves. Barnes et

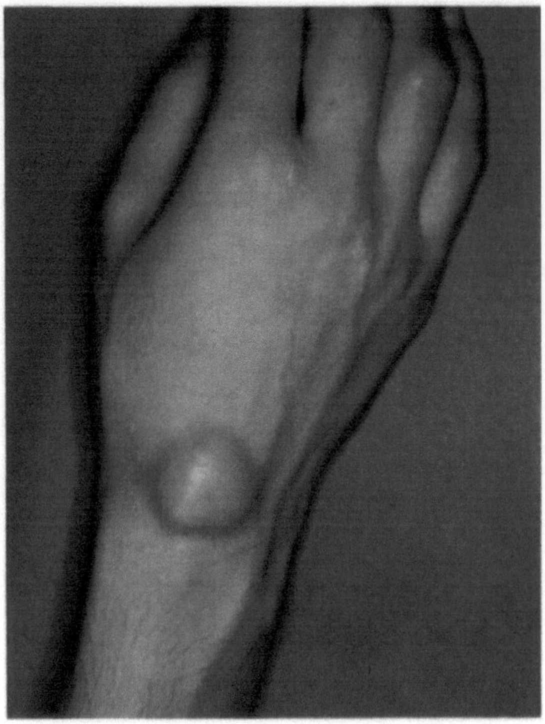

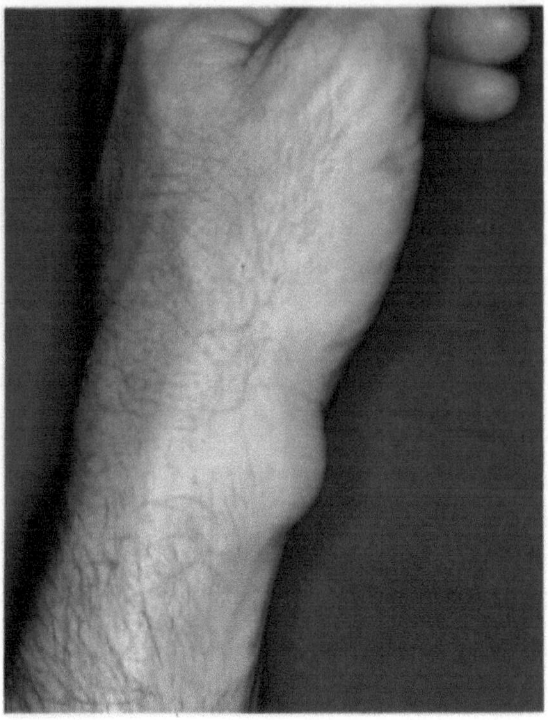

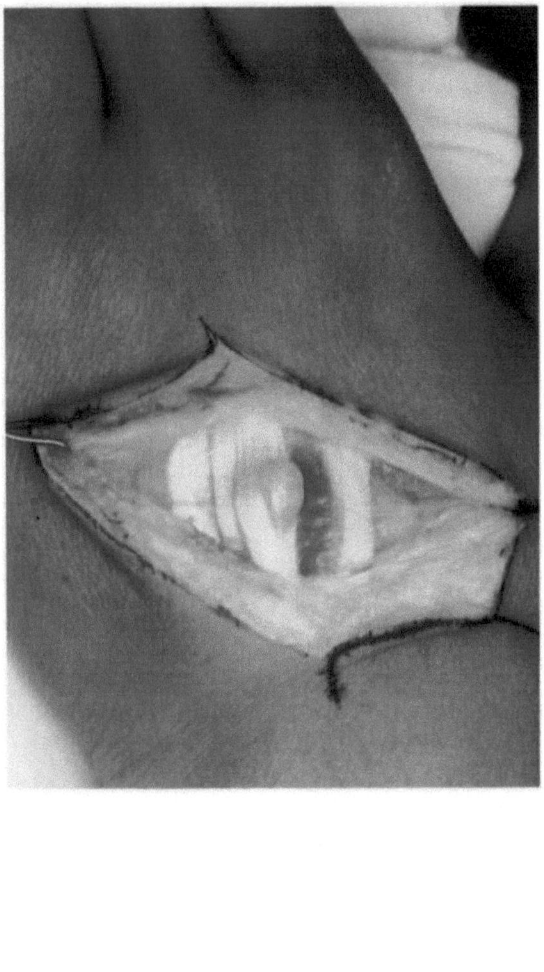

al. (1964) report the following reasons for seeking advice: unpleasant swelling, 80.5%; painful swelling, 17.9%; weakness, 1.2%; paraesthesiae, 0.4%. The ganglia on the dorsum of the wrist are the most common (Fig. 36): 70% (Holm and Pandey 1973), 61% (McEvedy 1962) or 60.3% (Calberg 1977). They form a visible and palpable mass, a rounded, regular swelling of the size of a pea to a walnut. Its generally firm and elastic consistence varies with the amount of contents; it may be hard and tense or soft and fluctuant. In all cases the cyst is mobile under the skin and rather fixed in depth. The appearance, colour and temperature of the overlying skin are normal. Most often, the cyst appears between the tendon of the extensor pollicis longus and the extensor tendons of the index, at the level of the scapholunare articulation (Calberg 1977, 70% of cases). More rarely, the cysts may appear in the anatomical snuffbox, between the extensor tendons of the fingers (Fig. 37) or between the extensor digiti

Fig. 36. Ganglion of dorsum of wrist *(above left)*

Fig. 37. Ganglion of extensor tendons of fingers *(left)*

Fig. 38. Ganglion at front of wrist *(above right)*

minimi and the extensor carpi ulnaris. A cyst originating dorsally may even emerge alongside the radial artery anteriorly.
Ganglia at the front of the wrist (Fig. 38) represent 20% of the synovial cysts of the hand and appear most often between the tendons of the brachioradialis and palmaris longus. They may also be situated between palmaris and brevis, between the flexor tendons of the fingers or, more medially, between the flexor carpi ulnaris and the common flexors. Clinically, these cysts resemble those on the dorsal aspect. However, because of the proximity of neurovascular structures, they may produce special symptoms: compression of the median nerve (Trevaskis et al. 1967) when the cyst lies between the two palmaris muscles, i.e. a carpal tunnel syndrome; compression of the ulnar nerve (Brooks 1952; Seddon 1952) by a cyst developing locally or, more rarely, arising from the radiocarpal joint and extending under the flexor tendons to reach the ulnar neurovascular bundle (Calberg 1977), i.e. an ulnar compartment syndrome; compression of the sensory branch of the radial nerve; compression of the radial artery in the groove of the pulse (Kelly 1973, cited by Calberg 1977) or, more often, of the radiopalmar branch.
With all ganglia the clinical site of the swelling may be at some distance from its true origin (Fig. 39). Thus, a lesion emerging in the groove of the pulse may derive from a cyst originating much more medially and extending between the flexor tendons and the palmaris longus.

Ganglia of the fingers represent 10% of synovial cysts of the hand and may affect any of the fingers. They appear most often on the flexor sheaths between the metacarpal neck and the shaft of the proximal phalanx (Fig. 40). The middle finger seems to be the most often affected (26 of Wales's 40 cases; Calberg 1977). They are sometimes difficult to detect as they are very small, palpable as a tiny, rounded swelling, tense and sensitive, opposite the palmar-digital crease. Cysts of the metacarpophalangeal pulley may be responsible for a trigger finger. Cysts on the dorsal aspect of the proximal interphalangeal joint extend under the extensor tendon and appear in the dorsolateral region of the finger. The mucoid cysts of the dorsal aspect of the distal interphalangeal joint are regarded by some as a simple variety of synovial cysts. However, they differ in certain features and will be considered in a separate chapter (see p. 60).

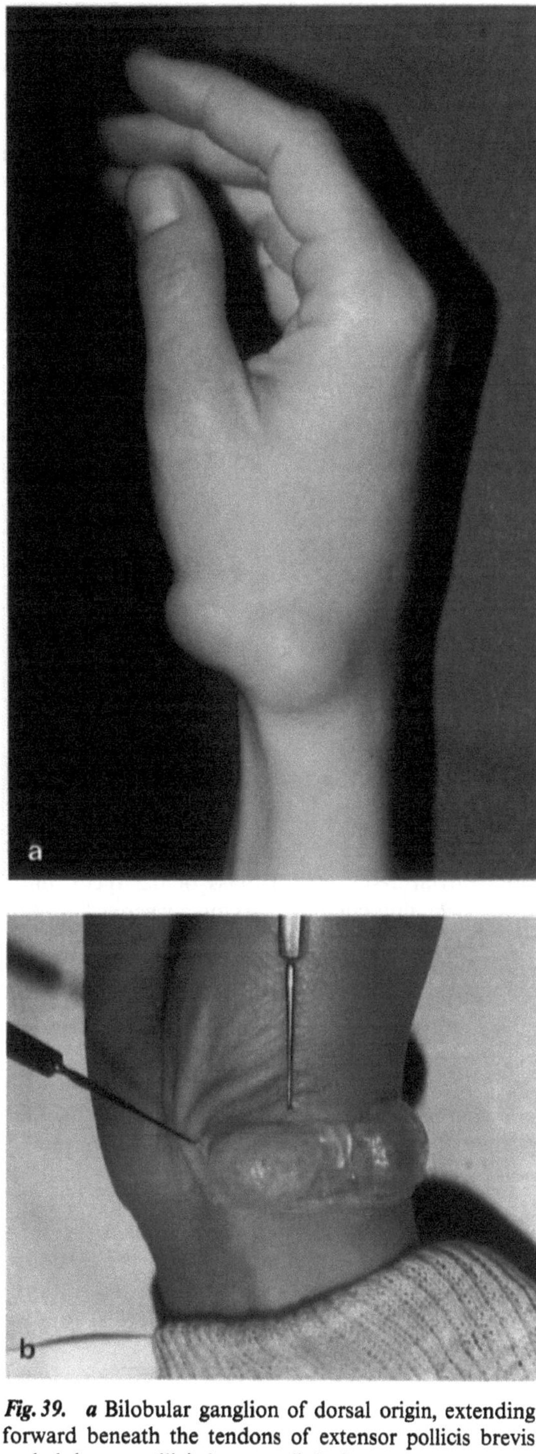

Fig. 39. ***a*** Bilobular ganglion of dorsal origin, extending forward beneath the tendons of extensor pollicis brevis and abductor pollicis longus: clinical aspect. ***b*** Same patient during operation

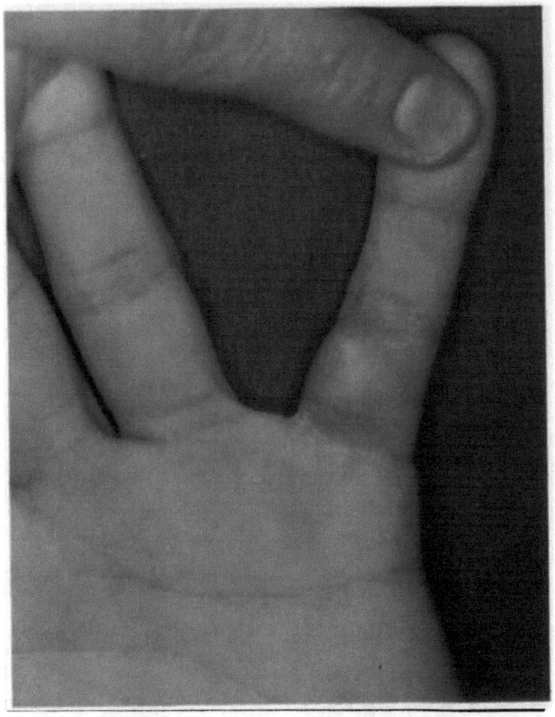

Fig. 40. Ganglion at base of 5th finger

Pathology

The synovial cyst forms a mass of 0.5-6 cm, unilocular or multilocular, containing a viscous "apple-jelly" fluid rich in mucopolysaccharides. The clinically evident swelling represents only part of the lesion, whose course is often complex and whose true origin may be remote from the point of emergence. Angelides and Wallace (1976) state that the dorsal tumours of the wrist arise at the intrinsic ligaments between the lunate and scaphoid, having detected at this site small translucent blebs, intra-articular cysts, constituting the origin of the lesion. Calberg (1977) warns that the morphology of synovial cysts is that of an iceberg, four-fifths immersed, twisting and slipping between the tendons like a "flow of lava" to appear at a distance, thus ensuring the failure of any minimal surgical procedure.

Delarue et al. (1955) in a study of the degenerative lesions of tendinous and aponeurotic tissues, provide a precise description:

Anatomically, these consist of unilocular or multilocular cystic masses, often divided into a series of small cavities by flattened septa with swellings corresponding to the vascular axes and containing a viscous substance. Histologically, these cavities appear to be embedded in tendinous tissue. The content consists of a poorly staining substance, though there is a very chromophile zone adjacent to the walls. Sometimes a histiocytic margin is seen. In the vicinity the tendinous fibers exhibit various changes. Some connective tissue bundles have diminished staining, accompanied by a loss of fibrillary structure and an irregular swelling of the collagen fibers. Sometimes the fibers completely disappear in an amorphous mass, hollowed out by small cavities. Finally, there are other bundles, notably those bounding the cavities, that strongly take up basic dyes.

Based on histochemical studies, Delarue et al. (1955) state that modifications of the basic substance occur continually and are the origin of the formation of the cavities filled with a viscous substance. Chronologically, there would seem to be a degeneration of the matrix producing differentiation of the fibres, accompanied by an accumulation of acid or neutral mucopolysaccharides, followed by a fibrinoid, mucoid or hyaline degeneration. Pierucci (cited by Iselin and Iselin 1967) regards synovial cysts as due to myxoid degeneration of connective tissue, surrounded by a fibrous capsule.

Aetiopathogenesis

The clinical and pathological data tend to support certain aetiopathogenic hypotheses with regard to synovial cysts. These tumours often appear spontaneously outside any pathological context and without known cause. Sometimes there is a history of initial trauma (contusion, continued or repetitive strain), and Calberg (1977) notes that the condition arises more readily in persons unused to hard physical work. The initial hypotheses regarded synovial cysts as hernias of the synovium through the capsular plane (Volkmann, Billroth 1882, cited by Carp and Stout 1928). The work of Carp and Stout (1928) modified this idea. They suggest, on the basis of surgical findings, that the cysts have initially no relation with the articulation and that the communication arises secondarily from irritation and then rupture of the capsular layer. McEvedy (1962) injected contrast medium into 40 synovial cysts and in no case found any spread into the joint. Andren and Eiken (1971), proceeding conversely, injected contrast medium into the wrist joint and looked for diffusion into the synovial cyst. The tumour was opacified in 40 of 59 cases, which verified the existence of a narrow tortuous channel linking it to the joint. The failure of the first experiment is accounted for by the mechanism of an anti-reflux valve permitting joint fluid to fill the cyst but preventing reverse flow. It is the persistence of this channel after incomplete surgical excision that explains recurrence of the cyst. Angelides and Wallace (1976) studied synovial cysts of the dorsum of the wrist under the operating microscope and did not find a constant communication; when it did exist, it was situated between the lunate and scaphoid on the intrinsic ligaments, which they recommended excising.

However, certain points remain unexplained: the high incidence at the scapholunare articulation, the greater number of women affected (⅔ of patients are female) and the age range (20-50 years), which differs from that of degenerative rheumatism. Many authors believe that the initial lesion is a degenerative process in the peritendinous and periarticular connective tissue adjacent to arthrosis. This is an attractive hypothesis, the more so as the wrist is the site of numerous and repeated mechanical constraints, and as the tendons, always in play, are particularly exposed to microtrauma. It should be noted that, in the setting of rheumatoid disease, synovial cysts are common, often multiple, and are situated on the dorsal aspect of the interphalangeal joints and the wrist. In support of a myxoid degeneration of connective and periarticular tissue, Razemon (1983), in a recent study, makes the following observations: the point of origin of the lesion is situated near the zones where the capsule is thick and resistant, whereas a hernia would tend to develop where the capsule is thin or absent; exploration reveals numerous microcysts in the vicinity of the main lesion, and it is their presence that explains the failure of nonsurgical measures and recurrence after inadequate excision; finally, examination of the endo-articular surface of the capsule does not reveal any canalicular orifice.

Treatment

The treatment of this common and benign condition is tricky. There are many methods, all saddled to a variable degree with the risk of recurrence, which is the only real complication of this disorder.

Conservatism may be justified, since spontaneous regression of a permanent nature is seen in about one-third of cases; more often, however, the cyst reappears or persists and becomes a painful nuisance. Compression of the cyst to produce its rupture is an old method; but it is possible in only half the cases, mainly at the back of the wrist and in the radial artery groove, and fails if the ganglion is too hard. Compression is exerted by means of a flat hard object (book, hammer, fingers) and causes sudden disappearance of the lesion. Though recurrence is very common with this method (50%-60% of cases), it is advocated by many authors who maintain that definitive cure can be obtained after the second or third rupture.

Other nonsurgical methods are more debatable. Radiotherapy is formally impermissible. Numerous studies have reported on aspiration of the cyst, with or without injection of various substances: tincture of iodine (Carp and Stout 1928), sodium morrhuate and ethanolamine (McEvedy 1930), hyaluronidase (Risolia, cited in Calberg 1977) and hydrocortisone (Posch 1966). Apart from Holm and Pandey (1973), who claim only 20% recurrences, the results obtained by this technique have not satisfied those who use them. The same applies to transfixion of the ganglion with a nylon thread tied loosely over the skin and moved about every other day, or the introduction under local anaesthesia of a Reverdin's needle to tear the cyst.

Surgical excision of the lesion using a pneumatic

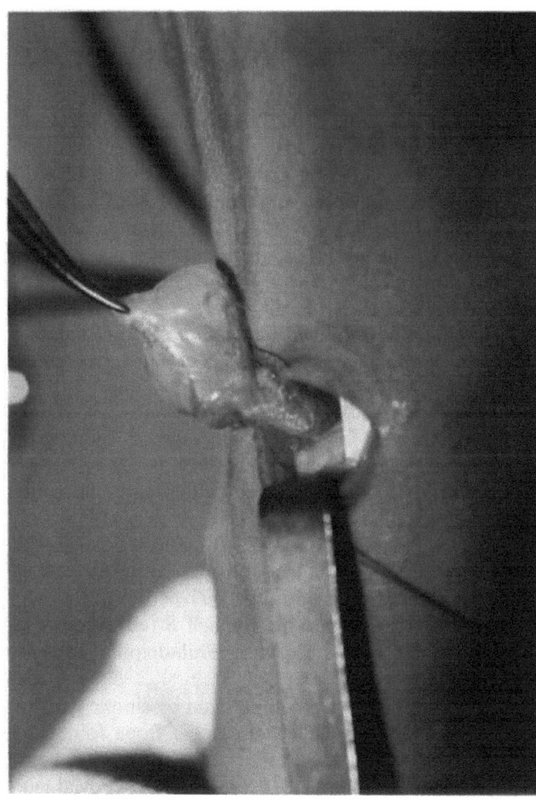

Fig. 41. Ganglion on dorsum of wrist: excision of ganglion and pedicle in their entirety

tourniquet and local-regional or general anaesthesia remains the best treatment. Some use it routinely some when other treatment attempts have failed, when the diagnosis is uncertain or nerve compression exists. It, too, has a risk of recurrence, in 5%-20% of cases). A transverse incision at the back of the wrist and an L-shaped one in the radial groove leaves an almost invisible scar and allows careful dissection with good exposure of the ganglion. Apart from the lesion itself, excision should include the adjacent tissue, especially that in contact with the joint capsule or tendon sheath. The object of the scrupulous dissection is to remove the cyst whole and without rupture (Fig. 41); it is unnecessary to suture the gap in the capsule, which does not cause any subsequent trouble. Razemon (1983) advises a more extensive procedure, combining excision of the pericystic cellular tissue with removal of a wide rim of capsule (2-4 cm diameter), exposing the bony structures, especially over the radius in dorsal ganglia of the wrist. Using this technique, he reports only 3 cases of recurrence in 180 operations, without any joint troubles in the form of instability.

Bibliography

Alter AH, Haber MH (1967) Hand tumors: A review. Hawaii Med J 26: 403-408

Andren L, Eiken O (1971) Arthrographic studies of wrist ganglions. J Bone Joint Surg [Am] 53: 299-302

Angelides AC, Wallace PF (1976) The dorsal ganglion of the wrist: Its pathogenesis, gross and microscopic anatomy, and surgical treatment. J Hand Surg 1: 228-235

Backstrom CG, Linell F, Osterberg G (1965) Cystic, myxomatous adventitial degeneration of the radial artery with development of ganglion in connective tissue. Acta Chir Scand 129: 447

Barnes WE, Larnsen RD, Posch JL (1964) Review of ganglia of the hand and wrist with analysis of surgical treatment. J Plast Reconstr Surg 34: 570-578

Betz J, Firket C (1963) Anatomie pathologique humaine. Desoer, Liège

Bourns HK, Sanerkin NG (1963) Mucoid lesions («Mucoid cysts») of the fingers and toes. Clinical features and pathogenesis. Br J Surg 50: 860-866

Boyes JH (1970) Tumors. In: Bunnell S (ed) Surgery of the hand, 5th edn. Lippincott, Philadelphia Toronto, pp 666-710

Brooks DM (1952) Nerve compression by simple ganglia, a review of thirteen collected cases. J Bone Joint Surg [Br] 34: 391

Bruner JM (1963) Treatment of «sesamoid» synovial ganglia of the hand by needle rupture. J Bone Joint Surg [Am] 45: 1689-1690

Butler ED, Hamill JP, Seipel RS, Delorimier AA (1960) Tumor of the hand. A ten-year survey and report of 437 cases. Am J Surg 32A: 457; 100: 293-302

Byrne JJ (1954) Hand surgery. Am J Surg 88: 470-472

Calberg G (1977) Les kystes synoviaux du poignet et de la main. Acta Orthop Belg 43/2: 212-230

Carl L, Nelson CL, Sawmiller S, Phalen GS (1972) Ganglions of the wrist and hand. J Bone Joint Surg [Am] 54/7: 1459-1464

Carp L, Stout AP (1928) A study of ganglia with especial reference to treatment. Surg Gynecol Obstet 47: 460-468

Constant E, Royer JR, Pollard RJ, Larsen RD, Posch JL (1969) Mucous cysts of the fingers. Plast Reconstr Surg 43: 241

Cooley SGE (1964) Tumors of the hand and forearm. In: Converse JM (ed) Reconstructive plastic surgery. Saunders, Philadelphia London, pp 1740-1786

Cornet L, Loubiere R (1974) Les tumeurs des parties molles de la main et des doigts. J Chir (Paris) 108/1, 2: 143-154

Coventry MB, Polley HF, Weiner AD (1959) Rheumatoid synovial cyst of the hip: Report of three cases. J Bone Joint Surg [Am] 41: 721

Dao L (1964) A new method of treatment of ganglions and synovial cysts. J Occup Med 6: 217-220

David G, Palmer DG (1969) Synovial cysts in rheumatoid disease. Ann Intern Med 70/1: 61-68

Delarue J, Laumonier P (1959) Anatomie pathologique. Flammarion, Paris

Delarue J, Mignot J, Payen J, Roussel A (1955) Les lésions dégénératives des tissus tendineux et aponévrotiques. Presse Med 63/30: 607-610

De Orsay RH, Mecray PM, Ferguson LK (1937) Pathology and treatment of ganglion. Am J Surg 36: 313

Derbyshire RC (1966) Observations on the treatment of ganglia. With a report on hydrocortisone. Am J Surg 112: 635-636

Diebold R, Camillerie P, Reynes C, Callard C (1978) Anatomie pathologique générale. Baillière

Dixon ASJ, Grant C (1964) Acute synovial rupture in rheumatoid arthritis. Clinical and experimental observations. Lancet I: 742

Eiken O (1971) Arthrographic studies of wrist ganglions. J Bone Joint Surg [Am] 53: 299

Galsford JC (1966) Tumors of the hand. Surg Clin North Am 40: 549-566

Glicenstein J (1977) Tumeurs des parties molles de la main. Ann Chir 31/12: 1087-1101

Haber MH, Alter AH, Wheelock MC (1965) Tumors of the hand. Surg Gynecol Obstet 121: 1073-1080

Holm PCA, Pandey SD (1973) Treatment of ganglia of the hand and wrist with aspiration and injection of hydrocortisone. Hand 5: 63-68

Hvid-Hansen O (1970) On the treatment of ganglia. Acta Chir Scand 136: 471-476

Iselin M, Iselin F (1967) Tumeurs de la main. Traité de chirurgie de la main. Flammarion, Paris, pp 739-769

Kendall TE, Robinson D, Masters F (1969) Primary ma-

lignant tumors of the hand. Plast Reconstr Surg 44/1: 37-40
Kleinert HV (1972) Etiology and treatment of the so-called mucous cyst of the finger. J Bone Joint Surg [Am] 54: 1455-1458
Lyle FM (1941) Radiation of ganglia of the wrist and hand. J Bone Joint Surg 23: 162
Mason ML (1937) Tumors of the hand. Surg Gynecol Obstet 64: 129-148
Matthews PH (1953) Ganglia of the flexor tendon sheaths in the hand. J Bone Joint Surg [Br] 55: 612-617
Matton G, Defloore E (1974) Benign tumors of the hand. Acta Orthop Belg 40: 172-198
McEvedy BV (1954) The simple ganglion. Lancet 266: 135-136
McEvedy BV (1962) Simple ganglia. Br J Surg 49: 586-594
McEvedy P (1930) Treatment of the simple ganglion by injections. Lancet II: 902
Nelson CL, Sawmiller S, Phalen GS (1972) Ganglions of the wrist and hand. J Bone Joint Surg [Am] 54: 1459-1464
Pack GT (1939) Symposium, tumors of the hands and feet. Surgery 5: 1
Posch JL (1956) Tumours of the hand. J Bone Joint Surg [Am] 38: 517-539
Posch JL (1966) Soft tissue tumors of the hand and surgery by J.E. Flynn. Williams & Wilkins, Baltimore, pp 999-1026
Razemon JP (1983) Traitement chirurgical des kystes dits synoviaux du poignet avec exérèse capsulaire. A propos de 300 observations. Ann Chir Main 2/3: 230-243
Richmond AD (1968) Carpal ganglion with ulnar nerve compression. J Bone Joint Surg [Br] 45: 513-515
Sames CP (1954) The simple ganglion. Lancet 266: 317
Sarpyener MA, Oscurumez O, Seyhan F (1968) Multiple ganglion of the tendon sheaths. J Bone Joint Surg [Am] 50: 985
Seddon HJ (1952) Carpal ganglion as a cause of paralysis of the deep branch of the ulnar nerve. J Bone Joint Surg [Br] 34: 386
Smith EB, Skipworth GB, Vonder Ploeg DE (1964) Longitudinal grooving of nails due to synovial cysts. Arch Dermatol 89: 364-366
Stack G (1960) Tumours of the hand. Br Med J I: 919-922, 517
Stack RE, Bianco AJ, MacCarty CS (1965) Compression of the common peroneal nerve by ganglion cysts. J Bone Joint Surg [Am] 47: 773
Stevenson TM (1966) Carpal tunnel syndrome. Proc R Soc Med 59: 824
Soren A (1966) Pathogenesis and treatment of ganglion. Clin Orthop 48: 173-179
Topho JK, Henriques U (1971) Ganglion on the wrist, a structure developed from the joint. Acta Orthop Scand 42: 244-250
Trevaskis AE, Tilly D, Marcks KM, Heffermann AH (1967) Loss of nerve function in the hand caused by ganglia. Plast Reconstr Surg 40: 39-97
Wallace PF (1973) The Ubiquitous Ganglion. Lancet I/794: 88-89
Weber B (1954) The simple ganglion. Lancet 266: 265
Woods JE, Murray JE, Vawter GF (1970) Hand tumors in children. Plast Reconstr Surg 46/2: 130-137
Zachariae L, Vibe Hansen H (1973) Ganglia. Recurrence rate elucidated by a follow-up of 347 operated cases. Acta Chir Scand 139: 625-628

Mucous Cysts

Mucous cysts, originally described by Hyde in 1883, are relatively uncommon benign tumours, also known as mucinous cysts of the fingers, synovial pseudocysts of the skin or periungual cysts. Although considered by some (Newmeyer et al. 1974) as a simple variety of synovial cysts, they exhibit a certain number of peculiarities that justify separate study.

Clinical Features

This lesion is situated on the dorsum of the distal finger phalanx between the terminal interphalangeal joint and the base of the nail, usually near the nail (Bourns and Sanerkin 1963). The cyst is rarely in the midline, more often at one side. It presents as a rounded or oval lesion of firm consistence, from 3 to 12 mm in diameter [average 6-7 mm, the size of a small pea (Fig. 42)]. Usually painless, it is translucent, vesicular, well-circumscribed and covered with normal or erythematous skin. Unlike the synovial cysts, it is adherent to the skin and remains mobile in relation to the underlying structures. It contains a yellow, clear, viscous fluid of gelatinous appearance. It develops slowly and may sometimes regress spontaneously or even disappear, but later recurrence is the rule. Often the spontaneous course is towards sinus formation; the overlying skin thins and ulcerates, allowing the gelatinous fluid to well out (Fig. 43). The cyst then subsides but reforms in 3-4 weeks, when it is less translucent. Mucoid cysts are seen mainly between the ages of 40 and 70, with a maximum around 50 years. According to the figures of Constant et al. (1969) two of every three cases occur in women (47 women in 69 patients). The lesion is usually solitary, but double or triple digital localisation is possible (Fig. 44). There is a marked preference for the right hand, with a greater incidence on the middle and index fingers. Very often there are associated ungual lesions in the form of longitudinal striations, sometimes preceding the development of the cyst. Finally, in 77% of the reported cases (Constant et al. 1969), there are changes suggestive of osteoarthrosis of the distal interphalangeal joint and Heberden's nodes.

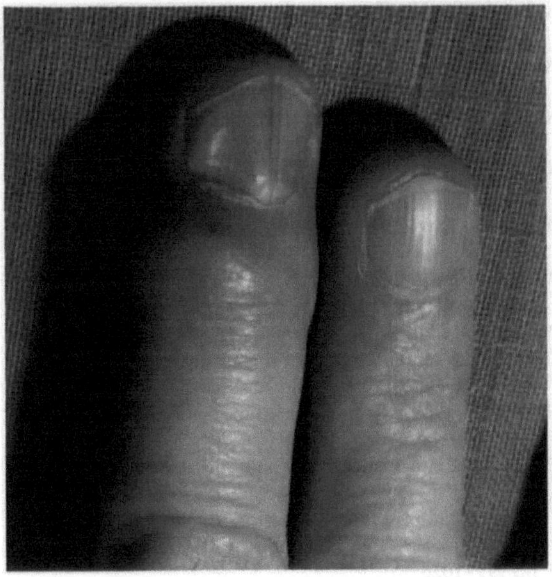

Fig. 42. Typical mucous cyst: note deformation of nail

Differential Diagnosis

A positive diagnosis of mucous cysts is usually easy in view of the site and clinical features of the lesion. Epidermal cysts are more likely to be situated on the palmar aspect of the fingers. Heberden's nodes are firm, painless, subcutaneous and composed of fibrous tissue, but they are quite frequently associated with mucous cysts. Synovial cysts themselves, apart from the nosological problems they pose, are usually mobile in relation to the superficial layers.

Histology

Histologically, mucoid cysts have special features. They present as a cavity hollowed out and developed in the dermis, resulting from the coalescence of numerous small cavities. Apart from the disappearance of the elastic fibres, a characteristic feature is the presence of a loose, amorphous, mucoid matrix, as well as an abundance of fibroblasts of various shapes. Above all, there exists no specific wall to

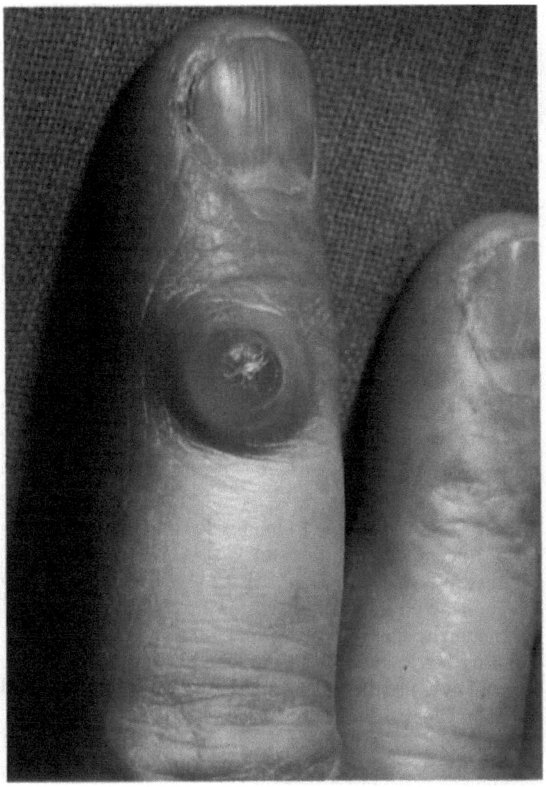

Fig. 43. Advanced mucous cyst about to break down

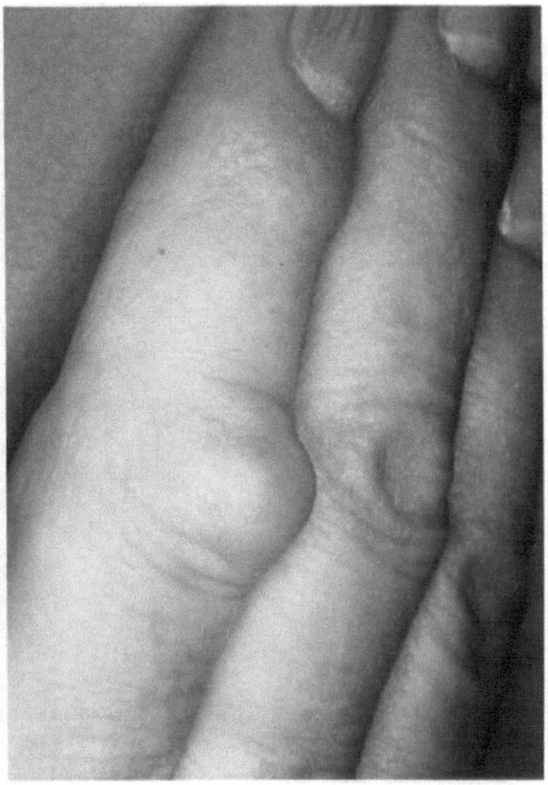

Fig. 44. Mucous cyst: rare location on dorsum of proximal interphalangeal joint

the lesion, at most a false capsule at the periphery indicative of compression of the connective tissue. Thus, this is not a true cyst but a pseudocyst. It is associated with epidermal changes of the nature of hyperkeratosis or atrophy. Inflammatory changes are moderate or even absent.

Aetiopathogenesis

This remains very debatable. Whereas Cooley (1964) considers that mucous cysts have the same structure as synovial cysts, Posch (1956) and Boyes (1970) make a sharp distinction in view of the absence of any proper wall. Some authors report a traumatic origin, but this is not at all regularly found. Gross (1937) and Woodburne (1947) hypothesise a mucoid degeneration of the dermis. King (1951), on the basis of the microscopic study of fibroblasts, considers that there is no degenerative process but an active metabolic secretion. This is supported by the work of Johnson et al. (1965), who allege a hypersecretion of hyaluronic acid by the fibroblasts, with diminution of the collagen framework. Bourns and Sanerkin (1963), from their observation of unicellular prolongations surrounded by mucoid material, ally these lesions to hidradenitis and attribute a positive role to the sweat glands. More recently, Newmeyer et al. (1968) showed clearly in 20 cases a constant communication between the cyst and the distal interphalangeal joint. A solution of dilute methylene blue injected into this joint was recovered from the cyst. Kleinert et al. (1972) report similar findings, confirmed at operation. Finally, Eaton et al. (1972) arrive at similar conclusions and accord periarticular exostoses a primary role, justifying their routine ablation during treatment of the cysts.

Treatment

This cannot be regarded as essential, given the benign nature of the tumour, which is tolerated well over long periods. However, treatment is often desired by the patient as the cyst may become painful, ulcerated or infected and constitutes a cosmetic blemish (Fig. 45). Numerous methods have been

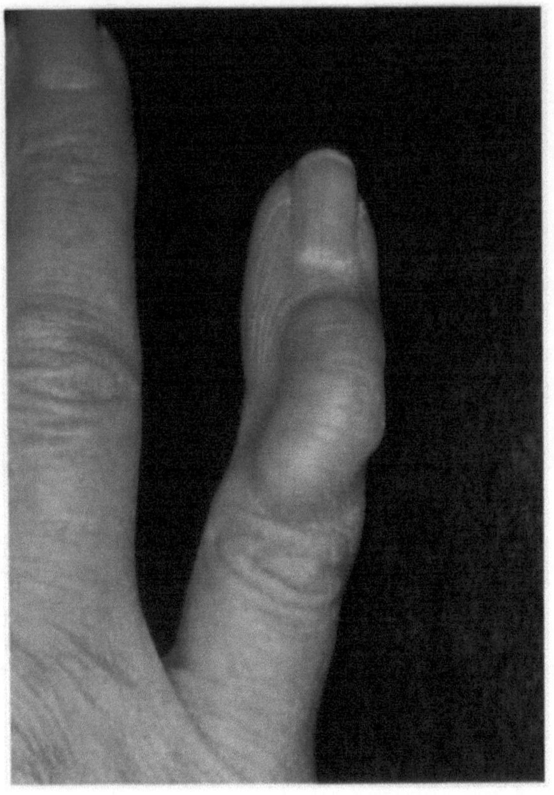

Fig. 45. Large mucous cyst with proximal extension

employed, all marked by a varying incidence of recurrence. Spontaneous disappearance of the cysts, or their rupture after injury or infection, is usually followed by recurrence. Radiotherapy, as proposed by Gross (1937) and Hultberg (cited by Arner et al. 1956), induces a cicatricial fibrous process, but it carries the risk of radiodermatitis and is formally to be prohibited. Puncture of the cyst followed by injection of hyaluronidase or cortisone is advised by many authors; two or three injections at intervals of 1 or 2 weeks are sometimes necessary to obliterate the cyst. The incidence of recurrence is considerable and justifies abandonment of this technique. The same applies to diathermy with curettage of the lesion or ablation of the cyst by simple incision.

Most authors are agreed on the following plan of treatment. Under local-regional anaesthesia or plexus block, with a pneumatic tourniquet and the aid of an operating microscope, the lesion is excised widely and completely, removing the related skin, the entire cyst and the pedicle connecting it to the joint. The dissection must be made carefully with due regard for the integrity of the nail bed or the termination of the extensor tendon. Direct skin suture is impossible in view of the loss of skin substance. Monitored cicatrisation may be used but requires attention for several weeks. Some authors use a rotation flap, but most often a full-thickness skin graft is employed. Kleinert et al. (1972) advise, in addition, synovectomy of the terminal joint and routine ablation of the osteophytes, followed by splint immobilisation for 3 weeks. Constant et al. (1969) report 25% recurrences in 51 patients treated by simple excision of the lesion, and 3% after wide excision and full-thickness skin graft.

Bibliography

Arner O, Lindholm A, Romanus R (1956) Mucous cysts of fingers. Acta Chir Scand 3: 314-321

Barnes WE, Larsen RD, Posch JL (1964) Review of ganglion of the hand and wrist with analysis of surgical treatment. Plast Reconstr Surg 34: 570

Bourns HK, Sanerkin NG (1963) Mucoid lesions (Mucoid cysts) of the fingers and toes. Clinical features and pathogenesis. Br J Surg 50/226: 860-866

Boyes JH (1970) Tumors. In: Bunnell S (ed) Surgery of the hand. 5th edn. Lippincott, Philadelphia Toronto, pp 666-710

Constant E, Roger JR, Pollard RJ, Larsen RD, Posch JL (1969) Mucous cysts of the fingers. Plast Reconstr Surg 43: 241-246

Cooley SGE (1964) Tumors of the hand and forearm. In: Converse JM (ed) Reconstructive plastic surgery. Saunders, Philadelphia London, pp 1740-1786

Eaton RG, Dobranski AI, Littler JW (1972) Marginal osteophyte excision in the treatment of mucous cyst. 27th Annual Meeting American Society Surgery Hand, Washington

Engle RB (1958) The treatment of myxomatous cutaneous cysts. Radiology 71: 93-95

Errikos C, Royer JR, Pollard RJ, Larsen RD, Posh JL (1969) Mucous cysts of the fingers. Plast Reconstr Surg 43/3: 241-246

Gross RE (1937) Recurring myxomatous cutaneous cysts of fingers and toes. Surg Gynecol Obstet 65: 289-302

Hyde JN (1883) Practical treatise on diseases of skin for use of students and practitioners. Lear, Philadelphia

Isaacson NH, McCarty D (1954) Recurring myxomatous cutaneous cyst. Surgery 35: 621-623

Johnson WC, Graham JH, Helwig EB (1965) Cutaneous myxoid cyst. A clinicopathological and histochemical study. JAMA 191/1: 15-20, 109-114

King ESJ (1951) Mucous cysts of skin and subcutaneous tissues. Aust NZ J Surg 21: 122-129

Kleinert H, Kutz J, Fishman J, McCraw L (1972) Etiology and treatment of the so-called mucous cyst of the finger. J Bone Joint Surg [Am] 54/7: 1455-1458

Newmeyer WL, Kilgore ES, Graham WP (1974) Mucous cysts: The dorsal distal interphalangeal joint ganglion. Plast Reconstr Surg 53/3: 313-315

Posch JL (1956) Tumors of the hand. J Bone Joint Surg [Am] 38: 517-539

Smith EB, Skipworth GB, Vander Ploeg DE (1964) Longitudinal grooving of nail due to synovial cysts. Arch Dermatol 89: 364-366

Woodburne AR (1947) Myxomatous degeneration cysts of skin and subcutaneous tissues. Arch Dermatol 56: 407-418

Giant-Cell Tumour

This, the most common of the tumours of the hand after the synovial cyst, still forms the subject of an unresolved terminological and aetiological debate. The first case reported by Chassaignac in 1852 was regarded as a malignant tumour of the tendon sheaths, as was then the case for all neoplasms containing giant cells. In 1857, Wells described their gross and microscopic features. The work of Heurtaux (1891) definitively established the benign nature of the condition, and the name of "myeloma of the tendon sheaths" was proposed in view of the presence of large cells analogous to myeloplaxes. Targett (1897) preferred to speak of "giant-cell tumours of the tendon sheaths" and Dor (1898), attributing a common origin to the two types of cells found (foam cells and giant cells) employed the term "myeloxanthoma". The list below gives the terms used to designate this affection, which was also, despite the work of Pinkus and Pick (1908, cited by Masson and Woolsto 1927), long considered as part of diffuse xanthomatosis.

Cancer of the tendon sheaths (Chassaignac 1852)
Xanthoma
Xanthogranuloma
Fibrous xanthoma
Myeloma of the tendon sheaths (Heurteaux 1891)
Myeloxanthoma (Dor 1898)
Giant-cell tumour (Targett 1897)
Villous arthritis (Dowd 1912)
Benign synovioma with giant cells
Pigmented villonodular synovitis (Jaffe et al. 1941)
Benign synovioma (Stewart 1948)

The term "pigmented nodular synovitis", introduced by Jaffe et al. in 1941, is currently that most widely employed in the English language literature. It was chosen by analogy with the disorder of the same name occurring in the knee and by the establishment of diffuse forms situated mainly at the knee but also at other articulations. It is a tumour common in the hand: 69 of 679 lesions (10%, Posch 1956); 52 of 394 tumours (13%, Bayes 1970). Reporting on 118 cases of giant-cell tumour, Jones et al. (1969) gives a maximum incidence between 40 and 60 years, with a range of 14-77 years. There was a clear female predominance (77 cases, 60%) with the following distribution by site: 91 in the fingers, 14 at the knee, 4 on the palm and dorsum of the hand, 4 at the wrist, 1 at the hip and 4 on the feet. Fyfe and MacFarlane (1980), in 51 cases of pigmented villonodular synovitis, likewise found a clear female predominance (2:1) over an age range of 20-60 years. The tumours were more often palmar (35 cases) than dorsal (16) and occurred mainly on the index (14) and medius (13), rarely at the thumb (2). The lesions developed most often opposite a distal interphalangeal (17) or metacarpophalangeal (13) articulation; more rarely, they occurred near a proximal interphalangeal joint (6), a carpometacarpal joint (3) or the wrist (2). Sherry and Anderson (1955), reporting on 34 cases of giant-cell tumour seen in patients between 18 and 70 years of age, found a very definite predominance in the fingers (30), the other cases constituting 1 in the palm and 3 in the foot.

Clinical Features

Xanthoma presents as a small swelling 5 mm-3 cm in length, lobulated, firm and laterally mobile; it is covered by normal skin and sometimes fixed in depth. It is painless, has usually existed for several months and grows progressively; sometimes, however, there is a rapid increase in size, often accompanied by pain. Usually solitary and adjacent to a distal interphalangeal joint (Fig. 46), it may be situated on any aspect of the finger and only rarely produces any marked disability (Fig. 47). A case of compression of the median nerve at the wrist was reported by Fyfe and MacFarlane (1980). Xanthoma may develop quite apart from any lipid abnormality, and none of the reported cases was associated with significant hypercholesteraemia.

Radiography

Radiographical examination reveals a tumour of the soft parts of the hand without intratumoral calcification. Erosion of the bone in contact with the tumour is common, as are degenerative lesions of the adjacent articulations.

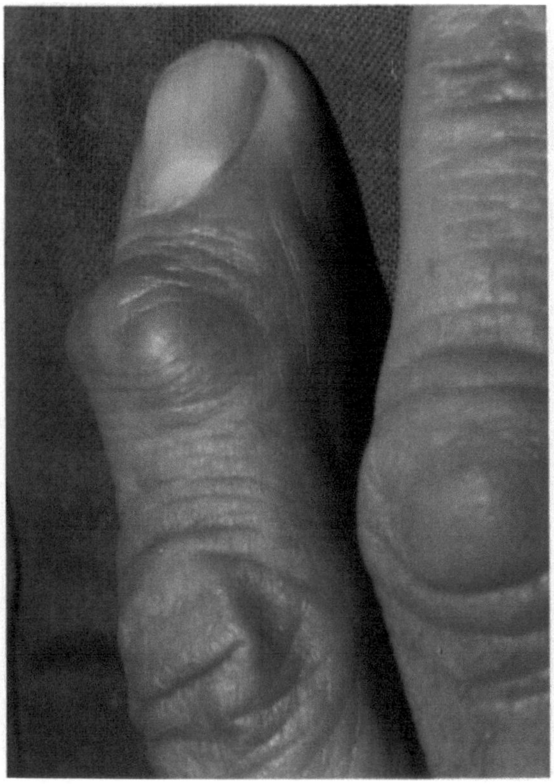

Fig. 46. Giant-cell tumour on dorsum of distal interphalangeal joint

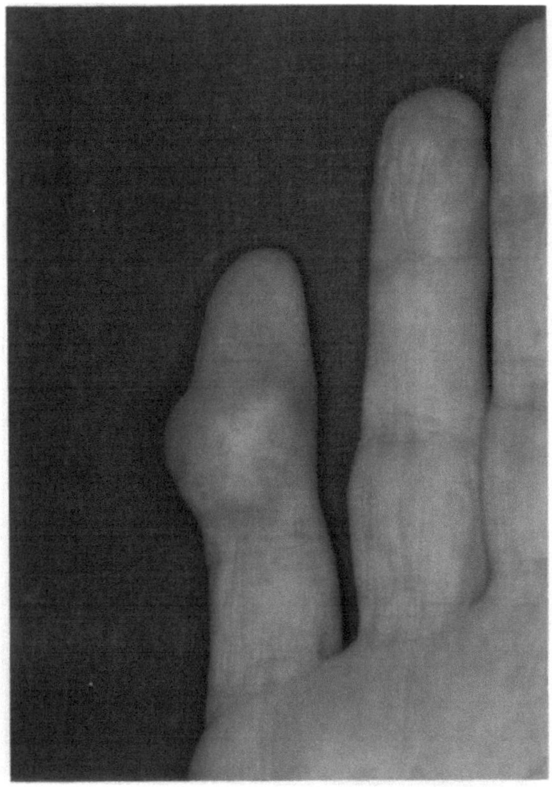

Fig. 47. Large giant-cell tumour on palmar aspect of finger

Pathology

Gross examination of the specimen often suffices for diagnosis. The lesion, which does not exceed 3 cm, is incompletely encapsulated, firm, fibrous and composed of lobulated nodules (Fig. 48). The grey-pink or buff colour of the tumour, with darker stripes, is suggestive, and sectioning shows fine trabeculations derived from the capsule. Finally, the tumour is often seen to adhere to the synovial capsule or the tendon sheath.

Microscopic investigation confirms that the lobulation observed is formed by spans of connective tissue compressed around the tumour and constituting an incomplete pseudo-capsule. Numerous cellular elements occur in a loose connective tissue rich in reticulin. The polyhedral cells with a large round nucleus and eosinophilic cytoplasm are macrophages; they contain brown-pigmented deposits of haemosiderin. The multinucleated giant cells (4-6 nuclei) resemble osteoclasts (myeloplaxes) or the cells found in foreign body reactions. Finally, the foam cells or xanthoma cells, usually mononucleated, derive from histiocytes and contain lipid inclusions. The distribution of the cell sheets within the connective tissue of the tumour is irregular, and in older lesions there is a decreased cellular reaction and an increase in the fibrous component. The extracellular tissue contains deposits of haemosiderin and fatty inclusions.

Aetiopathogenesis

The origin of xanthomata has given rise to numerous hypotheses. The metabolic cause proposed by some is not confirmable biochemically: the lesion does not occur in the context of a primary xanthomatosis, and there is no significant disturbance of lipid metabolism. Geschickter and Copeland (1949) found sesamoids frequently in relation to these tumours and regarded these accessory bones as the origin of the tumoral process. Young and Mudaceck (1954) allege that haemorrhage plays a part in the production of lesions of the synovial membrane. Fleissig (1913) suggested an inflammatory origin, an idea taken up by Jaffe et al. (1941)

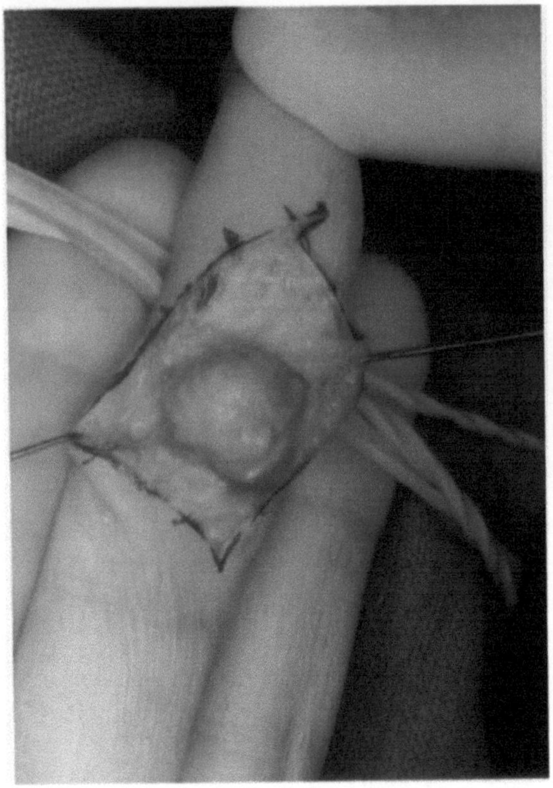

Fig. 48. Giant-cell tumour: gross appearance

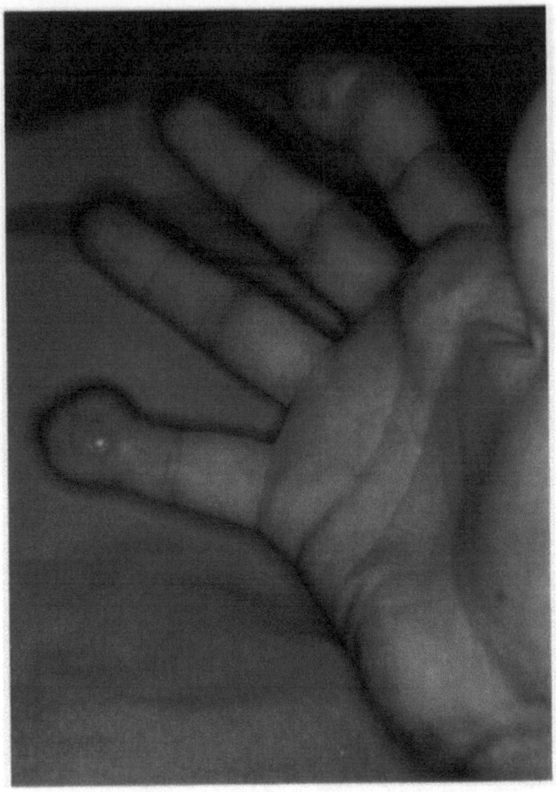

Fig. 49. Familial hypercholesterolaemic xanthomatosis: involvement of the hand

who considered that it was a reaction to trauma. They noted the structural analogy between fibrous xanthomata of the fingers and pigmented nodular synovitis of the knee and regarded these as two manifestations of the same disorder. Jones et al. (1969) pointed out the frequent association with degenerative traumatic lesions of the adjacent joints and the presence in the xanthomata of the same cells as in the synovium, arguing that the tumoral process is a reaction to the articular disorder (accumulation of histiocytes and giant cells). Experimentally, the injection of blood into the knee-joint of a dog produced a villous nodular hyperplasia, but no foam cells with lipid inclusions were found.

Sherry and Anderson (1955) state that thickening of the synovial membrane is the first lesion observed and is due to an increase in the size and number of cells containing pigmented inclusions probably derived from the blood. The villous processes would seem to result from a proliferation of synovial tissue composed of a loose, pigmented, connective tissue, identical with that found in the tendon sheath lesions. The tumour would then result from coalescence of these villosities, whose growth would induce the formation of a pseudo-capsule. Eisenstein (1968) studied the structure of the synovium and found two layers of cells, macrophages with numerous vacuoles superficially and fibroblasts deeply. These two types of cells can be recognised in electron microscopic studies of xanthomata. The giant cells would arise by fusion of the macrophages and the foam cells from the intracellular accumulation of lipids in the two types of cells. Reginato noted the frequency of xanthoma in rheumatoid disease, and Fyfe and MacFarlane (1980) saw this as due to a common chronic inflammatory origin. However, Hirohata (1968) thought that the inflammatory process was not primary, but secondary to a local accumulation of cholesterol. It may be concluded that, even if the precise aetiology is not clarified, most authors regard this disorder as the nodular and localised form of an inflammatory pathology that is probably traumatic and may also be expressed in a diffuse, multiple and intra-articular form.

Differential Diagnosis

Clinically, fibrous xanthomata are sometimes difficult to distinguish from other tumours of the soft parts of the hand (ganglion, epidermoid cyst, lipoma, fibroma, neurofibroma, sebaceous cyst). It is usually easier to exclude a rheumatoid nodule, the nodule of Dupuytren's disease or an inflammatory granuloma. Finally, it is useful to separate fibrous from "true" xanthomata, benign dysmetabolic lesions forming one component of a hereditary disorder with major hypercholesteraemia, i.e. the essentially familial hypercholesteraemic xanthomatosis. This comprises cutaneous lesions (tuberous or layered xanthomata, yellowish macules on the dorsum of the hands, xanthelasma of the eyelids), a corneal arc, tendon lesions (solitary or more often multiple tumours affecting the fingers; Fig. 49) and is associated with nodules at the elbows, knees, Achilles tendons and toes. The most characteristic histological features of the condition are the Touton giant cells, large cells with numerous nuclei arranged in central rings separating the peripherally vacuolated cytoplasm. These tumours are usually unpleasant, and their size may produce dysfunction. Complete excision is impossible as the tumour infiltrates the skin and deeper layers, without capsule or definite boundaries. The xanthomatous mass adheres to the tendons and surrounds the neurovascular bundles, which may be damaged if not previously isolated. Wide excision of the skin is often unavoidable. Local recurrences are as frequent as new lesions elsewhere. Nevertheless, an operation considerably improves the appearance and function of the affected fingers.

Histology

The histological features may lead to the erroneous diagnosis of rheumatoid synovitis. Byers et al. (1968) report four such mistakes, and it is best to exclude a synovial sarcoma in order to avoid an extensive and useless surgical procedure. Fibrous xanthoma is essentially a benign lesion which, even if it sometimes exhibits features of "invasion" of adjacent structures, never gives rise to metastases, never undergoes malignant transformation and never presents histological features characteristic of malignant tumours.

Treatment

This is entirely surgical. The operation must be scrupulous, as the tumour often presents clinically

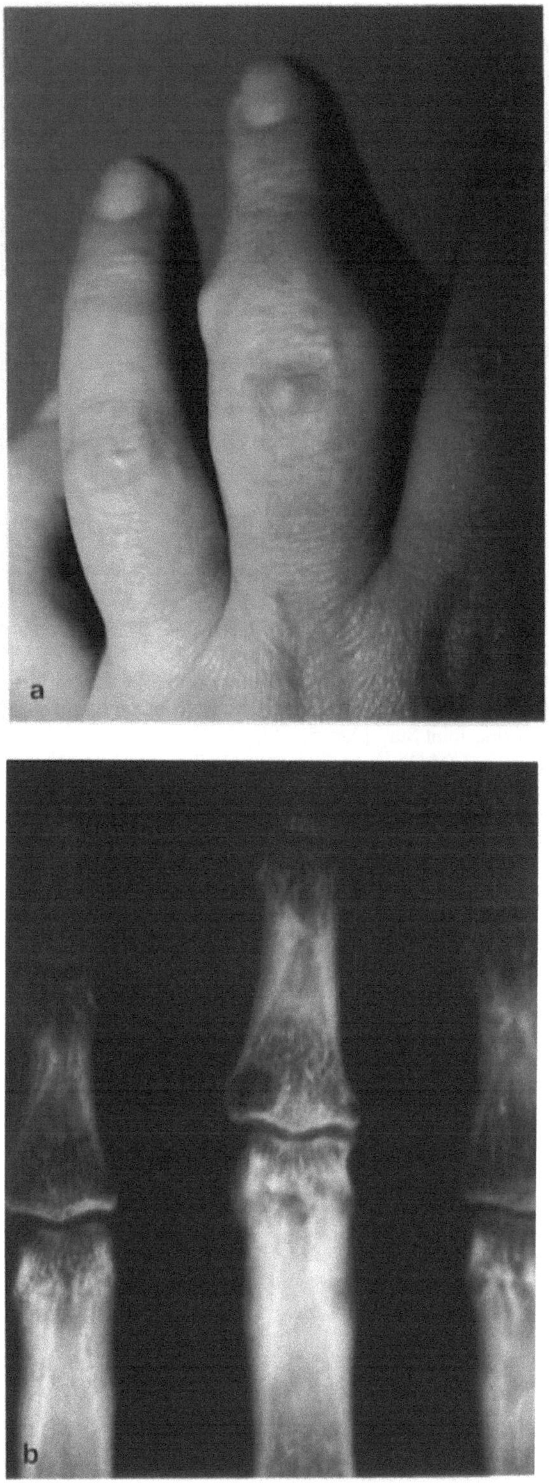

Fig. 50. ***a*** Giant-cell tumour: second recurrence after incomplete excision, clinical aspect. ***b*** Same patient: bony invasion during second recurrence

unsuspected extensions, particularly on the deep aspect of the tendon. Complete excision of the lesion is best done under local-regional anaesthesia and with a pneumatic tourniquet. If the excision is inadequate, it becomes the source of recurrences that may invade the adjacent structures (including the bone) and make subsequent operations more hazardous (Fig. 50a, b).

Bibliography

Boyes JH (1970) Tumors. In: Bunnell (ed) Surgery of the hand, 5th edn. Lippincott, Philadelphia

Byers PD, Cotton RE, Deacon OW, Lowy M, Nauman PH, Sissons HA, Thompson AP (1968) The diagnosis and treatment of pigmented villonodular synovitis. J Bone Joint Surg [Br] 50: 290

Chassaignac M (1852) Cancer de la gaine des tendons. Gazette des hopitaux civils et militaires, p 185

Dor L (1898) Relations des tumeurs à myéloplaxes et des xanthomes. Congrès Chir 12: 553

Dowd CN (1912) Villous arthritis of the knee. Ann Surg 56: 363

Eisenstein R (1968) Giant cell tumor of tendon sheath. Its hystogenesis as studied in the electron microscope. J Bone Joint Surg [Am] 50: 476

Fleissig J (1913) Über die bisher als Riesenzellensarkome bezeichneten Granulationsgeschwülste der Sehnenscheiden. Dtsch Z Chir 122: 239-265

Fyfe IS, MacFarlane A (1980) Pigmented villonodular synovitis of the hand. Hand 12/2: 179-188

Geschickter CF, Copeland MM (1949) Giant cell tumors of tendon sheath tumors of bone. Lippincott, Philadelphia, pp 353-362

Heurtaux MA (1891) Myélome des gaines tendineuses. Arch Gén Med 1: 40-54

Hirohata K (1968) Light microscopic and electron microscopic studies of individual cells in pigmented villonodular synovitis and bursitis. Kobe J Med Sci 14: 251-279

Jaffe HL, Lichtenstein L, Sutro CJ (1941) Pigmented villonodular synovitis, bursitis and tenosynovitis. Arch Pathol 31: 371

Jones FE, Soule EH, Coventry MB (1969) Fibrous xanthoma of synovium - a study of 118 cases. J Bone Joint Surg [Am] 51: 76-86

Martens M, Tanghe W, Muller JC, Willems D (1972) Pigmented villonodular synovitis of joints, tendons and bursae. Acta Orthop Belg 38/2: 233

Masson ML, Woolsto WH (1927) Isolated giant cell panthomatic tumors of the fingers and hand. Arch Surg 15: 499-529

Posch JL (1956) Tumors of the hand. J Bone Joint Surg [Am] 38: 517-539

Sherry JB, Anderson W (1955) The natural history of pigmented villonodular synovitis of tendon sheaths. J Bone Joint Surg 37: 1005

Stewart MJ (1948) Benign giant cell synovioma and its relation to «xanthoma». J Bone Joint Surg 303: 522

Targett JH (1897) Giant celled tumors of the integuments. Trans Pathol Soc Lond 48: 230-235

Wells S (1857) Tumor from flexor tendon of forefinger. Trans Pathol Soc Lond 6: 379-380

Young JM, Mudaceck AG (1954) Experimental production of pigmented villonodular synovitis in dogs. Am J Pathol 30: 799

Synovial Sarcoma

Stuer in 1893, Hardie and Salter in 1894 and Marsh in 1898 (cited by Mackensie 1966) described a malignant tumour of articular origin that was defined as a clinical and pathological entity by Smith in 1927. The term of synovial endothelioma used by Lejars and Rubens-Duval (1910) was replaced by that of synovialoma (Sabrazes et al. 1934) and then by the less ambiguous synovial sarcoma. Arising from some synovial component (articular, serous bursa, tendon sheath), this malignant tumour is rare, but its localisation in the feet and hands is common enough to constitute 10% of sarcomata of the soft parts of the extremities. Its slow growth, initially local and then extending distally over a varying period, is always associated with a poor prognosis.
There are numerous studies in the world literature:

Haagensen and Stout (1944): 104 cases, 3% in the hand
Pack and Ariel (1950): 60 cases
Wright (1952): 58 cases, 7% in the hand
Cadman et al. (1965): 134 cases, 14% in the hand
MacKensie (1966): 49 cases
Cameron et al. (1974): 39 cases, 6% in the hand

This tumour may arise at any age but is a particular feature of the young adult and is found more frequently in males (60%). The most common sites, in decreasing order, are the hip, knee, foot and hand. Cadman et al. (1965) found 71% of the lesions in the lower limb and 24% in the upper, of which 45% (14 cases) were on the hand and wrist. The fingers are very rarely affected.

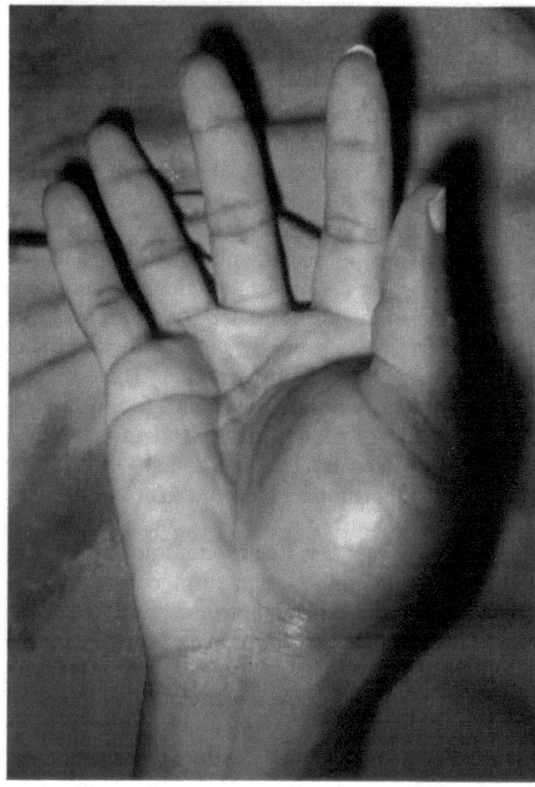

Fig. 51. Advanced synovial sarcoma of the thenar eminence. (Courtesy of Professor Génin)

Clinical Features

Clinically, synovial sarcoma presents most often as a solitary para-articular swelling, often small at the outset and enlarging only slowly, and also frequently neglected as it does not incommode the patient. This swelling, which may be superficial or deep, is usually firm, smooth, regular and painless, sometimes resembling a ganglion. It is covered by normal skin which is mobile over the tumour. Sometimes, however, there is a rapid course over a few months, producing a bulky tumour situated on the palm of the hand, wrist or finger, causing pain and dysfunction (Fig. 51). Again, the signs may be so trivial that the tumour is only detected in the course of clinical exploration for a change in general well-being or the discovery of a metastasis - nodal, pulmonary, bony, cerebral or cutaneous.

Radiology

Radiologically, it is not uncommon to find intratumoral calcification (40% of cases), while the destruction of adjacent bony structures is more the result of advanced or rapidly growing lesions.

Histology

Only histological examination of the lesion can confirm the diagnosis. It must be realised that there exist numerous structural variations, not only from one case to another but also in different parts of the same tumour, so that it is essential to examine the entire lesion. Macroscopically, synovial sarcoma forms a whitish mass of fibrous appearance, not encapsulated despite being apparently circumscribed. This appearance may be modified by changes due to necrosis or haemorrhage.

Microscopically, there are two different types of cells. There are the epithelioid cells, grouped in clumps, mutually adherent, and revealed by electron microscopy as having microvillosities and a cytoplasm rich in endoplasmic reticulum and ribosomes. There are also smaller pseudofibroblastic cells, separated by a ground substance rich in collagen and suggestive of fibrosarcoma. Thus, the tumour combines two different tissues found in variable proportion within a dense and more or less differentiated stroma. This multiplicity of possible histological appearances accounts for the diverse theories put forward.

Some authors consider that the tumour initially assumes an apparently benign form before the appearance of clear evidence of malignancy (such as mitoses and numerous atypical cells). MacKensie (1966) recognises three histological groups of synovial sarcoma: group A, displaying both types of cells; group B, in which pseudofibroblastic cells predominate; and group C with a predominantly pseudoepithelial structure. However, MacKensie (1966) admits that this classification does not correspond to a prognostic evaluation of these tumours, supporting the view of Cadman et al. (1965) that there is no parallelism between the histological type and the aggressivity of the lesion. Be this as it may, the presence of two types of cells is essential to the diagnosis, and it is only when there is a clear predominance of one type that confusion may arise - usually with a fibrosarcoma, but also with a metastatic epithelial tumour or even a rhabdomyosarcoma.

Treatment

The therapeutic approach to synovial sarcoma currently tends to uniformity in view of the gravity of the disease and the virtual ineffectiveness of the various methods used hitherto. Local excision confined to the tumour is quite unsuitable; the recurrence rate of over 90% and the long-term mortality of over 75% demand a more extensive procedure. Supplementary radiotherapy hardly improves the results.

Even wide excision well away from the tumour does not exclude recurrence. Only a radical procedure such as amputation, or perhaps disarticulation, afford any hope of eliminating this extensive tumour and its extensions among the aponeurotic planes. Some authors combine this with nodal eradication and supplementary radiotherapy, but the prognosis remains very gloomy with less than 50% of patients surviving long term. The indication for radical surgery is all the more justified because this extensive tumour has a very marked tendency to recurrence. This is very frequently seen after local excision, and its development, sometimes delayed for several years, is thought by most workers as considerably worsening the prognosis. Such recurrence is often accompanied by remote spread through the bloodstream or lymphatics, the most common metastases being observed in the lungs (81%), nodes (23%) and bones (20%).

Thus, this malignant tumour is quite formidable; the mean survival reported by Cadman et al. (1965) is between 6 and 7 years. The prognosis can be improved only by the earliest possible treatment, with rigorous and regular follow-up in every case.

Bibliography

Ariel IM, Pack GT (1963) Synovial sarcoma: Review of twenty-five cases. N Engl J Med 268: 1272-1275

Barland P, Novikoff AB et al (1962) Electron microscopy of the human synovial membrane. J Cell Biol 14: 207

Cade S (1962) Synovial sarcoma. J R Coll Surg Edinb 8: 1-51

Cadman NL, Soule EH, Kelly PJ (1965) Synovial sarcoma. An analysis of 134 tumors. Cancer 18: 613-627

Cameron HU, Kostuik JP et al (1974) A long term follow-up of synovial sarcoma. J Bone Joint Surg [Br] 56/4: 613-617

Chomette G (1958) Synovialomes malins chez l'enfant. Arch Anat Pathol A81: 227-233

Crampton RF (1957) Malignant synovioma of the hand. Br J Plast Surg 4: 56-62

Crocker DW, Stout AP (1959) Synovial sarcoma in children. Cancer 12: 1123-1133

Dishe FE, Darby AJ et al (1978) Malignant synovioma: Electron microscopical findings in three patients and review of the literature. J Pathol 124: 149-153

Enterline HT, Culberton JD et al (1960) Liposarcoma: A clinical and a pathological study of 53 cases. Cancer 13: 932-950

Fernandez BB, Hernandez FJ (1976) Poorly differentiated

synovial sarcoma. A light and electron microscopic study. Arch Pathol Med 100: 221
Fievez M, Haas F (1965) Les sarcomes de la synoviale. Ann Pathol 10/4: 351-362
Fragiadakis EG, Vatopoulos P (1973) Sarcoma of tendon sheath. Hand 5/1: 71-75
Gabbiani G, Kaye G, Lattes R et al (1971) Synovial sarcoma. Electron microscopic study of a typical case. Cancer 28: 1031
Grimley PM, Sokoloff DW (1973) Synovial giant cells in rheumatoid arthritis. Am J Pathol 49: 931
Haagensen CD, Stout AP (1944) Synovial sarcoma. Ann Surg 120: 826-842
Hand CR, McFarland GB (1970) Synovial sarcoma of the hand. J State Med Soc 122: 1-5
Hardie J, Salter SC (1894) Primary sarcoma of knee joint. Lancet I: 1619-1620
Kauffman SL, Stout AP (1959) Lipoblastic tumors of children. Cancer 12: 912-925
Kauffman SL, Stout AP (1961) Histiocytic tumors in children. Cancer 14: 469-482
Kauffman SL, Stout AP (1963) Extraskeletal osteogenic sarcomas and chondrosarcomas in children. Cancer 16: 432-439
King ESJ (1952) Tissue differentiation in malignant synovial tumors. J Bone Joint Surg [Br] 34: 97-115
Klein W, Huth F (1979) The ultrastructure of malignant synovioma. Beitr Pathol 153: 194
Lazarus JA, Marks MS (1943) Synovial sarcoma. With report of two cases. Surgery 13: 290-308
Lee SM, Steven IH et al (1974) Synovial sarcoma in children. Surg Gynecol Obstet 138: 143
Lejars F, Rubens Duval H (1910) Les sarcomes primitifs des synoviales articulaires. Rev Chir 41: 751-783
Luse SA (1960) A synovial sarcoma studied by electron microscopy. Cancer 13: 312-322
Mackensie DH (1966) Synovial sarcoma. Cancer 19: 169-180
Martens VE (1955) Unusual synovial tumors. JAMA 157: 888-890
Michon J (1983) Synovio-sarcomes des membres. Journ Chir Main II/3: 221-229
O'Brien P, Brasfield RD (1965) Hemangiopericytoma. Cancer 18: 249-252
Pack GT, Ariel IM (1950) Synovial sarcoma (malignant synovioma). Report of 60 cases. Surgery 28: 1047-1084
Raben M, Calabrese A, Higinbotham NL et al (1965) Malignant synovioma. AJR 93: 145-153
Reszel PA, Soule EH et al (1966) Liposarcoma of the extremities and limb girdles; a study of two hundred twenty two cases. J Bone Joint Surg [Am] 48: 229-244
Sabrazes J, Loubat E, De Grailly R, Magendie J (1934) Synovialosarcomes. Gaz Hebd Soc Méd Bordeaux 55: 754-762
Sherman RS, Chu FCH (1952) Roentgenographic study of the synovioma. AJR 67: 80-89
Shieber W, Grham P (1962) An experience with sarcomas of the soft tissues in adult. Surgery 52/2: 295-298
Smith LW (1927) Synoviomata. Am J Pathol 3: 355-364
Soule EH, Mahour GH et al (1968) Soft tissue sarcomas of infants and children. Mayo Clin Proc 43: 313
Stuer J (1893) Eine ungewöhnliche Geschwulst der Ellbogengelenksgegend. Medizinische Dissertation, Universität Würzburg
Taylor HB, Helwig EB (1962) Dermatofibrosarcoma protuberans. A study of 115 cases. Cancer 15: 717-725
Tillotson JF, McDonald JR et al (1951) Synovial sarcomata. J Bone Joint Surg [Am] 33: 459-473
Vidrine A, Welsh RA (1965) Hemangiopericytoma: Five cases. Surgery 56: 912-918
Wright CJE (1952) Malignant synovioma. J Pathol 64: 585-603

Epithelioid Sarcoma

Laskowski, in 1961, in a little-read Polish journal, gave the first description of "aponeurotic sarcomata", defining the clinical, histological and sequential characteristics of this malignant tumour. Before this, Berger (1938) had been intrigued by a particular form of synovial sarcoma of the wrist, and De Santo et al. (1941) had described polygonal and polyhedral tumoral cells analogous to those of the epithelium. Bliss and Reed (1968) reported four cases of "large cell" sarcoma. Enzinger (1970) made the major and most detailed study in a series of 62 cases, establishing the term of epithelioid sarcoma. Since then, this malignant tumour, whose deceptive clinical and histological features simulate a benign lesion, has been amply identified and recognised, partly thanks to progress in electron microscopy. However, many problems remain, particularly as regards its mode of evolution and its histopathological origin.

Epithelioid sarcoma is a relatively rare malignant tumour of the soft parts, characterised by the site and setting where it develops. It is a tumour of the young adult (mean age 23 years, range 10–60 years) with a predominance in males (3:1). Two major series confirm the special predilection of this tumour for the hand (dorsal aspect of the fingers, palm): 43% in Enzinger's (1970) series, whose 62 cases included 38 in the upper limb (fingers 13, hand 8, wrist 7, forearm 7, elbow 2, arm 1), 22 in the lower limb and 2 in the scalp; and 45% in the series of Prat et al. (1978), whose 22 cases included 15 in the upper limb (hand 10, wrist 1, forearm 2, elbow 1, axillary fossa 1) and 6 in the lower limb plus 1 in the neck. Most of these cases are therefore malignant tumours of the soft parts of the hand, making it a very common sarcoma of the upper limb below the elbow after fibrosarcoma and rhabdomyosarcoma and ahead of synovial sarcoma (Bryan et al. 1974, series of 101 sarcomata of the upper limb). Other much rarer sites of this tumour have been reported: penis, vulva, abdominal wall.

Clinical Features

The clinical appearance of epithelioid sarcoma of the hand is often so commonplace as not to worry the patient, thus delaying diagnosis. Prat et al. (1978) reported an average delay of 3 years before medical advice was sought. The lesion may be superficial, involving the dermis or subcutaneous areolar tissue, forming one or more, firm, regular, well-circumscribed nodules that are painless and of cystic appearance, with a diameter of usually less than 2.5 cm. These nodules grow progressively, lose their mobility and ulcerate through the skin after a varying period, producing in the worst cases an extensive ulcerated and infiltrated expanse. The initial site, however, may be deeper, in the form of a larger, poorly defined, hard, multinodular mass that is fixed and adherent to the tendons and aponeuroses and sometimes painful. Invasion of the adjacent structures may produce dysfunction, paraesthesiae, some muscle wasting or even joint stiffness.

Local progress is by contiguity, with bony involvement in advanced forms, while regional extension is by continued spread along the aponeuroses, tendon sheaths and neurovascular bundles.

Radiology

Radiological examination is usually negative, rarely revealing foci of microcalcification. Local spread may affect underlying bone with cortical erosion.

Histology

The diagnosis of epithelioid sarcoma rests entirely on histological evidence, the clinical features being neither specific nor even very disturbing. However, it must be considered as a possible diagnosis with such lesions of the hand in young persons. Grossly, the appearance is that of a whitish or yellowish tumour infiltrating the local structures and seeming to arise from a tendon sheath, the fascia, an aponeurosis or a neurovascular bundle.

Microscopic examination reveals the three characteristic features of the tumour: a nodular arrangement of the tumour cells with foci of central necrosis sometimes exhibiting microcalcification; large polygonal cells with eosinophilic cytoplasm and a central nucleus, analogous to epithelial cells – hence the name epithelioid; and fusiform cells arranged at the periphery of the nodules. These two types of cells are found in varying proportions; the fusiform cells are more numerous in early cases and the epithelioid cells in later cases, recurrences and metastases. Enzinger (1970) also insists on the existence of intermediate cells, as opposed to synovial sarcoma in which there is a sharp distinction between the two cell groups. Gabbiani et al. (1972) likewise confirm the presence of a less well differentiated cell group. Further, histological examination reveals some multinucleate cells, numerous mitoses, dense fibrous septa rich in collagen and a rather scanty reticulin network.

Differential Diagnosis

The evidence listed above suffices to refute the many other possible diagnoses, whether benign tumours (rheumatoid nodules, granuloma, fibromatosis, nodular tenosynovitis) or malignant (synovial sarcoma, fibrosarcoma, fibrous histiocytosarcoma, squamous-cell epithelioma). However, doubts often persist, and then the ultrastructural features revealed by electron microscopy are of great diagnostic value before embarking on treatment with serious implications. The essential characteristics, confirmed by numerous studies, are as follows: abundant cytoplasm, always containing microfilaments of varying amount and arrangement; granular endoplasmic reticulum, with lipid vesicles in some of the cells; irregular jagged nuclei; plasmic membranes bristling with microvillosities and filopodes; and intercellular connexions formed by complex interdigitations.

Development

Epithelioid sarcoma is marked by a course that is certainly always malignant but very variable in time. Indeed, while recurrence, solitary or multiple, is very common and affects 85% of cases after local excision, the appearance of regional or remote lesions may occur after several months or years, with a maximum of 34 years (Prat et al. 1978). The incidence of metastases is assessed between 30% (Enzinger 1970) and 60% (Prat et al. 1978), with a predilection for the lung and pleura, skin and scalp. Spread may also be via the lymphatics, and Enzinger (1970) reports six cases of nodal metastasis. The mean survival is 5 years, but there is a wide scatter from some months to several years, so an individual prognosis is difficult. There is, in fact, no correlation between the site and size of the tumour and survival.

In addition, many authors state that the degree of local spread cannot be established despite clinical, radiological, xerographic or lymphographic studies. Only histological evidence of the invasion of connective tissue structures, particularly the vascular sheaths, is always regarded as very sinister. Thus, epithelioid sarcoma poses problems of both criteria of cure and assessment of the effectiveness of treatment.

There exist two special clinical types, one revealed by cervical lymphadenopathy (Sugarbaker et al. 1981) and the other arising in patients with the Turner X0/XX syndrome (Males and Lain 1972).

Aetiopathogenesis

From this aspect the precise cellular origin of epithelioid sarcoma remains in dispute, despite the aid of ultrastructural studies. Various hypotheses have been advanced. Fischer and Branimir (1972) as well as Horvat et al. (1970) suggest a role for fibroblasts, in view of the presence of fusiform cells and the formation of collagen. The pseudo-epithelial arrangement and strongly eosinophilic cytoplasm of the polygonal cells suggest a possible histiocytin origin. Frable et al. (1973) report numerous analogies with fibrous histiocytosarcoma. However, Reymond et al. (1979) consider that the presence of microvillosities rules out any role for the histiocyte, even though this may have been suggested by appearances at optical microscopy. The argument most often advanced is that of a synovial origin. Gabbiani et al. (1972), taking into consideration the proximity of tendons and aponeuroses, the existence of two types of cells and the electron microscopic features, believe that epithelioid sarcoma is an atypical variant of synovial sarcoma. Patchefsky et al. (1977) advance the same hypothesis, based on the following arguments. Clinically, the site of the tumour, the frequency of recurrence and the existence of nodal metastases favour a synovial origin. Histologically, the synovial cell that may undergo fibroblastic transformation also exhibits microvillosities analoguous to those of the cells of epithelioid sarcoma. Thus, for many authors, epithelioid sarcoma appears as a variant of synovial sarcoma whose course may be somewhat slower.

Treatment

The currently advised treatment for this malignant tumour is essentially surgical. Tumoral resection alone is always inadequate, with a risk of virtually constant recurrence; wide excision is therefore advised by most workers with amputation in case of recurrence. Some even advise this radical procedure at the outset, once the diagnosis is made with certainty, the level of amputation being decided by the site of the tumour and the degree of lymphatic or vascular invasion. Peimer et al. (1977) perform axillary nodal eradication routinely, whether or not there is clinical spread. In cases of insufficient eradiation, complementary radiotherapy must be extended to the forearm. Chemotherapy, often used as a complement to surgery, may be necessary to reduce the volume of a tumour that cannot be extirpated. However it is difficult to settle on a line of treatment for a tumour which, though always malignant, has a variable and sometimes very slow course that hampers the assessment of the efficacy of treatment and the specification of criteria of cure.

Bibliography

Barland P, Novikoff AB, Hamerman D (1962) Electron microscopy of human synovial membrane. J Cell Biol 14: 207-220

Berger L (1938) Synovial sarcomas in serous bursae and tendon sheaths. Am J Cancer 34: 501-539

Black WC (1936) Synovioma of the hand. Report of a case. Am J Cancer 28: 281-284

Bloustein PA, Silverberger SG et al (1976) Epithelioid sarcoma. Cancer 38: 2380-2400

Bliss OB, Reed RJ (1968) Large cell sarcomas of tendon sheath, malignant giant cell tumors of tendon sheath. Am J Clin Pathol 49: 776-781

Boyes JG, Marroum MC (1978) Epithelioid sarcoma of hand and forearm. Hand 10: 302-305

Bryan RS, Soule EH et al (1974) Primary epithelioid sarcoma of the hand and forearm. J Bone Joint Surg [Am] 56: 458-465

Button M (1979) Epithelioid sarcoma: A case report. J Hand Surg 4: 368-371

Condamine JL, Louis B, Richter D (1984) Les sarcomes épithélioïdes à propos de 2 cas. Ann Chir Main 4/2: 197-209

De Santo DA, Tennant R, Rosahn PD (1941) Synovial sarcomas in joints, bursae and tendon sheaths. A clinical and pathological study of 16 cases. Surg Gynecol Obstet 72: 951-981

Enzinger FM (1970) Epithelioid sarcoma, a sarcoma simulating a granuloma or a carcinoma. Cancer 26: 1029-1041

Fischer ER, Branimir H (1972) The fibrocytic derivation of the so-called epithelioid sarcoma. Cancer 30: 1074-1081

Frable WJ, Kay S et al (1973) Epithelioid sarcoma. An electron microscopic study. Arch Pathol 95: 8-12

Gabbiani G, Kaye G et al (1971) Synovial sarcoma, electron microscopic study of a typical case. Cancer 28: 1031-1039

Gabbiani G, Yao-Shi FU et al (1972) Epithelioid sarcoma. A light and electron microscopic study suggesting a synovial origin. Cancer 30: 486-499

Hadju SJ (1979) Pathology of soft tissue tumors. Lea & Febiger, Philadelphia, pp 183-226

Hadju SJ, Shiu MH, Fortner JG (1977) Tendosynovial sarcoma. A clinicopathological study of 136 cases. Cancer 39: 1201-1217

Heppenstall RB, Yvars MF et al (1972) Epithelioid sarcoma, two case reports. J Bone Joint Surg [Am] 54: 802-806

Horvat B, Caines M, Fischer ER (1970) Ultrastructure of rhabdomyosarcoma. Am J Clin Pathol 53: 555-564

Jefferiss CD, Caldwell RA (1974) Epithelioid sarcoma. Case report of a young musician's hand. Hand 6: 208

Key JA (1929) Experimental arthritis. The reactions of joints to mild voutans. J Bone Joint Surg [Br] 11: 705

Laskowski J (1961) Sarcoma aponeuroticum. Nowotwory 11: 61-67

Laskowski J (1971) Sarcoma aponeuroticum. Pol Med J 10: 12

Linell F, Myhre-Jensen O et al (1973) Epithelioid sarcoma. Review and report of two cases. Acta Pathol Microbiol Immunol Scand [A] 236: 21-26

Mackenzie DH (1971) Two types of soft tissue sarcoma of uncertain histogenesis. Br J Cancer 25: 458-461

Males JL, Lain KC (1972) Epithelioid sarcoma in X0/XX Turner's syndrome. Arch Pathol 94: 214-216

Marmor L, Horner RL (1959) Metastasis to a phalanx simulating infection in a finger. Am J Surg 97: 236

Moore SW, Wheeler JE et al (1975) Epithelioid sarcoma masquerading as Peyronie's disease. Cancer 35: 1706-1710

Nelson FR, Crawford BE (1972) Epithelioid sarcoma. J Bone Joint Surg [Am] 54: 798-801

Patchefsky AS, Soriano R et al (1977) Epithelioid sarcoma, ultrastructural similarity to nodular synovitis. Cancer 39: 143-152

Peimer CA, Smith RJ et al (1977) Epithelioid sarcoma of the hand and wrist: Pattern of extension. J Hand Surg 2: 275-282

Pinol Aguade J, Ferrandiz C et al (1976) Sarcoma epithelioïde. Un tumor de miofibroblastos? Med Cutan Iber Lat Am 2: 91-104

Prat J, Woodruff J et al (1978) Epithelioid sarcoma: Analysis of 22 cases indication the prognostic, significance of vascular invasion and regional lymph node metastasis. Cancer 41: 1472-1487

Reymond JL, Stoebner P et al (1979) Sarcome épithélioïde, étude anatomo-clinique d'un cas. Ann Dermatol Venereol 106: 1013-1018

Santiago J, Feinermann LK, Lattes R (1972) Epithelioid sarcoma: A clinical and pathologic study of nine cases. Hum Pathol 3: 133-147

Saxe N, Botha JB (1977) Epithelioid sarcoma: A distinctive clinical présentation. Arch Dermatol 113: 1106-1108

Seemayer TA, Dionne PG, Tabah EJ (1974) Epithelioid sarcoma. Can J Surg 17: 37-42

Sennwald G et al (1984) Sarcome épithélioïde, analyse de l'attitude thérapeutique. Ann Chir Main 4/2: 257-267

Sergiou SP, Reid C (1980) Epithelioid sarcoma. Two case reports. Hand 12/3: 304-307

Smith RJ (1963) Involvement of the carpal bones with metastatic tumor. AJR 89: 1253

Soule EH, Enriquez P (1972) Atypical fibrous histiocytoma malignant fibrous histiocytoma, malignant histiocytoma and epithelioid sarcoma. A comparative study of 65 tumors. Cancer 30: 128-143

Stoebner P, Miech G et al (1973) Ultrastructure de la cellule mésothéliale néoplasique: Aspects typiques et variétés évolutives. Rev Cytol Clin 6: 15-21

Sugarbaker PH, Auda S et al (1981) Early distant metastases from epithelioid sarcoma of the hand. Cancer 48: 852-855

Sutton JS (1967) Ultrastructural aspects of in vitro development of monocytes into macrophages, epithelioid cells and multinucleated giant cells. Natl Cancer Inst Monogr 26: 71-141

Wanstrup J, Christensen HE (1966) Sarcoidosis. Ultrastructural investigations on epithelioid cell granulomas. Acta Pathol Microbiol Immunol Scand [A] 66: 169-185

Lipoma

Lipomata are the most common and very ordinary benign tumours of mesenchymal origin. They arise from adipose tissue and may be located anywhere in the body. However, there are certain preferential sites, and it is also noteworthy that, while lipomata are relatively common in the limbs, the extremities - hands and feet - are rarely affected. Related to tumours of the soft parts of the hand, the incidence of lipoma varies from 1.3% to 4%. Stout (1953) reported 28 cases in 1203 tumours, Haber et al. (1965) 32 in 2321. Posch (1956) 5 in 136, Butler et al. (1960) 6 in 437 and Phalen et al. (1971) 7 in 500. Compared with lipomata of the body as a whole, the incidence of lipomata of the hand varies from 0.5% to 1.5%, with those of the upper limb forming 11.5%-27% of all lipomata. Geschickter (1934) found only 6 lipomata of the hand in an overal body total of 390 lesions, while Barrile (1958), in a series of 476 lipomata, found 55 in the upper limb, of which only 3 were in the hand. Adair et al. (1932) in a series of 352 lipomata, reported 94 cases localised to the upper limb and only 2 in the hand. Finally, two studies dealing more particularly with lipomata of the upper limb report 15 tumours in the hand of a total of 112 (Phalen et al. 1971) and 13 of 51 (White and Hanna 1962).

Carroll and Doyle (1967) found lipomata three times more often in women, and Leffert (1972) noted also that these cases seen in young men usually arose in an obese setting. However, most authors report an almost equal sex distribution; for example, White and Hanna (1962) found 29 cases in men against 22 in women and Paarlberg et al. (1972) 33 cases in men against 26 in women. The same authors find no predilection for one hand over the other. Lipomata may be situated anywhere in the hand: wrist, dorsum, palm, thenar or hypothenar eminence, fingers. They may be cutaneous, intermuscular or periosteal. Apart from Mason (1939), who gives a greater incidence of dorsal lipomata, all authors note a marked palmar predominance. Of 59 cases described by Paarlberg et al. (1972), 28 were palmar (12 on the front of the fingers, 4 at carpal level, 9 on the thenar eminence and 3 on the hypothenar), 19 were on the dorsum and 11 at the wrist. Booher (1965) reported 36 cases of palmar lipomata, 5 dorsal and 7 digital. Finally, in the 13 cases of White and Hanna (1962), 5 were at the wrist, 6 in the palm and 2 in the fingers. It should be noted that lipomata of the back of the hand, wrist and fingers are usually small, whereas those of the palm tend to be bulky and multilocular (Fig. 52 a, b).

Mason (1939) distinguishes two groups of lipomata of the hand according to site: superficial or subcutaneous, and deep or subaponeurotic (subfascial). He describes endovaginal lipomata insinuating themselves into the sheath of one or more tendons, forming arborescent forms with many prolongations, and exovaginal lipomata, large lesions situated between the muscles and fibrous structures and producing little discomfort. Although Strauss (1931) reports a certain incidence of the endovaginal forms, Sullivan et al. (1956) and Paarlberg et al. (1972) found only one case each (Table 3).

Table 3. Incidence of lipomata of the hand

Reference	Tumours of soft parts of hand	Lipomata	Lipomata of upper limb	Lipomata of the hand
Stout (1953)	1203			28 (2.3%)
Posch (1956)	136			5 (4.0%)
Butler et al. (1960)	437			6 (1.3%)
Haber et al. (1965)	2321			32 (1.4%)
Phalen et al. (1971)	500			7 (1.4%)
		112		15 (13%)
Geschickter (1934)		390		6 (1.5%)
Barrile (1958)		476	55 (11.5%)	3 (0.6%)
			55	3 (5.4%)
Adair et al. (1932)		352	94 (27%)	2 (0.5%)
			94	2 (2%)
White and Hanna (1962)			51	13 (26%)

Clinical Features

The clinical signs are related mainly to the site and size of the tumours. These are often asymptomatic (80%-90% of cases), but they may, because of their size, constitute a mechanical obstacle to normal function of the hand (sensation of palmar fullness, discomfort on pressure, especially with bulky thenar lipomata). More rarely, they present with signs of nerve compression. It is interesting to relate this to the study of Barber et al. (1962) dealing with nerve compression due to benign tumours. Of 22 cases observed in the upper and lower limbs, 16 were due to lipomata and fibrolipomata and involved the radial (8 cases), ulnar (3), median (1) and sciatic (4) nerves. The other tumours found were ganglia, epidermal cysts and haemangiomata. In the hand compression of the median nerve may present as a carpal tunnel syndrome. Grantham (1966) reported a case of ulnar compartment syndrome which was mainly sensory, with deformation of the hypothenar eminence by the tumoral mass and a deficit in the ulnar nerve distribution. McFarland and Hoffer (1971) observed a case of ulnar compression with mainly motor features, e.g. paralysis of the intrinsic muscles. Finally, although they are often situated close to the arterio venous axes, no case of vascular compression due to a lipoma has been reported.

On examination, the tumour is often solitary and localised, rarely diffuse, painless, circumscribed, sometimes lobulated, and of soft and pseudocystic consistence (Fig. 53). It is mobile on both the deep and superficial layers, and there is usually no sign of inflammation or adenopathy. Although the clinical signs are often suggestive, the diagnosis of lipoma of the hand is rarely raised in view of its low incidence. Palpation of the tumour is particularly difficult in the deep forms due to interposition of the fibrous and aponeurotic palmar layers. Bunnell (cf. Boyes 1964) and Posch (1956) find that application of ice increases the consistence of the tumour, which is a useful sign.

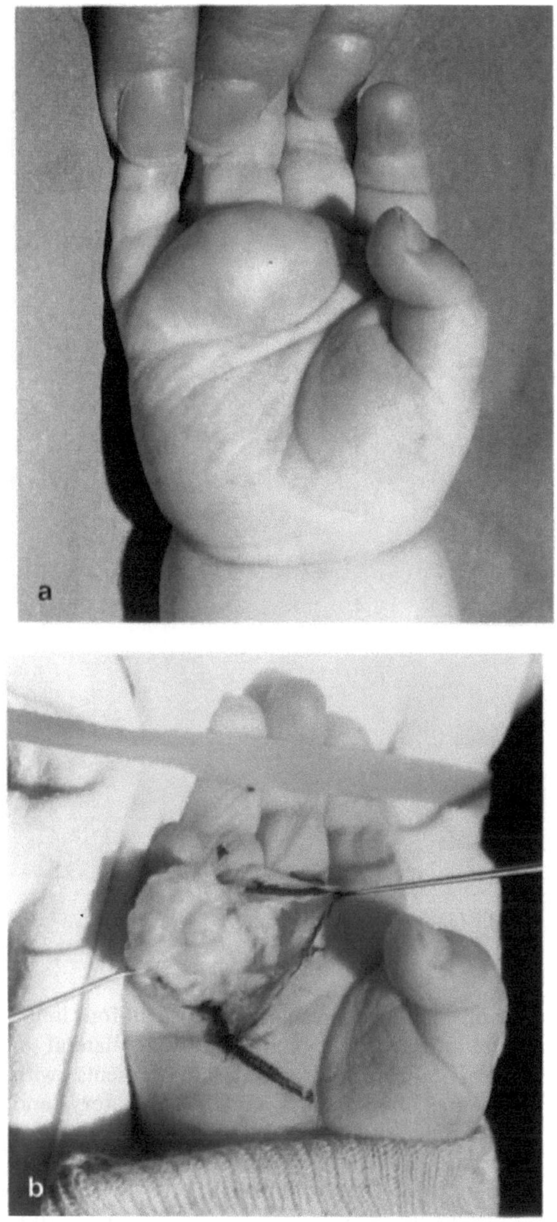

Fig. 52. ***a*** Large lipoma of palm in an infant: clinical appearance. ***b*** Same patient: gross appearance of lipoma

Clinical Types

Two clinical types of lipoma of the hand should be singled out.

1. Lipomata of the tendon sheaths were studied by Sullivan et al. (1956). Of the 45 cases reported up to then in the literature which were situated at the extremities of the upper and lower limbs, two different tumoral aspects were found: arborescent lipoma (26 cases) with numerous ramifications, described by Muller (1928), Virchow (1847, cited by Sullivan et al. 1956) and Billroth (1885) and lipoma simplex, without extensions (19 cases). These lesions are observed between the ages of 6 and 70 years, with a mean age of 45, and are of equal sex distribution. Tumours were found at the hand and wrist in 30 cases, at the feet and ankles in 9. Both a hand and a foot were

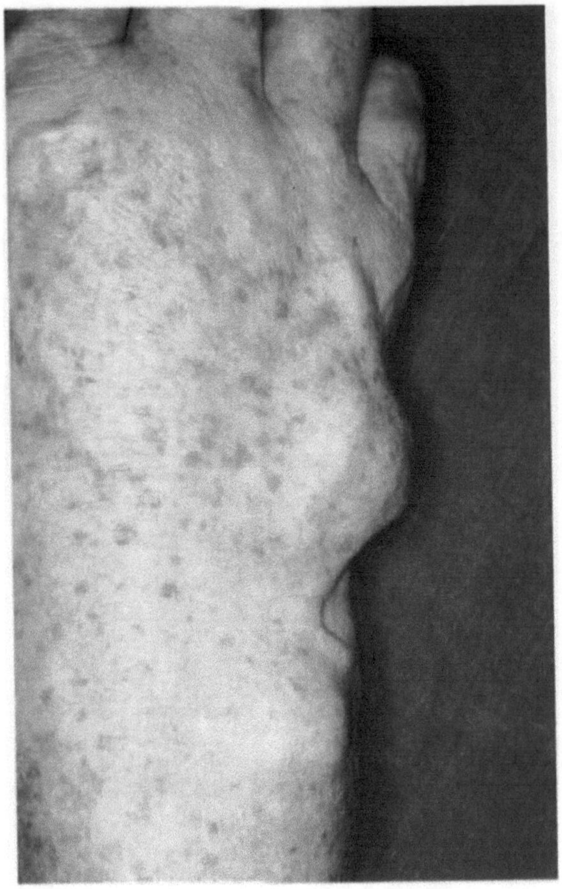

Fig. 53. Lipoma of back of hand

involved in 2 patients and in 1 case all four limbs were affected. The condition was bilateral in 14 cases. Clinically, the lesion presents with swelling, often painful and inflammatory, and most often situated in relation to the extensor tendons. One complication of these tumours was reported by Grandclaude et al. in 1933: spontaneous rupture of a tendon adjacent to an arborescent lipoma.

2. Lipoblastomoses of the infant are a rare and extensive form that may involve the whole of the palm in a few weeks. This is a benign tumour composed of embryonic lipoblasts and can be completely excised. The cases reported are few. Vellios et al. (1958) studied the histological and cytological features of this tumour in the context of a thoraco-axillary lipoblastomosis. Van Meurs (1987) reported a case requiring five operations because of recurrences and Cox (1954), a case of a cervical tumour causing death by asphyxia.

Differential Diagnosis

Diagnosis is most often made only during surgical exploration. Owing to the rarity of the tumour, other diagnoses may be suspected, particularly ganglia. The radiological appearance may be a guide as it shows the particular opacity of the fatty tissue, especially when the tumour lies deep and adjacent to muscular masses. In some cases bony erosion has been noted next to the lipoma, and Stein (1959) reports total destruction of a distal phalanx.

Histology

Histologically, lipomata are solitary or multiple benign tumours, rounded or multilobular, and may be located in any part of the body although they are usually observed in regions normally occupied by fatty tissue. The tumours are sharply circumscribed and generally encapsulated; they consist of adult fat cells of normal appearance arranged in lobules, separated by fibrous septa (Fig. 54). In addition to their abundance of capillary vessels, there frequently exists a proliferation of fibrous or myxomatous tissue, so producing fibrolipomata or myxolipomata. Within the tumour the presence of necrotic foci or disseminated calcification is sometimes seen.

Aetiology

The aetiology of lipomata remains unknown. Most authors agree in thinking that they probably do not arise from mature fat cells but rather from primitive mesenchymal cells. However, Ewing (1940) notes certain predisposing factors - heredity, endocrine pathology and disordered lipid metabolism. Leffert (1972) found 11% with endocrine abnormalities in 122 lipomata: 5 thyroid disorders, 7 cases of diabetes and 1 acromegaly.

Treatment

Surgery for asymptomatic lipomata of the hand is not essential, but as the tumour is rare, and there is a risk of diagnostic error, excision is recommended in all cases. This is obviously necessary when there are clinical signs of nerve compression or dysfunction. The operation should be performed under general or local/regional anaesthesia with a pneumatic tourniquet to ensure a bloodless field. The incision is made over the tumour with due regard for the principles of hand surgery. The local structures must be defined and safeguarded. The tumour is usually well-encapsulated, and its excision is simple. Any point of bony implantation should be sought for to exclude a lipoma originating from the periosteum (Quenu and Schwartz 1907, cited by Moulouguet 1946). In some cases surgical excision may prove difficult, especially with the arborescent forms, when the lesion is near a joint, or when the lipoma extends from the front to the back of the hand or extends along an interosseous muscle. Deep lipomata of the thenar eminence in the region of the adductor of the thumb raise the problem of safeguarding the motor branch of the ulnar nerve (risk of injury of the motor branch to the first interosseous).

The diffuse forms are also troublesome to treat since there is no definite plane of cleavage in the hand and fingers. It is a matter of actually scraping out under the dermis a lipomatous mass adherent to the adjacent structures. With lipomata of the tendon sheaths it is necessary to incise the distended sheath and excise the tumour with all its villosities and prolongations.

Postoperative Course

This is usually satisfactory, with the regression and ultimate disappearance of the signs of nerve compression. Recurrence is rare. If it happens, it is due to incomplete excision, as may appear with arborescent forms with numerous prolongations or along a dorsopalmar path.

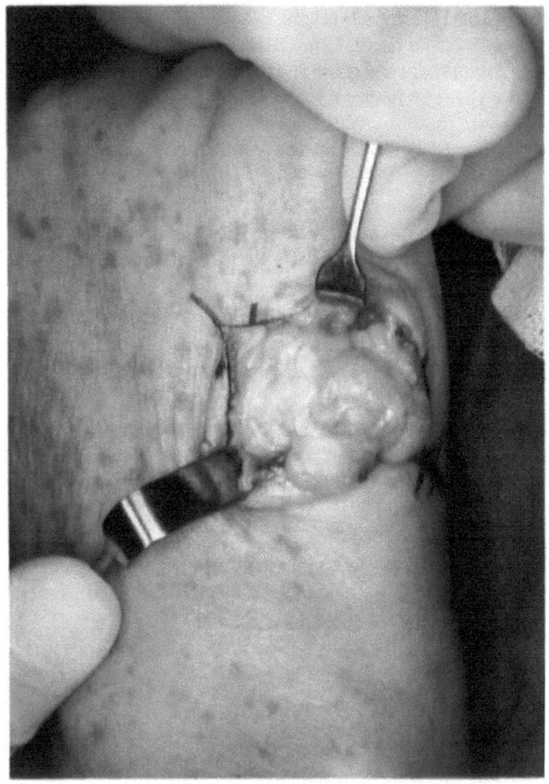

Fig. 54. Gross appearance of a lipoma

Bibliography

Adair FE, Pack GT, Farrior JH (1932) Lipomas. Am J Cancer 16: 1104-1120

Aldredge WM, Halpert B (1948) Lipoma of the thenar. Surgery 24: 853-854

Armingeat J (1938) Lipome périostique enserrant en demi-cercle le radius gauche, à l'union du tiers moyen et du tiers supérieur. Bull Mem Soc Chir 30: 291-295

Auvray M (1926) Lipome de la paume de la main. Bull Mem Soc Chir 52: 47-48

Barber KW, Bianco AJ Jr, Soule EH, Maccarty CS (1962) Benign extraneural soft-tissue tumors of the extremities causing compression of nerves. J Bone Joint Surg [Am] 44: 98-104

Barrile NM (1958) Gran lipoma palmar subaponeurotico. Pressa Med Argentina 45: 318-320

Bartlett E (1930) Periosteal lipoma. Report of two cases. Arch Surg 21: 1015-1022

Bick EM (1936) Lipoma of the extremities. Ann Surg 104: 139-143

Billroth T (1885) General surgical pathology and therapeutics. Appleton, New York

Booher RJ (1965) Lipoblastic tumors of the hands and feet: Review of the literature and report of thirty-three cases. J Bone Joint Surg [Am] 47: 727-740

Bosch DT, Bernhard WG (1950) Lipoma of the palm. Am J Clin Pathol 20: 262

Bowen TL, Stone KH (1966) Posterior interosseous nerve paralysis caused by a ganglion at the elbow. J Bone Joint Surg [Br] 48: 774

Boyes JH (ed) (1964) Bunnell's surgery of the hand. Lippincott, Philadelphia

Butler ED, Hamill JP, Seipel RS, De Lorimier AA (1960) Tumors of the hand. A ten-year survey and report of 437 cases. Am J Surg 100: 293-302

Campbell CS, Wulf RF (1954) Lipoma producing a lesion of the deep branch of the radial nerve. J Neurosurg 11: 310-311

Capener N (1966) The vulnerability of the posterior interosseous nerve of the forearm. J Bone Joint Surg [Br] 48: 770

Carroll RE, Doyle JR (1967) Lipoma of the hand. J Bone Joint Surg [Am] 49: 581

Carvolo JE, Dahlin DC (1953) Lipoma involving bone and simulating malignant bone tumor. Report of a case. Mayo Clin Proc 28: 361-363

Charache H (1942) Tumors of tendon sheaths. Arch Surg 44: 1038-1052

Clifford RH, Kelly AP (1959) Diagnosis and treatment of tumors of the hand. Clin Orthop 13: 204–208
Cox RW (1954) Hibernoma: The lipoma of immature adipose tissue. J Pathol 68: 511–518
Dickson AB, Ayres WW, Mason MW, Miller WR (1951) Lipoma of bone of intraosseous origin. J Bone Joint Surg [Am] 33: 257–259
Dupont C, Cloutier GE, Prevost Y, Dion MA (1965) Ulnar tunnel syndrome at the wrist. A report of four cases of ulnar-nerve compression at the wrist. J Bone Joint Surg [Am] 47: 757–761
Ewing J (1940) Neoplastic diseases - a treatise on tumors. Saunders, Philadelphia
Fairbank HAT (1953) A parosteal lipoma. J Bone Joint Surg [Br] 35: 589
Fenning JB (1965) Deep ulnar-nerve paralysis resulting from an anatomical abnormality. A case report. J Bone Joint Surg [Am] 47: 1381–1382
Fleming RJ, Meyer A, Garcia A (1962) Parosteal lipoma. AJR 87: 1075–1084
Gay JR, Love JG (1947) Diagnosis and treatment of tardy paralysis of the ulnar nerve. Based on a study of 100 cases. J Bone Joint Surg 29: 1087–1097
Geschickter CF (1934) Lipoid tumors. Am J Cancer 21: 617–641
Geschickter CF, Lewis D (1934) Tumors of tendon sheaths, joints and bursae. Am J Cancer 22: 96–126
Gold AM, Oppenheim A (1954) Deep intermuscular lipoma of an extremity. A case report. J Bone Joint Surg [Am] 36: 146–148
Grandclaude C, Razemon P, Bizard G (1933) Angio-lipome arborescent de la gaine tendineuse du long extenseur du pouce ayant déterminé la rupture du tendon. Ann Anat Pathol 10: 444–446
Grantham SA (1966) Ulnar compression in the loge De Guyon. JAMA 197: 509–510
Gross S, Wood C (1953) Hibernoma. Cancer 6: 159–163
Guyon F (1961) Note sur une disposition anatomique propre à la face antérieure de la région du poignet et non-encore décrite. Bull Soc Anat 36: 184–186
Haber MH, Alter AH, Wheelock MC (1965) Tumors of the hand. Surg Gynecol Obstet 121: 1073–1080
Hayes JR, Mulholland RC, O'Connor BT (1969) Compression of the deep palmar branch of the ulnar nerve. A case report and anatomical study. J Bone Joint Surg [Br] 51: 469–472
Hustead AP, Mulder DW, MacCarty CS (1955) Nontraumatic, progressive paralysis of the deep radial (posterior interosseous) nerve. Arch Neurol 79: 269–274
Kauffman SL, Stout AP (1959) Lipoblastic tumors of children. Cancer 12: 912–925
Lagache G, Guiot Y (1955) Lipoma profond de la main. Lille Chir 10: 36–37
Leffert RD (1972) Lipomas of the upper extremity. J Bone Joint Surg [Am] 54/6: 1262–1266
Lewis D, Geschickter CF (1935) Diffuse lipoma of the right upper extremity. Proban A and B yielded by bioassay of fat. Ann Surg 102: 154–158
Martin A, Grenier J (1922) Un cas de lipome du doigt. Paris Med 43: 303–304
Mason ML (1939) Tumors of the hand. Surg Gynecol Obstet 64: 129–148
McEnery ET Jr, Schmitz RL, Nelson PA (1959) Palmar lipoma: Report of a case. Arch Surg 79: 699–700
McFarland GB, Hoffer MM (1971) Paralysis of the intrinsic muscles of the hand secondary to lipoma in Guyon's tunnel. J Bone Joint Surg [Am] 53: 375–376
Moon N, Marmor L (1964) Parosteal lipoma of the proximal part of the radius. J Bone Joint Surg [Am] 46: 608–614
Mouchet A (1926) Discussion of Auvray M: Lipome de la paume de la main. Bull Mem Soc Chir 52: 47–48
Moulonguet P (1946) Lipomes profonds des membres. Lipomes périostés. Mem Acad Chir 72: 301–303
Mueller MC, Robbins JL (1960) Intramedullary lipoma of bone. Report of a case. J Bone Joint Surg [Am] 42: 517–520
Mulholland RC (1966) Nontraumatic progressive paralysis of posterior interosseous nerve. J Bone Joint Surg [Br] 48: 781
Muller P (1928) Lipome pré-synovial palmaire d'un doigt. Bull Mem Soc Chir 20: 390–392
Paarlberg D, Linscheid R, Soule EH (1972) Lipomas of the hand. Including a case of lipoblastomatosis in a child. Mayo Clin Proc 47: 121–124
Petit-Dutaillis D, Godlewski S, Dry J (1960) Paralysie radiale progressive. Exérèse chirurgicale d'un lipome périosté comprimant la branche postérieure du nerf radial. Guérison. Rev Neurol (Paris) 102: 267–271
Phalen GS, Kendrick JT, Rodriguez JM (1971) Lipomas of the upper extremity: A series of fifteen tumors in hand and wrist and six tumors causing nerve compression. Am J Surg 121: 298–306
Posch JL (1956) Tumors of the hand. J Bone Joint Surg [Am] 38: 517–540
Richmond DA (1953) Lipoma causing a posterior interosseous nerve lesion. J Bone Joint Surg [Br] 35: 83
Rodriguez JM, Phalen GS (1970) Lipomas in the hand and wrist: Diagnosis and treatment. Cleve Clin Q 37: 201–205
Rosen H (1959) Parosteal lipoma. Bull Hosp Jt Dis Orthop Inst 20: 96–102
Rowland SA (1967) Lipofibroma of the median nerve. J Bone Joint Surg [Am] 49/3: 581
Samuel E (1947) The radiological diagnosis of lipomata. Br J Radiol 20: 55–57
Schmitz RL, Keeley JL (1957) Lipomas of the hand. Surgery 42: 696–700
Seddon HJ (1952) Carpal tunnel ganglion as a cause of paralysis of deep branch of the ulnar nerve. J Bone Joint Surg [Br] 34: 386–390
Sharrard WJ (1966) Posterior interosseous neuritis. J Bone Joint Surg [Br] 48: 777
Shea JD, McClain EJ (1969) Ulnar-nerve compression syndromes at and below the wrist. J Bone Joint Surg [Am] 51: 1095–1103

Smith T (1868) Fatty tumour growing from the neck of the radius. Trans Pathol Soc Lond 19: 344-345

Smith WE, Fienberg R (1957) Intraosseous lipoma of bone. Cancer 10: 1151-1152

Stein AH (1959) Benign neoplastic and non neoplastic destructive lesions in the long bones of the hand. Surg Gynecol Obstet 109: 189-197

Stout AP (1953) Tumors of the soft tissues. In: Atlas of tumor pathology. Armed Forces Institute of Pathology, Washington

Strauss FH (1931) Deep lipomas of the hand. Ann Surg 94: 269-273

Sugar S, Murphy BM (1955) Liposarcoma of the foot. Case report. J Michigan State Med Soc 54: 468-469

Sullivan CR, Dahlin DC, Bryan RS (1956) Lipoma of the tendon sheath. J Bone Joint Surg [Am] 38/6: 1275-1280

Sutherland JC, Callahan WP, Campbell GL (1952) Hibernoma: A tumor of brown fat. Cancer 5: 364-368

Templeton FE (1937) Roentgen diagnosis of lipomata. AJR 37: 210-216

Van Meurs DP (1947) The transformation of an embryonic lipoma to a common lipoma. Br J Surg 34: 282-284

Vellios F, Baez J, Shumacker HB (1958) Lipoblastomatosis: A tumor of fetal fat different from hibernoma. Am J Pathol 34/6: 1149-1155

Wells HG (1940) Adipose tissue, a neglected subject. JAMA 114: 2177-2183; 2284-2289

White JR (1924) Arborescent lipomata of tendon sheaths. A report of two cases. Surg Gynecol Obstet 38: 489-490

White WL, Hanna DC (1962) Troublesome lipomata of the upper extremity. J Bone Joint Surg [Am] 44/7: 1353-1359

Wu KT, Jordan FR, Eckert C (1974) Lipoma, a cause of paralysis of deep radial (posterior interosseous) nerve: Report of a case and review of the literature. Surgery 75/5: 790-795

Liposarcoma

Liposarcomata are malignant tumours composed of fat cells. They have been known since the time of Virchow (1863) and have been the object of numerous reviews (Moulonguet and Cazal 1960; Pollosson, cited in Sawhney and McDonald 1975; Stout 1944). The histological diagnostic criteria are strict. A sarcoma arising in fatty tissue is not necessarily a liposarcoma, and the presence of fatty droplets in the fibroblastic cytoplasm does not definitely indicate the lipoblastic nature of the tumour. To define a liposarcoma, apart from the elaboration of fatty inclusions by the tumour cells, it is necessary to demonstrate the presence of the histological features of young lipocytes and marked differentiation, at least in places, into fatty vesicles.

This tumour, which may be located anywhere, superficial (hip, groin, popliteal fossa) or deep (retroperitoneal), is the third most frequently occurring sarcoma of the soft parts according to Pack and Pierson (1954). It develops most often in men around an average age of 50 years. It is rarely localised to the extremities and then is three times more common in the lower than in the upper limb; in the latter it is more often proximal than distal. Only exceptionally is the hand affected by liposarcoma, and some major series of liposarcomata (Enzinger and Winslow 1962, 103 cases; Enterline et al. 1960, 53; Groff and Horvitz 1964, 17) include no lesions of the hand; this is also true in the work of Kendal et al. (1969) and Haber et al. (1965). Thus, the reports in the literature are very rare: Stout (1944), 1 case; Booher (1965), 1 case on the palm; Pack and Pierson (1954), 1 lesion at the inner border of the hand among 105 liposarcomata; Posch (1956), 1 case between the thumb and index; Reszel and Soule Coventry (1966), 1 case in 222 liposarcomata; Shiu et al. (1975), 1 case in 77 liposarcomata; Sawhney and McDonald (1975), 1 case on the left middle finger. The last authors refer in their article to the astonishing series of Brasfield and Das Gupta (cited by Sawhney and McDonald 1975) of 13 liposarcomata of the hand but with serious reservations about the strictness of the histological criteria observed.

Clinically, there is a swelling of variable consistence, soft or firmer, without pain or evidence of inflammation and without sensorimotor deficit or special dysfunction. Radiological assessment is often negative; there is sometimes a clear image demarcated from the adjacent soft tissues or a heterogeneous appearance with alternating opaque and clearer regions.

Aetiopathogenesis

The history given by these patients often refers to previous injury, but most authors regard this as purely coincidental. Sometimes there is an associated multiple lipomatosis, as in the case reported by Sawhney and McDonald (1975). Sternberg (1952) and Wright (1948) postulate the malignant transformation of a long-standing lipoma. This has not been generally accepted, and Enterline et al. (1960), in particular, presume that liposarcomatous degeneration can only be considered in the case of an encapsulated tumour exhibiting localised sarcomatous foci, the remainder being strictly identical with normal adipose tissue.

Histology

Grossly, the colour and consistence of a liposarcoma are variable; it is soft or firm, white, yellowish or pink, with necrotic and haemorrhagic foci giving the variegated appearance of a malignant tumour. Microscopically, different types of cells are found, though with great variations from one tumour to another (Enterline et al. 1960). These include small, star-shaped or fusiform myxoid cells, with an oval nucleus and a vacuolated cytoplasm containing numerous mucoid inclusions; immature lipocytes, round cells with vacuolated cytoplasm; "signet cells", with a large vacuole occupying almost the entire cytoplasm and an elongated compressed nucleus; mature lipocytes, adult adipose cells; "round cells" with intracytoplasmic fat droplets; and multinucleate giant cells.

The varying proportions of these different cells and the nature of the ground substance have led to different classifications. Stout (1944) identified four groups: group I, a well-differentiated myxoid lesion; group II, a poorly differentiated myxoid tumour with pseudo-fibrosarcomatous streaks; group III, adenomatous with round cells and characteristically, large adipose cells; and group IV, a mixed version of groups II and III. Chomette's (1963) classification also includes four groups: myxoid liposarcoma, round-cell liposarcoma, differentiated liposarcoma simulating an ordinary lipoma and polymorphous liposarcoma with numerous cell types. Finally, Enterline et al. (1960) distinguish the following five categories: well-differentiated myxoid tumour, poorly differentiated myxoid lesion, "lipoma-like" lesion, mixed myxoid tumour and non-myxoid type.

The histological examination is of primary importance in two respects: (1) diagnostic, making it possible to exclude a benign tumour such as a ganglion, chronic synovitis or neurofibroma, or a malignant tumour such as a xanthofibrosarcoma, nodular fasciitis or myxosarcoma; and (2) prognostic, since there seems to be a marked difference in the course of the differentiated and anaplastic types. Thus, in a global study of liposarcomata independent of site, Enzinger and Winslow (1962) found 50% of cases had recurrences and Enterline et al. (1960) 31% with metastases (mainly pulmonary, rarely nodal), principally arising in the non-myxoid types. Hence the prognosis seems markedly different for a grade I tumour with 75% survival (Reszel and Soule Coventry 1966) and for a grade IV where the survival rate falls to 32%. Likewise, most authors refer to the importance of site in the prognosis of liposarcoma, a distal site being more amenable to radical procedures.

Treatment

The treatment of liposarcomata is always surgical, based on wide resection with amputation if necessary. The rare cases reported in the hand are associated with relative stability, and a recurrence is probably due to an inadequate procedure. Radiosensitivity of liposarcomata is noted by Pack and Pierson (1954) for the myxoid forms, but Enterline et al. (1960) find it ineffective in the non-myxoid types. Most authors agree in suggesting combined radiation and surgical treatment. Some suggest preoperative irradiation to shrink the tumour bulk and facilitate surgical excision, while others use postoperative radiation routinely in view of the tendency of the tumour to recur.

Clinical Types

The clinical variants of liposarcoma include liposarcomatous degeneration in von Recklinghausen's disease, though no lesion of this type has yet been reported in the hand. There is also primary liposarcoma of bone, a tumour of metaphyseal origin, most often situated in the femur but with one case in the finger reported by Stewart in 1931.

Clinically nonspecific, this lesion may be manifested radiologically by an apparently benign intraosseous cavity or the appearances of multiple osteolysis with gaps in the cortex and destruction of bone.

Bibliography

Booher RJ (1965) Lipoblastic tumors of the hands and feet. J Bone Joint Surg [Am] 47: 727

Chomette G (1963) Les liposarcomes. Arch Anat Pathol A113

Dreyfuss U, Ben Arieh JY et al (1978) Liposarcoma a rare complication in neurofibromatosis, case report. Plast Reconstr Surg 61/2: 287-289

Edland RW (1968) Liposarcoma. A retrospective study of fifteen cases. A review of the literature and a discussion of radiosensitivity. Am J Röntgenol 103/4: 778-789

Enterline HTJD, Culberson DB et al (1960) Liposarcoma, a clinical and pathological study of 53 cases. Cancer 13: 932

Enzinger FM, Winslow DJ (1962) Liposarcoma. A study of 103 cases. Virchows Arch [A] 335: 367

Groff DB, Horvitz A (1964) Liposarcomas. Am J Surg 107: 725

Haber MH, Alter AH et al (1965) Tumors of the hand. Surg Gynecol Obstet 121: 1073

Holtz F (1958) Liposarcomas. Cancer 11: 1103

Kauffman SL, Stout AP (1959) Lipoblastic tumors of children. Cancer 12: 912

Kendal TE, Robinson DW et al (1969) Primary malignant tumors of the hand. Plast Reconstr Surg 44: 37

Mandard JC, Mandar AM et al (1973) Les liposarcomes primitifs de l'os à propos de 5 cas. Ann Pathol 18/3: 329-346

Moulonguet P, Cazal JF (1960) La dégénérescence sarcomateuse des lipomes. Ann Pathol 5/1: 70-85

Pack GT, Pierson JC (1954) Liposarcoma, a study of 105 cases. Surgery 36: 687

Phelan JT, Perez-Mesa C (1962) Liposarcoma of the superficial soft tissue. Surg Gynecol Obstet 115: 609

Posch JL (1956) Tumors fo the hand. J Bone Joint Surg [Am] 38: 517-539

Reszel PAEH, Soule Coventry MB (1966) Liposarcoma of the extremities and limb girdles, a study of two hundred seventy two cases. J Bone Joint Surg [Am] 48: 229

Roujeau J, Galian P (1963) Les liposarcomatoses à pro-

pos de l'observation. Arch Anat Cytol Pathol 11: 157-160
Sawhney KK, McDonald JM (1975) Liposarcoma of the hand. Am Surg 47: 117-119
Sawyer KC, Sawyer AE et al (1968) The unpredictable fatty tumor. Arch Surg 96: 773
Shiu MH et al (1975) Results of surgical and radiation therapy in the treatment of liposarcoma arising in the extremity. AJR 123: 577-582
Soule EH (1957) Tumors. Lipomatous tumors: Classification, pathology and diagnosis. In instructional course lectures. Am Acad Orthop Surg 14: 311-320
Spittle MF, Neuton KS et al (1970) Liposarcoma, a review of 60 cases. Br J Cancer 24: 696-704
Starkloff GB, Saxton JA et al (1951) Liposarcoma of an extremity associated with multiple subcutaneous lipomas. Ann Surg 133: 261
Sternberg SS (1952) Liposarcoma arising within a subcutaneous lipoma. Cancer 15: 975
Stewart FW (1931) Primary liposarcoma of bone. Am J Pathol 7: 87-93
Stout AP (1944) Liposarcoma, the malignant tumor of lipoblasts. Ann Surg 119: 86
Virchow R (1863) Die krankhaften Geschwülste, Bd 1. Hirschwald, Berlin, S 369-434
Wright CJE (1948) Liposarcoma arising in a simple lipoma. J Pathol 60: 483

Myxoma

In 1863 Virchow used the term myxoma to designate a tumour that he was to describe precisely in 1871 as having the macroscopic and histological characteristics of Wharton's jelly of the umbilical cord. The numerous cases reported subsequently by different authors led to the definition of a heterogeneous group of tumours having in common a predominance of myxomatous tissue but capable of a benign or malignant course (myxoma, myxosarcoma, fibromyxoma, lipomyxoma, mesenchymoma). Not until 1948 did Stout establish a new classification in which the myxoma constituted a tumoral entity with well-defined macroscopic and histological features and course. Myxoma is now seen as a benign tumour of mesenchymal origin composed of star-shaped and fusiform cells contained in a poorly vascularised, mucoid, oedematous stroma. This tumour may arise from different tissues (soft tissues, muscles, bone) and is also characterised by purely local growth and the absence of remote metastases.

The largest series, that of Stout (1948), deals with 143 extracardiac myxomata found in various parts of the body, since mesenchymal tissue is widely spread (genitourinary tract, bone, subcutaneous areolar tissue, mesentery, muscle). It emerges that myxomata occur equally frequently in males and females and may do so at any age, the average being 55-60 years. The cases reported in the hand are very rare and can be situated in the subcutaneous areolar tissue, the aponeurotic structures or the muscles. They are seen mainly in the fingers, more rarely in the palm. Attar (1956) found 2 digital lesions in 21 cases of myxomata.

Ireland et al. (1973) found 5 myxomata of the hand (4 of the fingers and 1 of the thenar eminence) in 58 cases. The first cases situated in the palm were reported by Keoshian et al. (1964), Andreef and Paneva-Holevitch (1967) and Tolhurst (1973).

Clinical Features

Clinically, myxoma presents as a usually solitary, painless tumour that is well-circumscribed, rounded, more or less deep and adherent neither to the skin nor to the skeleton. When it is intramuscular, it is mobile at rest and fixed during muscular contraction. Its size may produce functional disturbance (Keoshian et al. 1964) or compress adjacent nerves to produce paraesthesiae (Tolhurst 1973) or an actual median nerve compression syndrome with paraesthesiae and muscular atrophy (Ireland et al. 1973).

Complementary examinations do not supply unequivocal information. Radiological examination with soft X-rays shows a fairly well circumscribed shadow which is sometimes the site of calcifications (Milgram and Preininger 1973). A skeletal survey occasionally reveals signs of fibrous dysplasia; this association with myxoma is a peculiar feature that we will discuss later.

Histology

In the absence of specific clinical and paraclinical signs, the diagnosis and precise identification of myxoma depend mainly on the macroscopic and histological criteria.

Macroscopically, myxoma presents as a greyish, lumpy, irregular, multilobular tumour of soft gelatinous consistence, with an appearance on section of oedema fluid and without haemorrhagic or necrotic areas. There is no true capsule around the tumour but a pseudo-capsule caused by crowding of the peripheral tissues compressed by the tumoral growth. Myxomata cannot therefore be enucleated and are integrated with the structures they invade; they are infiltrating tumours sending extensions between the muscle bundles and the aponeurotic layers.

Microscopically, the appearance is the same as that seen with Wharton's jelly of the umbilical cord and which exists outside embryonic life only in the vitreous substance of the eye. The amorphous and viscous mucoid matrix constitutes virtually the entire tumour; it stains with polychrome blue, mucicarmine and thionine. It is poorly vascularised and contains only sparse cellular elements and occasional fibres buried in the ground substance. Collagen fibres are seen at the periphery of the tumour, since they constitute the pseudo-capsule, or in the tumour itself as fibrous septa in intramuscular myxomata. Two types of cells may be distinguished: fusiform cells with an ovoid nucleus and acidophilic cytoplasm, either solitary or interconnected by long, slender cytoplasmic extensions; and stellate or ramifying cells, few in number, with a spherical nucleus and exhibiting numerous intracytoplasmic vacuoles which give the cells a foamy appearance. The nature of these stellate cells is obscure; they could be undifferentiated mesenchymal cells containing glycogen and mucin. The fusiform cells are probably secreting fibroblasts. Finally, the oedematous nature of the matrix could be due to its abundance of hyaluronic acid, a mucopolysaccharide acid with a great capacity for fixation of water in the intercellular spaces. There are never any mitoses or nuclear deformations, and the histological features are those of a benign tumour.

All this fulfils Stout's (1948) criteria for specification of this tumour: star-shaped or fusiform cells in a loose matrix with collagen fibres, poor vascularisation and absence of chondroblasts, lipoblasts, rhabdomyoblasts or other mesenchymal elements. This classification, which is not always observed, permits the diagnosis of myxoma as such. However, this type is very rare; the tumours more commonly met with, although myxomatous substance predominates in them, also contain other tissues of mesenchymal origin, in particular adipose tissue (lipomyxoma) or fibrous tissue (fibromyxoma). The presence of histological features of malignancy characterises myxosarcoma, a tumour outside the extent of ordinary myxomata.

Aetiopathogenesis

Numerous aetiological hypotheses have been advanced. Some authors (Willis 1953) maintain that myxomata do not form a tumoral entity but are rather the result of a degenerative process arising in some other tumour, whose origin may sometimes be detected at histological examination (chondromyxoma, lipomyxoma, fibromyxoma). Others propose an embryonic origin. Ewing (1941) sees the myxomata as dysembryomata derived from a proliferation of mesenchymal rests that have preserved their developmental potential. Similarly, Herbut (1952) and Senn (1900) speak of embryonic remnants as yet undifferentiated into fibrous tissue. On the other hand, Stout (1948) and Ackerman (1953) argue that this is an entirely specific tumour formed from well-differentiated cells. The resemblance of its particular ground substance to Wharton's jelly could be due to the persistence of slowly evolving embryological elements. Other authors (Thibonneau and Kress 1954) are still strongly in favour of metaplasia in view of the frequent association with other connective tissues. Enzinger (1965) and Ireland et al. (1973) regard the stellate cells as degenerate fibroblasts, incapable of forming collagen while producing mucopolysaccharides rich in hyaluronic acid. Mazabraud et al. (1967), noting the frequent association of myxoma and fibrous dysplasia of bone, suggest a common origin for both disorders. Finally, there is the viral hypothesis put forward by Glazunov and Puckov (1960).

Clinical Types

One particular type of myxoma reported in 1926 by Henshen Volke has formed the subject of numerous studies (Krogius 1928; Vehlinger 1940; Braunwarth 1953, cited by Mazabraud et al. 1967). This type, combining myxomata of the musculotendinous junctions and aponeuroses with fibrous dysplasia of bone has been particularly reviewed by Mazabraud et al. (1967). Of the nine cases reported, seven had multiple myxomata; the sex ratio was four males to five females. Clinically, the bony manifestations usually precede the appearance of the musculotendinous tumour(s) by several years. Radiologically, the bony lesions are those of rarefaction and sclerosis; histologically, some of the bony foci are identical to the lesions seen in the musculotendinous system. An abundant and richly vascularised mesh contains numerous spans of metaplastic bone of varying shape and abnormal structure, without any trace of chondroid or cartilaginous tissue. The electron microscopic study of a muscular myxoma associated with a fibrous dysplasia (Jaffe-Lichtenstein) shows structural analogies that suggest a dysplastic origin common to both disorders (Leung et al. 1971). Wirth et al. (1971) see the muscular myxoma as an extraskeletal localisation of fibrous dysplasia. Finally, for Mazabraud et al. (1967) this lesional association indicates a global dysplasia of the locomotor system in its two specialised forms - bony and musculotendinous-aponeurotic.

Differential Diagnosis

This remains controversial. An essential feature in the course of these tumours is the high percentage of recurrences, leading to repeated excisional interventions. This extraordinary tendency to recurrence (Uradel, cited in Stout 1966, found 8 recurrences in 11 tumours) is due to the difficulty of enucleation, the laboriousness of removal and the invasion of adjacent structures. Patel and Tubiana (1947) consider that the recurrences are often of a different histological nature and exhibit signs of malignancy. Mitchell et al. (1962) see recurrence as virtually constant and of an increasingly malignant and threatening nature, while other authors (Fenster 1933) even believe that myxoma and myxosarcoma are synonymous and can give rise to remote metastases. This malignant transformation of extracardiac myxomata is far from being generally accepted, however, and has been formally rejected by some.

Thus Stout (1948), adding to his three histological criteria the evolutional one of absence of remote metastases, considers that this is always a benign tumour developing strictly locally. Attar (1956), too, states categorically that myxosarcoma does not exist and that the discovery of a metastasis during the course of a tumour must lead to reconsideration of the initial diagnosis of myxoma. Sponsel et al. (1952) retain the two terms of myxoma and myxosarcoma but attribute specific characters to each. Myxosarcoma, in particular, exhibits areas of haemorrhage and necrosis on section. Histologically, the nuclei are the site of numerous mitoses, and it is these types that recur most often and give rise to remote metastases (of nine myxosarcomata, seven recurrences required amputation of a limb and three had metastases; of six myxomata, three of them digital, no recurrences were seen after local excision; Sponsel et al. 1952).

Treatment

This is entirely surgical. Total excision of the tumour is the ideal, but extirpation is difficult, and the infiltration of adjacent structures impairs the outcome. However, incomplete excision predisposes to routine recurrence and reoperation. Histological examination of the operative specimen will determine the quality of the excision. The best results are to be hoped for with lesions with little spread that are seen early. When adjacent structures are widely invaded, and when their sacrifice would seriously compromise the function of the hand, the decision to amputate may be unavoidable.

Bibliography

Ackermann L (1953) Surgical pathology. Mosby, St. Louis

Andreef I, Paneva-Holevitch E (1967) Myxoma der Hohlhand. Beitr Orthop Traumatol 14: 454-456

Attar S (1956) Myxoma; clinicopathologic study. Am J Surg 91: 755-760

Backstrom CG, Linell F, Ostberg G (1965) Cystic, myxomatous adventitial degeneration of the radial artery with development of ganglion in connective tissue. Report of two cases. Acta Chir Scand 129: 447-451

Bauer WH, Harell A (1954) Myxoma of bone. J Bone Joint Surg [Am] 36: 263-266

Chardot C, Parache R (1967) Les myxomes purs des coulées conjonctives des membres et des parois du tronc. A propos de 19 observations. Bull Cancer (Paris) 54: 297-318

Chaves E, Nobrega C, Oliveira AM (1971) Congenital myxoma in childhood: A report of two cases. Cancer 28: 239-243

Damian CR, Soule EH, Ivins JC (1973) Myxoma of somatic soft tissues. A report of 58 patients, 3 with multiple tumors, and fibrous dysplasia of bone. Mayo Clin Proc 48: 401-410

Dutz W, Stout AP (1961) The myxoma in childhood. Cancer 14: 629-635

Enzinger FM (1965) Intramuscular myxoma. A review and follow-up study of 34 cases. Am J Clin Pathol 43: 104-113

Ewing J (1941) Neoplastic diseases, 4th edn. Saunders, Philadelphia, p 182

Fenster E (1933) Primäres malignes Myxom des Herzens mit Metastasen. Z Pathol 45: 565-570

Filarski SA, Doyle WF, Mutz SB (1974) Traumatic myxomatous degeneration of a peripheral nerve. Hand 6/3: 229-232

Fine G, Morales A, Horn RC Jr (1968) Cardiac myxoma: A morphologic and histogenetic appraisal. Cancer 22: 1156-1162

Glazunov MF, Puckov JG (1960) Human muscular myxoma and intracellular inclusions. Vopr Onkol 6: 11-27

Herbut P (1952) Urological pathology. Lea & Febiger, Philadelphia, p 256

Hodson JJ, Prout RES (1968) Chemical and histochemical characterization of muco-polysaccharides in a jaw myxoma. J Clin Pathol 21: 582-589

Ireland DCR, Soule EH, Juins JC (1973) Myxoma of somatic soft tissues. Mayo Clin Proc 48: 401-410

Keoshian LA, Lower RR, Eversole SL, Nelsen TS (1964) Myxoma of the palm. A case report. Plast Reconstr Surg 34: 66-70

Kindblom LG, Stener B, Angervall L (1974) Intramuscular myxoma. Cancer 34/5: 1737-1744

Leak LV, Caulfield JB, Burke JF, McKhann CF (1967) Electron microscopic studies on a human fibromyxosarcoma. Cancer Res 27: 261-285

Lejeune E (1972) Dysplasie fibreuse et myxomes des tissus mous. Rev Rhum Mal Osteoartic 39: 281-288

Leung TK, Vauzelle IL, Patricot LM, Lejeune E, Queneau

P (1971) Etude ultrastructurale et cytochimique d'un myxome musculaire associé à une dysplasie fibreuse. Ann Pathol 16/4: 417-428

Mazabraud A, Girard J (1957) Un cas particulier de dysplasie fibreuse à localisations osseuses et tendineuses. Rev Thum Mal Osteoartic 24: 2-8

Mazabraud A, Semat P, Roze R (1967) A propos de l'association de fibromyxomes des tissus mous à la dysplasie fibreuse des os. Presse Med 75/4: 2223-2228

Mazabraud A, Totyl L, Roze R, Semat P (1969) Un nouveau cas d'association de dysplasie fibreuse des os à un myxome des tissus mous. Sem Hôp Paris 45: 863-868

Milgram JW, Preininger RE (1973) Calcifying myxoma. J Bone Joint Surg [Am] 55/2: 401-405

Mitchell MM, Paradire C, Massot JP (1962) Les myxomes. Presse Méd 70/38: 1773-1776

Patel JP, Tubiana R (1947) Tumeurs primitives rétropéritonéales. Ann Pathol 2: 182-198

Perou ML, Kolis JA, Zaeske EV (1967) Myxoma of the toe: Case report. Cancer 20: 1030-1034

Rosin RD (1973) Myxomes intramusculaires. Br J Surg 60/2: 122-124

Salmon M, Musinger M (1959) Sur les myxomes, à propos d'un cas de myxome des muscles du mollet. Marseille Chir 2: 282-295

Salmon M, Mignard P, Gonggryp N, Garbe P (1976) Les myxomes des muscles. A propos de 3 observations. S.O.F.C.O.T. XLVIII Reunion annuelle, pp 209-217

Senn O (1900) Pathology and surgical treatment of tumors. Saunders, Philadelphia

Sponsel KH, McDonald JR, Ghormley RK (1952) Myxoma and myxosarcoma of the soft tissues of the extremities. J Bone Joint Surg [Am] 34: 820-826

Stout AP (1948) Myxoma, the tumor of primitive mesenchyme. Ann Surg 127: 706-719

Stout AP (1966) Myxoma. In tumors of the soft tissues. Atlas of tumor pathology, 2nd ser, fasc 1. Armed Forces Institute of Pathology, Washington, pp 35-37

Thibonneau J, Kress O (1954) Myxosarcomas. Am J Cancer 8: 267

Tighe JR (1963) The histological demonstration of mucopolysaccharides in connective tissue tumors. J Pathol Bacteriol 86: 141-149

Tolhurst DE (1973) Myxoma of the palm. Hand 5/3: 260-262

Virchow R (1863) Die krankhaften Geschwülste, Bd 1. Hirschwald, Berlin, S 369-434

Virchow R (1871) Die Cellularpathologie in ihrer Begründung und pathologische Gewebelehre, 4. Aufl. Hirschwald, Berlin, S 563

Willis RA (1953) Pathology of tumors, 2nd edn. Butterworth, London, p 653

Wirth WA, Leavitt D, Enzinger FM (1971) Multiple intramuscular myxomas. Cancer 27: 1167-1173

Fibromata and Fibromatoses

Classification of fibrous connective tissue tumours was for long restricted to benign encapsulated fibromata on the one hand and malignant metastasising fibrosarcomata on the other. This overdogmatic distinction has been broadened and complicated by more precise histological and cytological studies and the recognition of a number of intermediate presentations.

Benign tumours of fibrous connective tissue are characterised by proliferation of fibroblasts and histiocytes, mesenchymal cells with macrophagic capacity. Depending on which cell type predominates, this definition places fibroma and histiocytoma as extremes with histiocytofibroma as an intermediate form. The group of fibromatoses comprises a heterogeneous collection of lesions, distinguished from fibroma by their size, extent, number and location of their components, which are preferentially more deeply seated. As regards prognosis, the pseudosarcomatous types that invade local structures without remote dissemination lie between benign and malignant tumours.

Numerous classifications have been proposed by different authors (Stout 1954; Enzinger 1965a; Pinol-Aguade et al. 1971; McKenzie et al. 1970; Weschler 1972; Fleischmajer et al. 1973; Folana 1974; Barrière et al. 1975; Allen 1977). In a study based on 44 juvenile fibromatoses, Stout (1954) distinguished in this particular category: desmoid fibromatoses or desmoid tumours composed of pseudo-cicatricial tissue; palmar and plantar fibromatosis affecting essentially the aponeuroses, of the nature of Dupuytren's contracture, and rare in children; the juvenile aponeurotic fibromatosis described by Keasbey; keloid fibromatosis; diffuse muscular fibromatosis associated with interstitial fibroblast proliferation plus connective tissue fibres producing localised thickening of the muscle; and congenital, multiple, generalised fibromatosis invading the superficial tissues, viscera and bones. Stout mentions the difficulty, sometimes even the impossibility, of distinguishing between fibromatoses and differentiated fibrosarcomata. However, a certain number of features common to the fibromatoses may be maintained: a more fibrous than cellular proliferation of fascicular structure forming long wavy bundles; higher collagen density; narrow, regular, fusiform cells parallel to the fibres, with greater cellularity in the fibrosarcomata than the fibromatoses; few mitoses and never any nuclear deformities; no necrosis or haemorrhage; and a poorly circumscribed tumour which is infiltrative but hardly destructive.

Mackensie (1964) suggests an arrangement extending from inflammatory to tumoral pathology and from benign to malignant types: cicatricial fibroses - hypertrophic scars, keloids; fibromata, histiocytofibromata, histiocytomata; nodular pseudosarcomatous fasciitis; the fibromatoses proper-desmoid form or desmoid tumour or invasive fibroma, palmar or plantar form juvenile aponeurotic form or calcifying aponeurotic fibroma, keloid form diffuse muscular form congenital multiple generalised form; differentiated fibrosarcomata; and undifferentiated fibrosarcomata.

Degos' classification (1953a) allows for the date of appearance of the lesions and the difficulties of histological diagnosis:

1. Congenital and juvenile fibromatoses:
   a) Congenital solitary fibromatosis or fibrous hamartoma of the newborn (Enzinger 1969)
   b) Generalised multinodular fibromatosis (Mauclaire and Legry 1915)
   c) Congenital fibromatosis of the sternocleidomastoid or fibromatosis colli or congenital torticollis
   d) Infantile digital fibromatosis or benign juvenile digital fibromatosis of Reye
   e) Diffuse muscular fibromatosis (Stout 1948)
   f) Juvenile palmar-plantar fibromatosis or calcified juvenile aponeurotic fibromatosis of Keasbey
   g) Juvenile hyaline fibromatosis
2. Late onset fibromatoses:
   a) Palmar fibromatosis or Dupuytren's disease
   b) Plantar fibromatosis or Ledderhose's disease
   c) Fibromatosis of the corpora cavernosa or Peyronie's disease
   d) Elastofibroma of the back
3. Pseudosarcomatous fibromatoses:
   a) Atypical cephalic fibroxanthoma of the elderly

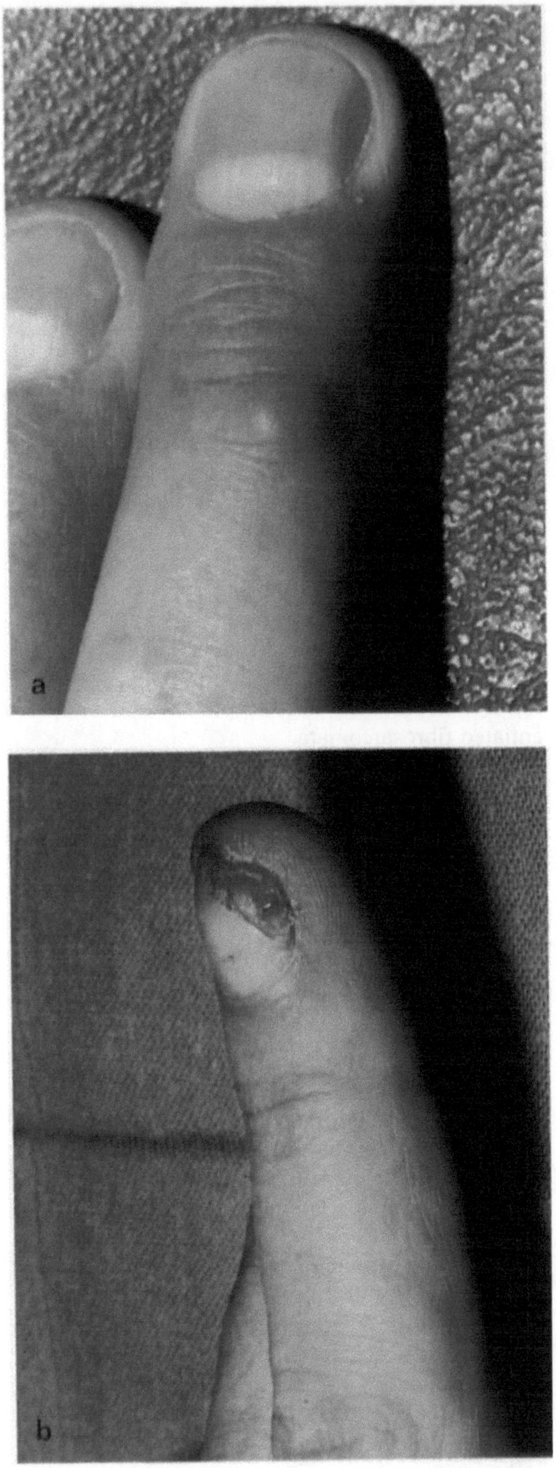

Fig. 55. ***a*** Digital fibroma. ***b*** Fibroma: rare subungual localisation

b) Nodular fasciitis or subcutaneous pseudosarcomatous fibromatosis
c) Pseudosarcomatous muscular fibromatosis, including Enzinger's infantile aggressive fibromatosis, Kern's proliferative myositis (cited by Weschler 1972), Munchmeyer's progressive myositis ossificans (cited by Weschler 1972), polypoid pseudosarcoma (Stout 1948; Lattes 1964)

A synthesis of the different reported classifications and certain intermediate forms described separately leads to the following overview of the different forms in the soft tissues of the hand:

1. Fibroma, histiocytofibroma and histiocytoma
2. Fibromatoses:
   a) Juvenile forms: juvenile benign digital fibromatosis of Reye, palmar-plantar aponeurotic fibromatosis of Keasbey
   b) Adult forms: Dupuytren's disease, acquired digital fibrokeratosis or acral fibrokeratoma, Koenen's tumours, phalangeal pads (knuckle-pads)
3. Pseudosarcomatous forms:
   a) Nodular fasciitis
   b) Desmoid tumours

Fibroma, Histiocytofibroma and Histiocytoma

Fibroma is a rare tumour in the hand, accounting for only 0.5% in the series of Haber et al. (1965) and 1% in Stack's (1960) series. It occurs mainly in adults and presents as a lesion a few millimeters in diameter, indolent embedded in the dermis and often adherent to the deeper layers (Fig. 55 a, b).
This white or yellowish tumour is well-encapsulated and characterised by an intradermal proliferation respecting the papillary body and formed of more or less abundant fusiform fibroblastic cells, associated with a hyperplastic reaction in the epidermis. Treatment consists of surgical excision limited to this (always benign) tumour.
Histiocytofibroma or dermatofibroma is the most common type of skin fibroma. This benign tumour, still called fibroma simplex (Unna 1896) or fibroma *en pastille* (Civatte 1967) occurs mainly in adults and more often in women. Clinically, there is an erythematous nodule 1-2 cm in diameter, often situated in the hand and very frequently in the fingers. It is a firm lesion, more or less adherent to the skin and deep layers, often painless and consisting

of a flattened lozenge-shaped nodule within the skin (Fig. 56). It remains unchanged for a long period but may sometimes regress spontaneously and rarely ulcerates. Grossly, the lesion is yellow or light brown (Fig. 57) but may sometimes be very pigmented and simulate a melanoma. Histologically, the tumour is not encapsulated and exhibits proliferation of fibroblasts and histiocytes in varying proportions. There are also giant cells containing inclusions of lipids and haemosiderin as well as inflammatory residues. The tumour is always benign. Surgical treatment is not routine and is indicated only in cases with pain or dysfunction, where the diagnosis is doubtful or for aesthetic reasons (Fig. 58).

The term *histiocytofibroma* tends more and more to link the generality of fibromata, histiocytomata and intermediate forms, distinguished only by histological variations indicative of the developing nature of the lesions.

Fibromatoses

Juvenile Fibromatoses

Benign Juvenile Fibromatosis of the Fingers (Reye)

Reviewing a series of fibrous tumours, some of which were wrongly regarded as fibrosarcomata, Reye in 1965 described a new benign pathological entity exhibiting progressive clinical features and original microscopic peculiarities. Since then, numerous authors have confirmed the specific nature of this disorder, which remains rare (Shapiro 1969; Pinol-Aguade et al. 1971; Mackensie 1964; Degos et al. 1975). Juvenile benign fibromatosis occurs in the very young child, sometimes from birth and usually within the first 3 years. It affects both sexes equally, generally as multiple lesions and only rarely as solitary. The disease is mainly located in the fingers, but can also affect the toes or both digits simultaneously. The lesions are preferentially situated on the lateral and dorsal aspects of the fingers, most often at the distal phalanges, and may extend towards the nail, which is elevated without destruction, or towards the palmar region. All the fingers may be involved, both of the right or left hand; both hands may be affected.

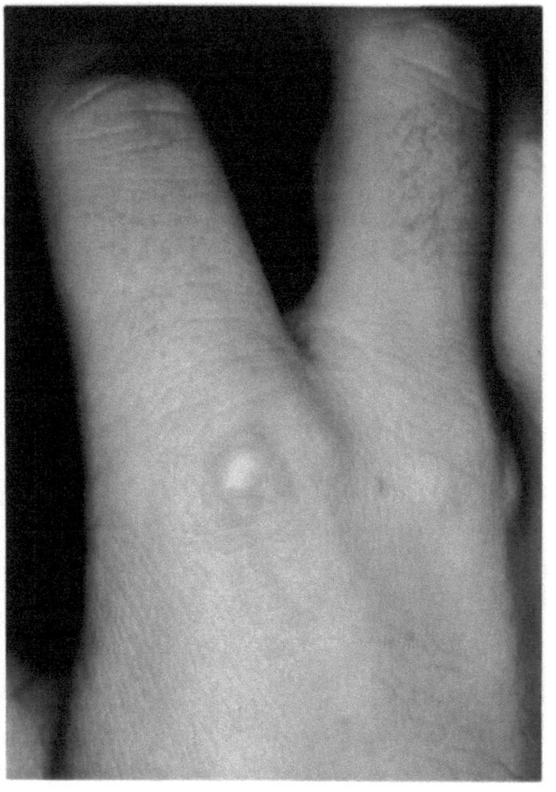

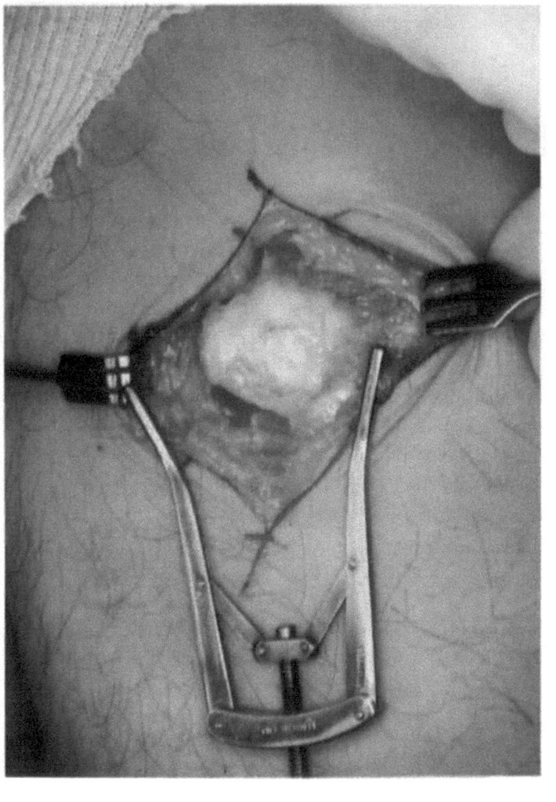

Fig. 56. Histiocytofibroma of dorsum of hand *(above)*

Fig. 57. Fibrous tumour: operative appearance *(right)*

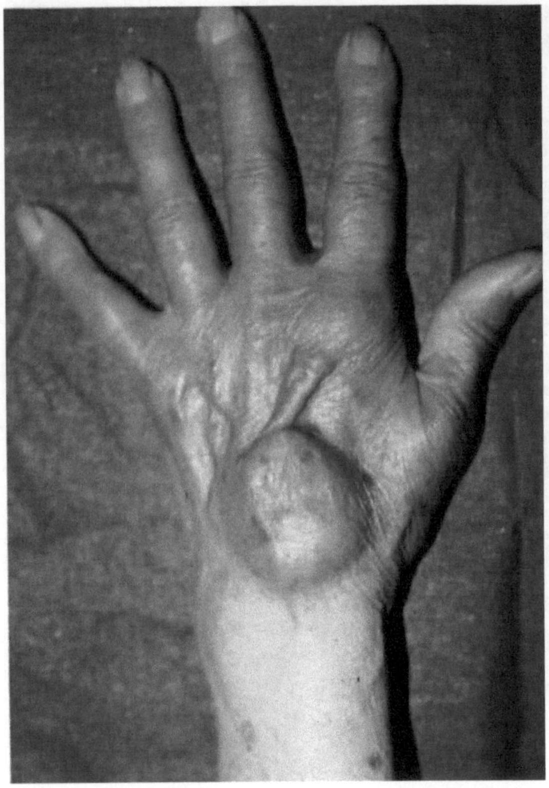

Fig. 58. Large subcutaneous fibroma in an elderly woman

Clinical Features

The lesions present as intradermal nodules 1 cm in diameter, firm, elevated, hemispherical, painless, smooth-surfaced and pink or of a normal skin colour. More rarely, they constitute circumscribed bulges associated with a deeper situation of the lesion. Fine telangiectases may be visible in the region of the nodules. The evolution of these tumoral structures is usually slow, although a sudden increase in growth is sometimes observed. Occasionally, there may be repercussions on the function of the finger (Beckett and Jacobs 1977).

Radiography

Radiological surveys usually indicate the integrity of the phalanges and adjacent joints. However, numerous instances of bony anomalies have been reported. Bloem et al. (1974) record two cases, one with deformity of the 3rd and 4th metacarpals, the other with shortening of the 3rd and 4th metacarpals and corresponding metatarsals. Lakhanpal et al. (1978) note deformity and shortening of the four outer metacarpals, shortening of the 3rd metatarsal and bilateral absence of the 2nd metatarsal; thus these authors surmise that there may be other pathological associations to be included in the same syndrome. Pinol-Aguade et al. (1971) report a case of associated and stable, bilateral, plantar fibromatosis. Further, Bloem et al. (1974), in two of their patients, note disorders of skin pigmentation on the forehead and temples and hypertrophy of the lower lip. The whole matter remains unsettled, but familial polyfibromatosis has never been noted, and neither has antecedent trauma.

Development

In general, the lesions tend to persist or enlarge, and new nodules may appear. Mackensie (1964), however, report a case of spontaneous regression of a nodule. Even after surgical intervention recurrence is common and justifies Reye's initial description of "recurrent fibrous digital tumours of childhood". The prognosis is always good, as the tumour is benign, hardly limits finger function and never gives rise to metastases.

Histology

The predominant feature under optical microscopy is connective tissue proliferation of average cell density under a normal or flattened epidermis, with abundant collagen of normal structure. The elastic fibres, however, are rare and may be more or less degenerate. These anomalies are situated throughout almost the entire dermis, while the basal membrane is respected. The vessels are numerous, with appearances of endothelial engorgement, and inflammatory elements are rare. The fibroblasts seen are morphologically normal, without atypical forms, abundant mitoses or nuclear deformation. The main and essential feature is the presence of intracytoplasmic inclusions, sometimes visible with conventional stains (H & E) but better shown by the ferric haematoxylin of methyl pyronine and by orange acridine in the ultraviolet. These inclusions, which have not been detected by every worker in optic microscopy, do not seem to contain DNA (Feulgen negative). They were first observed by Reye (1965), then by Shapiro (1969) and studied under electron microscopy by Mackensie (1964) and by Pinol-Aguade et al. (1971). The ultrastructural study of benign juvenile fibromatosis of the fingers has revealed the following features:

- The connective tissue is composed of unchanged collagen fibres arranged in thick bundles.
- The elastic tissue is scanty or even totally absent.
- The vessels, nerve structures and sweat glands appear normal.
- Inflammatory cells are present: macrophages and mastocytes.
- There are numerous fibroblasts without morphological changes, but their endoplasmic reticulum is very developed, with the presence of marked saccular dilatations. Mitochondria and centrioles are seen.
- Certain fibroblasts (2%, Pinol-Aguade et al. 1971) contain bulky, dense intracytoplasmic inclusions, not bounded by membranes and generally located near the nucleus. These inclusions consist of fibrillary clumps that are parallel or interlaced, with vesicular granulations; they are not found in any other type of fibromatosis and are considered by Battifora and Hines (1971) as specific to this lesion (though they have not been detected by Bloem et al. 1974 or by Lakhanpal et al. 1978).

Electron microscopic analysis permits differentiation of two types of inclusions: some are rounded, well-circumscribed, large, often situated near the nucleus and with a central, amorphous, finely granular appearance; the others are smaller, irregular, poorly circumscribed and intracytoplasmic. Finally, Battifora and Hines (1971) also describe other dense and smaller bodies around all these inclusions, usually solitary, rarely grouped in clumps.

Aetiopathogenesis

Numerous hypotheses have been advanced. The frequent multiplicity of the lesions and their occasionally observed, spontaneous involution argue against a tumoral nature. Their pace of development and the existence of intracytoplasmic inclusions are in favour of a viral origin, but this aetiology has not been confirmed by electron microscopic studies. In support of it, Pohjampelto et al. (1967) have isolated an ultrafiltrate that induces agglutination of human fibroblasts in culture, but no other experimental study in depth has so far really established a viral origin. Mehregan (1966) considers the inclusions to result from the intracytoplasmic accumulation of a product of abnormal metabolism synthesised by certain abnormal fibroblasts. Mackensie (1964), noting that these inclusions are analogous to those seen in elastofibroma by Winkelmann and Sams (1969) and in fibroblasts in the course of elastogenesis by Pinol-Aguade et al. (1971) considers that there is an isolated and localised degeneration, an "elastodysplasia", a malformation of elastic tissue that may persist or disappear.

Treatment

The therapeutic approaches proposed by various authors relate to the benign nature of the affection, its site and its potential for advance and recurrence. Shapiro (1969) advises surgical ablation of all the nodules with a 2-mm margin at the periphery of the lesion and extending to the periosteum in depth. However, in multiple forms the scrupulous observance of this approach may bring about damage and functional impairment disproportionate to the benign nature of the condition. In view of the possible involution of the lesions, Mackensie (1964) advises refraining from treatment. Bean (1969) suggests biopsy to confirm the diagnosis of the lesion and its benign nature before deciding to abstain from all surgical procedures. However, surgery is essential when there is functional impairment, in order to free a joint or a tendon or even to preserve a nail. In such cases, excision of the nodule should aim to be complete in order to avoid recurrence.

Juvenile Aponeurotic Fibromatosis (Keasbey)

Described by Keasbey in 1953 and still called palmar-plantar fibromatosis, this condition has a specific anatomoclinical nature. It occurs mainly in children and is marked by proliferation of connective tissue elements with foci of calcification and a certain degree of cartilaginous differentiation. Owing to its speed of progress, Kermarec and Ayraud (1968) suggest replacing "juvenile aponeurotic fibroma" by "recurrent tenofibroma". It is a rare unilateral lesion, of equal sex distribution, which may appear from birth, occurs chiefly in children and adolescents, and is situated primarily on the palm of the hand (especially the thenar and hypothenar eminences) and the sole of the foot (Fig. 59). Certain cases, however, have been seen in older subjects (Lichtenstein and Goldman 1964a, four cases above the age of 30) and at other sites (Rios-Dalenz et al. 1965, on the back). Allen and Enzinger (1970) report the following distribution of 19 cases; wrist 3, palm 10, index 3, middle finger 3. Distribution is equal on both sides and the depth of the lesion is variable (subcutaneous, tendinous, deep, juxta-articular).

The terminology proposed by Keasbey is not accepted by every author, as this is not just a benign

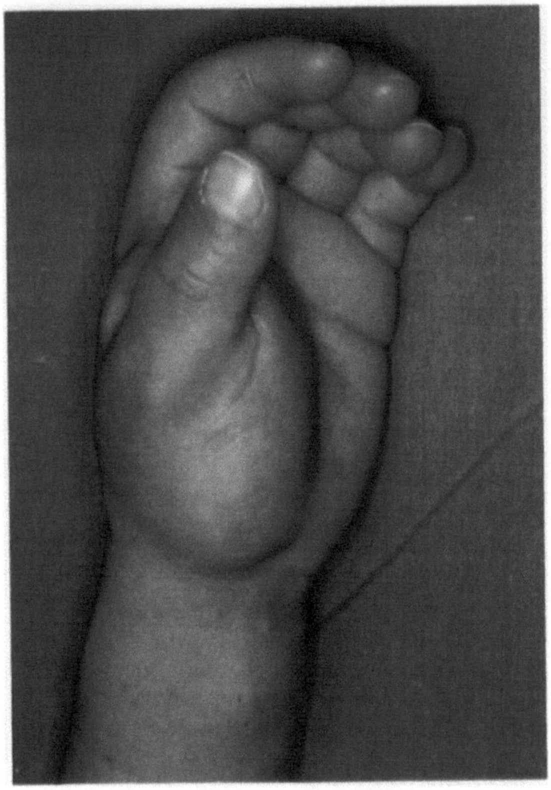

Fig. 59. Juvenile aponeurotic fibromatosis of the thenar eminence

fibroblastic tumour, and it also differs from the extraskeletal cartilaginous tumours. In view of the various constituents found at histology, Des Marchais et al. (1972) have proposed the term "calcifying cartilaginous fibromatosis". Lichtenstein and Goldman (1964a), considering that a process is involved analogous to that of fibromatosis and developing on a basis of cartilaginous tissue, suggest the name "cartilaginous analogue of fibromatosis".

Clinical Features

The condition is usually asymptomatic and painless and presents as a solitary tumour. It is one firm nodule, 1-3 cm in diameter, mobile in relation to the skin and more or less fixed on the deeper layers. It is poorly circumscribed, situated in the subcutaneous, aponeurotic or tendinous tissue and always independent of bony structures. The tumour arises spontaneously, unrelated to injury or local irritative cause. Its course is slow and occupies several months or years. The nodule may at the most handicap the function of a finger but does not produce sensory deficit when located in relation to a digital collateral nerve.

Radiological examination may suggest the diagnosis if it reveals calcifications in the form of very fine intratumoral granulations.

Pathology

Two types of appearance of the tumour may be distinguished: a ramifying, poorly limited type, often found in children under the age of 6 years, and a nodular type seen in older patients, though Keasbey (1953) argues that these are successive stages in the course of the same lesion. Histologically, the tumour is ill-defined peripherally and surrounded by a pseudo-capsule infiltrating the fat and adjacent muscle. It is composed of fibrous tissue, made up of wavy collagen fibres arranged in packed bundles. There is a fibroblastic proliferation whose appearance is quite specific, since it is seen to consist essentially of oval or rounded nuclei oriented in the same direction while the cytoplasm and cell membranes are barely visible. Punctiform or linear calcifications, sometimes detectable by the naked eye, are arranged irregularly in the collagen stroma. These calcifications are surrounded by special cells, very analogous to chondrocytes and forms intermediate between fibroblasts and chondrocytes. Often, there can also be noted multinucleated cells resembling osteoclasts. Finally, very frequently there are areas of chondroid or osseous metaplasia.

Study under polarised light, as proposed by Kermarec and Ayraud (1968), shows arrangement of the collagen fibrils in bundles, with strict interfibrillary parallelism. This appearance is comparable to that seen in tendons and aponeuroses and differs from that of an ordinary fibroma. Rather, it favours Keasbey's concept in terms of a neoformation of a teninous or aponeurotic nature.

Development

The spontaneous course of this tumour is towards slow progressive growth. It is a benign tumour not exhibiting local invasion, malignant change or remote metastases. The outlook after surgical excision is dominated by the major risk of recurrence, of the order of 50%. This appears between 1 month and 12 years after ablation and is not necessarily related to the surgical procedure; indeed, Allen and Enzinger (1970) note that recurrence appears even after wide excision and report one case of "cure" with 5-years' regression even though the ablation was judged inadequate. Finally, despite the fact that some authors consider the recurrence rate to be independent of the age of the patient, Allen and Enzinger report seven recurrences in seven cases observed in children under the age of 5 years.

Differential Diagnosis

This is always difficult and must be based on the opinion of competent pathologists. Numerous errors have been reported, the tumour having been regarded as an ossifying fibroma, a fibrosarcoma, a lipofibrosarcoma or a chondroblastoma. What is important is to be able to exclude a malignant tumour because of the therapeutic implications. The abundant cellularity and nuclear hyperchromatism of the juvenile aponeurotic fibromatoses should be recognised and not allowed routinely to arouse suspicion of a malignant tumour. In practice a deep palmar tumour in a young child should suggest the possibility of a juvenile aponeurotic fibromatosis.

Treatment

Unlike malignant tumours this disorder does not need to be treated by a radical surgical procedure. A limited, non-damaging excision is advised by most authors. The treatment must be conservative and respect the adjacent vascular, nervous and tendinous structures. The fear of recurrence should not lead to mutilating procedures, and it is cardinal to respect the function of a finger affected by this benign tumour. If there is a recurrence, surgical excision is not justified since the tumour will stabilise.

Adult Fibromatoses

Dupuytren's Contracture

This was first described by Baron de Dupuytren in 1832 and is evidenced by progressive flexion contracture of one or more fingers. This lesion is associated with fibrous contracture of the palmar aponeurosis and is really outside the context of tumours of the hand; it will therefore be reviewed only briefly.

Clinical Features

Dupuytren's contracture is more common in males than in females and occurs mainly after the age of 40 years. It is marked by the presence of two types of lesions, which may be situated in the palm or on the palmar aspect of the fingers: modules, usually located in the distal palm and at the proximal phalanx, and longitudinal bands indicative of the aponeurotic contracture and responsible for flexion of the fingers. Their appearance is progressive, being initially limited to solitary nodules and then slowly developing to bring about contraction of the fingers over several years. This evolution may have its exacerbations, often excited by injury. Though the condition is more often found at the ulnar fingers, it frequently affects the radial digits more severely. The extent of the deformity can be calculated using the coding established by Tubiana and Michon in 1961, which allows for the type of lesion, its site and the severity of the contracture.

Associated Lesions. Three types of so-called ectopic lesions may be associated, with the involvement of the hands: dorsal knuckle-pads (usually at the proximal interphalangeal joint), plantar nodules (Ledderhose's disease) and involvement of the corpora cavernosa (Peyronie's disease[1]). Certain other disorders may also be associated with Dupuytren's contracture: diabetes, alcoholism and epilepsy, the effect of which on the genesis and gravity of the disorder is variously assessed. Finally, one should note that the contracture is often hereditary.

Histology

The nodules are richly cellular (fibroblasts), and the bands have numerous collagen fibres and only a few cells. Gabbiani and Majno (1972) show by electron microscopic study that the fibroblasts contain contractile structures (myofilaments) similar to those of smooth muscle and which are also found in granulation tissue; they named these cells myofibroblasts. Brickley-Parsons et al. (1981), followed by Bailey et al. (1977), demonstrate that these fibroblasts produce a special type of collagen (type III), not only in the macroscopically affected areas but also in the unaffected remainder of the palmar aponeurosis.

Aetiology

The aetiology of Dupuytren's contracture still remains unsolved. The role of trauma has not been certainly established, and a hereditary factor is not constant. The part played by associated disorders has seen rather variably assessed. At the histologi-

[1] Translator's note: this is usually called "Peyronie's disease" in the UK and USA.

cal level, Hueston (1969) contests a purely aponeurotic origin of the disease and insists on the role of the dermis.

Differential Diagnosis

Dupuytren's contracture may pose difficult diagnostic problems in two types of circumstances: forms that are nodular from the outset, suggesting an epidermal cyst of the palm or a localised fibromatosis, and ectopic localisations on the back of the fingers (Hueston 1969), which may simulate an epidermal or mucous cyst.

Acquired Fibrokeratoma

Acquired digital fibrokeratoma, described by Bart et al. in 1968, constitutes a distinct anatomical and clinical entity within the general context of fibrous tumours. Though not yet specified as such, this lesion had already been recognised and reported by Moncorps (1931), Hare (1954), Samman (1965) and Steel (1965). Since a digital localisation is not exclusive, though certainly the most common, the term "acquired fibrokeratoma" seems preferable to designate this lesion. It is a relatively uncommon fibroepithelial tumour, always acquired, that only rarely recurs after surgical treatment and never undergoes malignant degeneration. Its mode of origin remains obscure; a traumatic origin is postulated by some authors, and there is no familial or hereditary tendency.

Clinical Features

Acral fibrokeratoma usually occurs in adults between the ages of 30 and 60 years, with a marked male predominance. In the hand it is almost exclusively localised to the fingers. The tumour is preferentially situated opposite the proximal and distal interphalangeal joints on their lateral and dorsal aspects. The palm and back of the hand and the palmar aspect of the fingers are rarely the site of the lesion. On the other hand, fibrokeratoma is relatively common at the circumference of the nail and sometimes even under the nail, which it elevates and deforms.

Acquired fibrokeratoma initially presents as a single, isolated nodule which is projecting, hemispherical or elongated, firm, well-circumscribed and painless. It never produces any functional disturbance. It grows progressively to reach its definitive size of 3-15 mm in a few months and then remains stable. The existence of a peripheral hyperkeratotic rim is characteristic and confirms the diagnosis of acquired fibrokeratoma; this feature is constant, even when the clinical appearance is different as in the pedunculated and flattened forms.

Histology

This is a fibroepithelial tumour with certain specific features. In the epidermis there are major hyperkeratotic lesions with a thick stratum corneum and evidence of granulomatosis, acanthosis and papillomatosis, with hyperplasia of the malpighian cells. The basal membrane is normal. Finally, the epidermis is very thick at the central part of the tumour and becomes progressively thinner as it rejoins the adjacent normal epidermis. In the dermis, there is a vertical connective-tissue axis composed of interlacing collagen fibres interspersed with numerous fibroblastic cells and some inflammatory cells. This connective tissue is in continuity with the underlying connective tissue.

Differential Diagnosis

This lesion may be confused with a number of other lesions: nodular fibroma (which does not, however, exhibit a peripheral hyperkeratotic rim), Koenen's tumours, that is, manifestations of Bourneville's tuberous sclerosis (which have a different clinical appearance, are usually multiple with a tendency to recurrence and are associated, in particular, with other manifestations of this grave disorder), phalangeal pads (which are less circumscribed and generally appear in adolescence and cause functional handicap), mucous cysts (which are quite different and contain clear viscous fluid), a possible supernumerary finger and, finally, a verruca, which can be excluded by histological examination.

Treatment

The treatment of fibrokeratoma is simple and consists of excision of the tumour, usually under local anaesthesia, followed by direct suture. However, such a procedure is not always indicated, for it is mainly cosmetic considerations that are decisive. Fibrokeratoma of this acquired type does not usually recur after surgical excision.

Koenen's Tumours

Bourneville's tuberous sclerosis is a rare hereditary condition, transmitted in irregular dominant fashion and not sex linked. It is a disorder in the general setting of the phakomatoses or neuro-ectodermal dysplasias, with an incidence of 1 per 100000 births. It combines: neuropsychiatric disorders (mental retardation, epilepsy, absence of focal neurological signs and intracranial calcifications); skin and mucosal lesions such as Pringle's sebaceous adenomata, achromatic patches and Koenen's tumours; ocular features dominated by retinal phakoma, a benign, yellowish, flattened, asymptomatic, retinal tumour; and diverse visceral malformations and tumours (renal, pulmonary, cardiac). The course of this disorder in its complete form is chronic and progressive, leading to death from intercurrent infection, cerebral tumour or status epilepticus. However, there are numerous incomplete forms, compatible with a relatively normal life.
Koenen's tumours were mentioned by Kohte in 1903 (after Martin 1981) and described by Koenen in 1932. It is, in fact, a rare manifestation of this disorder, which was found in only 17% of cases by Dwyer (1971). Unlike sebaceous adenoma and the achromatic patches, it is delayed and does not appear until after puberty. However, it is regarded as pathognomonic of Bourneville's tuberous sclerosis.
Koenen's tumours are small, grey or pink, fleshy proliferations. They are firm, ovoid or rounded or like a "grain of wheat" and have a diameter of 2-10 mm. They are periungual lesions - sessile or pedunculated, regular and well-circumscribed, attached to the outer lip of the nail fold and presenting a horny distal tip (Fig. 60). They are also painless and usually multiple, sometimes two or three on the same nail but with a common base, or they may occur on several nails of the hands and feet (Fig. 61). They recur after surgical excision. There is often an associated longitudinal groove in the nail. Colomb et al. (1976) "consider that Koenen's tumours have some prognostic importance insofar as the neurological involvement is less severe when the ungual lesions are more numerous.
Histologically, Koenen's tumours have a fibrous component and a vascular component, neither being specific. They are considered by Bogaert and Pallas (1958) and by Kissel et al. (1962) as angiofibromata and are indeed composed of a loose connective tissue, rich in dilated irregular capillaries and with only a few fibroblasts. Disappearance of the dermal crests is a feature of the histological diagnosis. It should be said that recently Martin

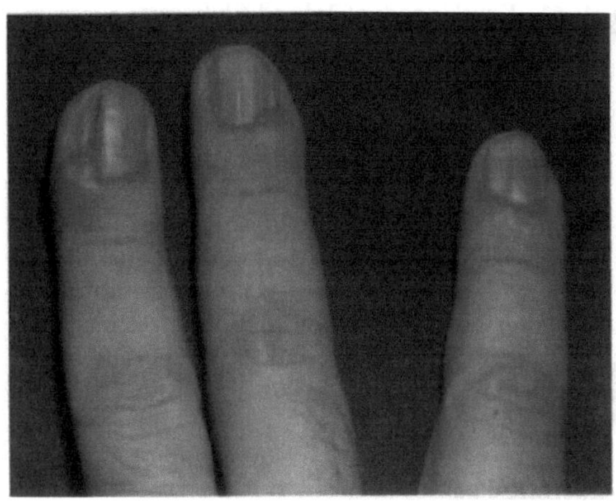

Fig. 60. Koenen's tumours of the fingers

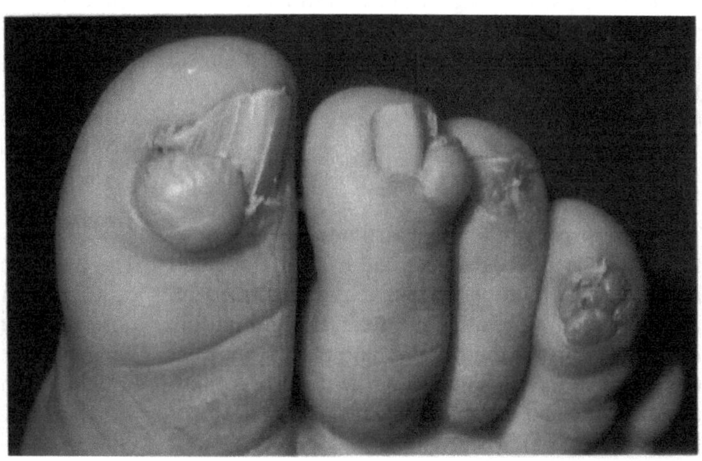

Fig. 61. Koenen's tumours of the toes

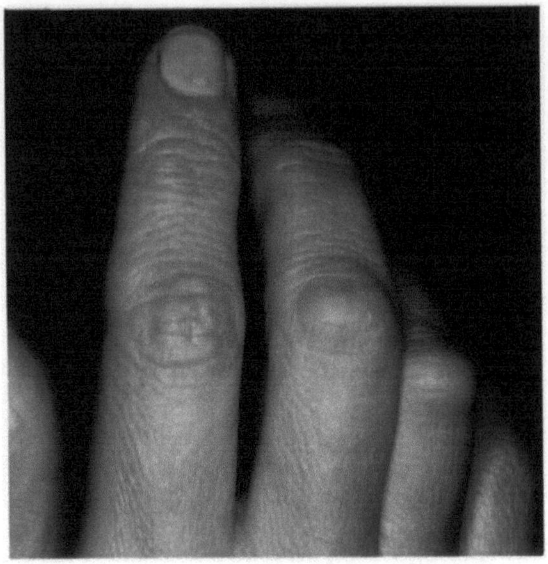

Fig. 62. Knuckle-pads of 3rd and 4th fingers

(1981), on the basis of a solitary Koenen's tumour without other manifestation of Bourneville's tuberous sclerosis, has questioned whether this histologically nonspecific lesion is truly pathognomonic of phakomatosis. He suggests that Koenen's tumours may constitute a specific and autonomous disease entity with an entirely benign prognosis, not recurring after surgical excision and quite distinct from the *formes frustes* of Bourneville's tuberous sclerosis.

Knuckle-Pads

The knuckle-pads of the English language literature (French, *les coussinets des phalanges;* Ramos and Silva 1956, *pulvillus digiti*) are a fibromatosis described by Garrod in 1893, consisting of small tumours situated on the back of the interphalangeal joints. The French designation is due to Touraine (1942).

Clinical Features

These are small swellings, sometimes barely visible, firm, circumscribed, painless and oval or elongated in shape. The multiple lesions may be situated on all the fingers, more often on the index and middle ones, rarely on the thumb. They are placed opposite the proximal interphalangeal or the metacarpophalangeal joints and are more evident when the finger is flexed (Fig. 62). They appear in adolescents and adults of both sexes, are sometimes a nuisance and are always displeasing. As mentioned by Ramos and Silva (1956), this lesion has often been reproduced in works of art. This is particularly the case with Michelangelo, being found on the right hand of the "David" in Florence (thumb and index) the right hand of the "Sleeping Slave" (distal interphalangeal joint of the 2nd, 3rd and 4th fingers), the right hand of "Julian de Medici" (Florence) and the left hand of "Moses" (Rome). Lorenzo the Magnificent in a painting by Vasari also exhibits such a lesion.

Histology

Knuckle-pads of the phalanges are characterised by fibrosis of the dermis and hyperkeratosis of the epidermis. The resulting appearance is that of a fibroma with hyperplasia of the connective tissue and the presence of fibroblasts and collagen bundles. There is also hypertrophy of the vessel walls and the nervous structures.

Aetiopathogenesis

Knuckle-pads were considered initially and until much later as simple callosities (tylosis, heloderma) and regarded by many as secondary to repeated microtraumas. However, Touraine (1942) saw them as only one feature of hereditary polyfibromatosis, and hence their frequent association with Dupuytren's contracture, the existence of other associated lesions (keloid, fibroma, induration of the corpora cavernosa) and the occurrence of such lesions in other members of the same family.

Differential Diagnosis

Knuckle-pads must be distinguished from hyperkeratotic callosities of traumatic origin in which epidermal hyperplasia is the main feature, and from Heberden's nodes, which are osteoarticular and not cutaneous and are situated mainly at the distal interphalangeal joints.

Treatment

The treatment of phalangeal knuckle-pads is surgical, if at all indicated, as medical measures predispose to recurrence. The skin and abnormal underlying tissue are excised, with a full-thickness skin graft to cover the defect.

Pseudosarcomatous Fibromatosis

Nodular Fasciitis

In 1955 Konwaler et al. described under the name of subcutaneous pseudosarcomatous fibromatosis a benign disorder of connective tissue, characterised by its speed of development and histological features. The essential difficulty with this lesion is to make a precise diagnosis of its benign nature, for it may be erroneously considered malignant and so provoke an extensive surgical procedure quite out of proportion to its true gravity. This uncommon and little-known affection produces subcutaneous nodules based on the superficial or deep fascia, growing rapidly and tending to infiltrate. Although there is no evidence of an inflammatory origin, Shuman (1966) prefers to use the term "nodular fasciitis" because of the elective anatomical site of the lesion and the absence of recurrences (unlike the fibromatoses arising from any fibrous structure, which have frequent recurrences).

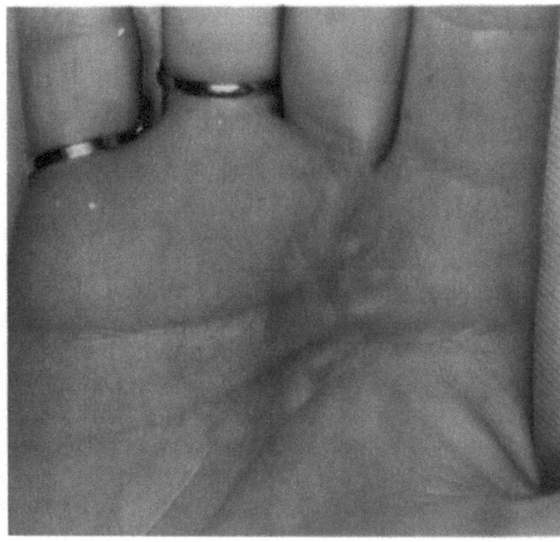

Fig. 63. Pseudosarcomatous fibromatosis: multinodular form

Clinical Features

This is a single and solitary tumour, firm, sensitive and sometimes spontaneously painful, mobile in relation to the skin and more or less fixed in depth. It grows rapidly, reaching a diameter of 1.5-2 cm in less than a month. Rarely, several nodules co-exist which are of more supple consistency and whose evolution may be slower (Fig. 63). This disorder may appear at any age (range 3-67 years, mean 38) and has a male predominance (45 males in the 65 cases reported by Price et al. 1961). It may arise from any fascia and is more often seen in the forearm than the hand or wrist.
Grossly, the lesion is solitary, rounded or oval, grey or whitish or even pink (Fig. 64). On section, besides the presence of adipose tissue and fibrous structures, gelatinous strands of myxoid appearance are found distributed irregularly throughout the nodule.

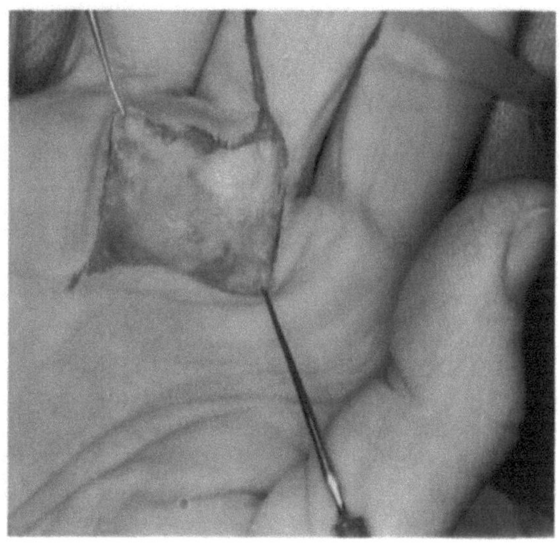

Fig. 64. Pseudosarcomatous fibromatosis: gross appearance

Histology

Histological examination of the tumour is sometimes difficult to interpret and may be a source of grave diagnostic errors. Within an amorphous myxoid matrix there is a proliferation of young fibroblasts, stellate or fusiform cells that may have intracytoplasmic inclusions, granules or vacuoles. The large ovoid nucleus may exhibit mitosis but without anomalies. Other cells may be observed - lymphocytes, macrophages, giant cells - analogous to those described by Stewart and Foot (1951) in desmoid tumours, as well as undifferentiated mesenchymal cells. There is also a peritumoral pseudo-capsule. Hutter et al. (1962b) describe in the nodular fasciitides a particular arrangement consisting of a major central vascularisation in which the fibroblasts are numerous and arranged in compact bundles, while at the periphery they are disposed in columns. The surrounding matrix is loose and of an oedematous nature, with irregularly distributed fibroblasts. These authors also note a considerable distribution of collagen and the presence of multinucleate giant cells.

Price et al. (1961) distinguished three histological types of nodular fasciitis, using an arbitrary classification not particularly correlated with the anatomical and clinical features.

In type I the matrix is abundant, of myxoid appearance and situated mainly at the centre of the tumour, the fibroblasts are distributed peripherally, multinucleate cells are frequent, and there is major vascularisation.

In type II there is less ground substance, cell distribution is more regular with hypercellular strands and collagen fibres present as well as lymphocytes, macrophages and giant cells, and the vascularisation is normal.

Type III resembles type II, but is much richer in collagen. In all cases the tumour is non-encapsulated and may in places exhibit features of infiltration of adjacent structures (muscles, nerves). The presence of major vascularisation and of cell mitoses accounts for the difficulty in pathological diagnosis.

Errors are indeed frequent. In a series of cases clinically diagnosed as nodular fasciitis and reported by Price et al. (1961), the postoperative diagnoses after histological examination were as follows: in one group 20 diagnoses of malignant tumours, including 14 fibrosarcomata, 1 myxosarcoma, 3 liposarcomata and 3 unidentified sarcomata; in another group 24 diagnoses of benign tumours, including only 10 cases of nodular fasciitis, 4 fibromata, 4 neurofibromata, 2 neurilemmomata, 1 myxoma, 1 haemangioma and 2 unidentified benign tumoral reactions. Thus, given the difficulty of diagnosis, it is especially dangerous to overlook the benign nature of a tumour since an unduly severe surgical procedure may lead to unnecessary mutilation.

If necessary, examination of the specimen should be entrusted to several histologists in order to confirm the absence of any evidence of malignancy.

Treatment

This should follow that of a benign tumour and be limited to surgical excision of the lesion. Cure is routine, leaving a functionally normal hand, and long-term follow-up confirms the absence of metastases or local invasion. Not only is there no recurrence but some multinodular forms have regressed spontaneously, without operation, after a few years.

Desmoid Tumour

The invasive fibromata of Lecène and Delamare (1908) or the desmoid tumours of the English-language literature are rare fibrous lesions arising from connective tissue, fasciae and aponeuroses. Always benign, they are marked by their striking tendency to local spread and infiltration of adjacent structures and by their frequency of recurrence. Although first recognised in the abdominal wall, which is the most common site and where it was first described in 1832 by MacFarlane as originating from the rectus aponeurosis, this is in fact an ubiquitous tumour. Extra-abdominal lesions were reported in 1923 by Nichols.

According to Pack and Ehrlich (1944), invasive fibromata constitute 0.03% of all tumours; Stout (1967) says that they form 1.3% of all fibromatoses. The extra-abdominal forms represent only 20% of desmoid tumours and are apportioned by Das Gupta et al. (1969) in their series of 72 cases as follows: 28% in the upper limb, 22% in the lower limb, 11% in the head and neck and 21% on the trunk. In the limbs this tumour is mainly located at the proximal end; distal lesions are exceptional. Desmoid tumours occur in young adults (average age 30 years) and affect women twice as often as men. Among the suggested aetiopathogenetic factors, apart from the influence of trauma and scarring, a very definite hormonal influence has been noted, in particular their common occurrence during or just after pregnancy, puberty or oestrogenic treatment. Cases of desmoid tumours reported in the hand are extremely rare (less than 10) and situated at the extensor tendons, lumbrical and hypothenar muscles or bony structures.

Clinical Features

This is generally a solitary, deep-seated tumour that is painless, firm and regular. As it progresses and if situated in relation to a muscle or neurovascular channel, it may produce a syndrome of nerve compression with distal paraesthesis. If large enough, it can sometimes cause limitation of movements. The general condition of the patient is always unaffected. Radiologically (using soft rays) one sees a well-defined tumoral opacity that is regular and uniform.

Histology

The diagnosis of the lesion depends on the histological findings. Grossly, this is a firm, whitish, dense tumour, never encapsulated, infiltrating the adjacent structures and adhering to the periosteum or neurovascular axes without any plane of cleavage. Microscopic examination shows fibroblastic cells and collagen fibres in varying proportions. Sometimes the appearance approximates to that of an ordinary fibroma. Often, however, the characteristics are those of the fibromatoses, i.e. fascicular fibrous tissue composed of fibroblasts with a considerable production of collagen. In certain regions the paucity of cells and the arrangement of the fibres produce a pseudotendinous appearance.

Differential Diagnosis

The most tricky differential diagnosis is from the so-called low-grade fibrosarcoma, Stout's stage I. The presence of rare but normal mitoses and of normal nuclei and nucleoli are in favour of desmoid fibroma. Jaffe (1958) emphasises distinctive cellular features, especially the size of the fibroblasts, long, fusiform cells which are twice as large in well-differentiated fibrosarcomata. It should be noted that, in the event of recurrence, the histological appearance remains unchanged and that nodal invasion never occurs.

Clinical Types

Apart from the double or triple forms localised in the same limb (and considered by Meary et al. (1978) as manifestations of one and the same tumour with local extension), attention should be drawn to the intraosseous lesion, certainly rare but very specific and also reported in the hand. This develops very progressively, sometimes painfully, and is characterised radiologically by its rounded cystic appearance, possibly preserving a fine honeycomb trabeculation, and extending peripherally with effraction of the cortex, suggesting a malignant tumour. Again, the diagnosis is made on the histological features, which are identical with those of desmoid tumours of the soft parts, whereas the clinical and radiological findings taken by themselves may suggest an intraosseous cyst, a giant-cell tumour or a haemangioma. The treatment usually advised consists of eradication of the tumour by thorough curettage if the cortex is intact, or by bony resection followed by cortico-cancellous grafting when the lesion is more invasive.

We should also mention Gardner's syndrome, which combines diffuse colorectal polyposis and multiple mesodermal tumours, the later including myxomata, leiomyomata and desmoid fibromata.

Development

Whatever its site, the prognosis of this fibromatosis is typically characterised by the frequency of recurrence, even though this has been variously assessed by different authors. The course is slow in all cases, and metastases or nodal involvement never occur. Enzinger and Shiraki (1967) and Meary et al. (1978) have reported cases of stabilisation and even of spontaneous regression.

Treatment

These various types of clinical course raise the problem of the limits of surgical treatment, especially in the hand. Meary et al. (1978) advise early excision, which must be precise and rigorous, sacrificing any skin, muscle, aponeurosis, bony cortex or segment adherent to the tumour. Bastian et al. (1977) suggest immediate histological examination of the wound margins to determine the limits of excision. Ramsey (1955), on the basis that the recurrences are not more aggressive, prefers an incomplete resection when the tumour invades a major neurovascular axis. All authors are agreed in rejecting the idea of amputation, which can only be exceptionally indicated for a benign tumour. Finally, the role of radiotherapy is controversial; many find it ineffective, although Wara et al. (1977) report some degree of radiosensitivity.

Bibliography

Ahlqvist J, Pohjanpelto P, Hjelt-Hurme K (1967) Recurring digital fibrous tumor of childhood. Clinical and morphological aspects of a case. Acta Pathol Microbiol Immunol Scand [A] 70: 291-296, 98

Allen PW (1972) Recurring digital fibrous tumors of childhood. Pathology 4: 215-223

Allen PW (1977) The fibromatoses. A clinico-pathologic classification based on 140 cases. Am J Surg Pathol 1: 255-270

Allen PW, Enzinger FM (1970) Juvenile aponeurotic fibroma. Cancer 26: 857-867

Allisson JR Jr, Allisson JR Jr (1966) Knuckle pads. Arch Dermatol 93: 311-316

Bailey AJ, Sims TJ, Gabbiani G et al. (1977) Collagen of Dupuytren's disease. Clin Sci 53: 499

Barber HM, Galasko CSB (1973) Multicentric extra-abdominal desmoid tumors. Journal Bone Joint Surg [Br] 55/4: 858-863

Barriere H, Bouhour MT (1973) Fibromatose digitale juvenile de Reye. Bull Soc Fr Dermatol Syphiligr 80/2: 161

Barriere H, Litoux P, Bureau B, Welin J, Stalder JF (1975) Fibromatose de l'enfant. Essai de classification, à propos d'une observation de fibromatose ossifiante étendue. Ann Dermatol Venereol 102/1: 5-20

Bart RS, Andrade R, Kopf AW, Leider M (1968) Acquired digital fibrokeratoma. Arch Dermatol 97: 120-129

Bastian D, Simon PH, Perotin J (1977) Deux cas de fibromes desmoïdes extra-abdominaux. Ann Chir 31/11: 961-965

Battifora H, Hines JR (1971) Recurrent digital fibromas of childhood, an electron microscope study. Cancer 27: 1530-1536

Bean SF (1969) Infantile digital fibroma. Arch Dermatol 100: 124

Beckett JH, Jacobs AH (1977) Recurring digital fibrous tumors of childhood: A review. Pediatrics 59: 401-406

Bloem JJ, Vuzevski VD, Huffstadt AJC (1974) Recurring digital fibroma of infancy. J Bone Joint Surg [Br] 56: 746-751

Bogaert VL, Pallas JE (1958) Etude sur la sclérose tubéreuse de Bourneville à forme cerebelleuse. Rev Neurol 98: 673-689

Bonerandi JJ, Follana J, Migozzi B, Temime P (1976) Fibromatose digital infantile. Ann Dermatol Venereol 103: 161-168

Bong SL (1983) Desmoid tumor of the hand. Case report and literature review. J Hand Surg 8/1: 95-97

Booher RJ, McPeak CT (1959) Juvenile aponeurotic fibroma. Surgery 46: 924-13

Bories-Jalabert M (1976) Etude du fibro-kératome acquis. Thèse Medecine, Toulouse

Brickley-Parsons D, Glimcher MJ, Smith RJ et al. (1981) Biochemical changes in the collagen of the palmar fascia in patients with Dypuytren's disease. J Bone Joint Surg [Am] 63: 787

Brion M (1958) Les mains dans la peinture. Michel, Paris

Broder MS, Leonidas JC, Mitty HA (1973) Pseudosarcomatous fasciitis: An unusual cause of soft tissues calcification. Radiology 107: 173-174

Brown A (1969) Juvenile aponeurotic fibroma. Proc R Soc Med 62: 327

Burry AF, Kerr JF, Pope JH (1970) Recurring digital fibrous tumour of childhood, an electron microscopic and virological study. Pathology 2: 287-291

Chabeau G (1972) Aspect de fibro-kératome digital avec déformation unguéale posttraumatique. Bull Soc Fr Dermatol Syphiligr 79: 431

Chaouat Y, Aron-Brunetiere R, Favres B, Binet O, Ginet C, Aubart D (1980) Une nouvelle entité: Le rhumatisme fibroblastique à propos d'une observation. Rev Rhum Mal Osteoartic 47/5: 345-351

Civatte J (1967) Dermatofibrosarcome de Darier Ferrand. Histopathologie cutanée. Flammarion, Paris, pp 225-226

Collard JJ, Debas E, Bourlond A (1974) Sclérose tubéreuse de Bourneville: Valeur cannelure unguéale. Arch Belg Dermatol 30/4: 289-290

Colomb D, Racouchot J, Jeune R (1976) Les lésions des ongles de la sclérose tubéreuse de Bourneville isolées et associées aux tumeurs de Koenen. Ann Dermatol Venereol 103/4: 431-437

Cordier G, Garnier H, Barbier J, Calmettes C (1966) La polypose recto-colique généralisée associée à des lésions ectodermiques et mésodermiques. A propos d'un cas de syndrome de Gardner. Presse Med 74: 767-772

Coskey RJ, Nabai H, Rahbari H (1979) Recurring digital fibrous tumor of childhood. Cutis 23: 359-362

Coste F, Marceron L, Tissier M (1943) 2 cas de coussinets des phalanges. Bull Soc Fr Dermatol Syphiligr 50: 203

Culberson JD, Enterline HT (1960) Pseudosarcomatous fisciitis: A distinctive clinical-pathologic entity. Report of five cases. Ann Surg 151/2: 235-240

Das Gupta RK, Brasfield RD, O'Hara J (1969) Extra-abdominal desmoids: A clinicopathological study. Ann Surg 170: 109-121

Debard A, Richardet JM, Brissaud HE (1979) Sclérose tubéreuse de Bourneville chez l'enfant, étude de 30 cas. Sem Hop Paris 55/21-22: 1065-1072

Degos R (1953a) Les fibromes. Dermatologie, Flammarion, Paris, pp 805-810

Degos R (1953b) Dermatologie, fibromes, histiocytomes, fibromatoses. Collection Médico-chirurgicale, Flammarion, Paris, pp 805-808

Degos R (1953c) Dermatologie épiloïa. Collection Médico-chirurgicale, Flammarion, Paris, pp 784-786

Degos R, Civatte J, Bellaich S, Laregue M, Bonvalet D (1971) Fibromatose juvénile des doigts. Bull Soc Fr Dermatol Syphiligr 78: 6-8

Degos R, Civatte J, Bellaich S, Laregue M, Bonvalet D (1975) Fibrome juvénile des doigts après 8 ans d'évolution. Bull Soc Fr Dermatol Syphiligr 82: 233

Dehner LP, Askin FB (1976) Tumors of fibrous tissue origin in childhood. Cancer 38: 888-900

Des Marchais J, Paquin G, Guy R (1972) Fibromatose cartilagineuse calcifiante (un cas de fibrome aponévrotique juvénile). Union Med Clan 101: 247

Desmons F, Bombart M (1977) A propos de la fibromatose nodulaire congénitale (d'après 6 observations). Ann Dermatol Vénéreol 104: 355-360

Duperrat B, Debeyre J (1943) Coussinets fibreux des phalanges. Bull Soc Fr Dermatol Syphiligr 50: 203

Dupin J (1966) Un cas de fibrome multiple sous-aponévrotique envahissant. Arch Fr Pédiatr 23: 1100-1101

Dupuytren G (1832) Leçons orales de Clinique Chirurgicale, vol 1. Baillière, Paris, pp 2-24

Dwyer JM (1971) Pulmonary tuberous sclerosis. Q J Med 40: 115-125

Enzinger FM (1965a) Fibrous tumors of infancy. In tumors of bone and soft tissue. Year Book Médical Publishers, Chicago, pp 375-379

Enzinger RM, Shiraki M (1967) Musculo aponevrotic fibromatosis of the shoulder girdle. Cancer 20: 1131-1140

Enzinger FM, Lattes R, Torloni H (1969) Histological typing of soft tissue tumors in international histological classification of tumors, n°3. World Health Organization, Geneva

Fievez M, Mandard AM (1966) Les fibromatoses de l'enfant. A propos de 22 cas. Ann Pathol 11/1: 83-99

Fleischmajer R, Norwich A, Reeves I (1973) Juvenile fibromatoses. Arch Dermatol 107: 574-579

Folana P (1974) Tumeurs fibreuses et fibromatoses. Thèse, Paris

Gabbiani G, Majno G (1972) Dupuytren's contracture: Fibroblast contraction? An ultrastructural study. Am J Pathol 66: 131

Galinski AW (1972) Recurrent digital fibroma of childhood. A case report. J Am Pediatr Assoc 62: 114-117

Garot D (1978) Manifestations pulmonaires de la sclérose tubéreuse de Bourneville. Thèse, Créteil

Garrod AE (1893) On an unusual form of nodule upon the joints of the finger. St Barth Hosp Rep 29: 157-161

Garrod AE (1904) Concerning pads upon the finger joints. Br Med J II: 8

Gianotti F, Ermacora E, Magrin U (1977) Fibromatose juvenile hyaline. Ann Dermatol Vénéreol 104: 152-153

Goldman RL (1970) The cartilage analogue of fibromatosis aponeurotic fibroma. Cancer 26: 1325-1331

Gonatas NK (1961) Extra abdominal desmoid tumors. Arch Pathol 71: 214-222

Hardouin JP, Marche C, Simonpoli C (1963) Les tumeurs desmoides (ou fibromes envahissant de Lecene et Delamare) à propos de 2 observations. Sem Hôp Paris 39/33: 1583-1591

Hare PJ (1954) Rudimentary polydactyly. Br J Dermatol 66: 402

Hare PJ, Smith PAS (1969) Acquired (digital) fibrokeratomas. Br J Dermatol 81: 667-670

Heiple KG, Perrin E, Aikawa M (1972) Congenital generalized fibromatosis. A case limited to osseous lesions. J Bone Joint Surg [Am] 54: 663-669

Hemric JR, Allen HB (1979) Acquired digital fibrokeratoma. Cutis 23: 304-306

Herman PS, Datnow B (1974) Acquired (digital) fibrokeratomas. Acta Dermatol Venereol 54: 73-76

Hill DR et al. (1973) Radiation therapy of desmoid tumors. AJR 117: 84-89

Hozumi Y (1971) Case of desmoid tumor of the first dorsal interosseous muscle. Orthop Surg (Tokyo) 32: 985-986

Hueston JT (1969) The control of recurrent Dupuytren's contracture by skin replacement. Br J Plast Surg 22: 152

Hutter RVP, Foote FW, Francis KC, Higinbotham NL (1962a) Parosteal fasciitis; Self limited benign process that simulates malignant neoplasm. Am J Surg 104: 800-807

Hutter RVP, Stewart FW, Foot FW (1962b) Fasciitis a report of 70 cases with follow up proving the benignity of the lesion. Cancer 15/5: 993-995

Iwasaki H, Tsuneyoski M, Ensoji M (1974) Infantile digital fibromatosis. Histo-pathological and electron microscopic, study with a review of the literature. Acta Pathol Jpn 24: 717-732

Jaffe HL (1958) Tumors and tumorous condition of the bones and joints. Lea & Febiger, Philadelphia

Jansion P, Caillau M, Calop JP, Chalopin R (1942) Cous-sinets des phalanges. Bull Soc Dermatol Syphiligr 49: 482

Jensen AR, Martin LW, Longino LA (1957) Digital neurofibrosarcoma in infancy. J Pediatr 51: 566-570

Jones HW (1923) Two cases of knuckle pads. Br Med J 1: 759

Jung EG (1974) Symptômes cutanés de la sclérose tubéreuse de Bourneville. J Génét Hum 28: 218-222

Kapiloff B, Prior JT (1952) Fibromatosis in children. Plast Reconstr Surg 10: 276-282

Kauffman SL, Stout AP (1965) Congenital mesenchymal tumors. Cancer 18: 461-476

Kay S (1960) Subcutaneous pseudosarcomatous fibromatosis. Report of 4 cases. Am J Clin Pathol 33/5: 433-438

Keasbey LE (1953) Juvenile aponeurotic fibroma (calcifying fibroma) a distinctive tumor arising in the palms and soles of young children. Cancer 6: 338-346

Keasbey LE, Panselau HA (1961) The aponeurotic fibroma. Clin Orthop 19: 115-131

Keller RB, Bauz-Giangreco A (1975) Juvenile aponeurotic fibroma. Report of three cases and a review of the literature. Clin Orthop 106: 198

Kermarec J, Ayraud N (1968) Fibrome aponévrotique juvénile. Forme anatomo-clinique spéciale de fibromatose pseudo-sarcomateuse. Arch Pathol 16/1: 43-47

Kint A, Verlinden L (1972) Acquired digital fibrokeratoma. Arch Belg Dermatol 28: 387

Kissel P, Dureux JB, Shmitt J (1962) Les phacomatoses, considérations générales, intérêts spéculatifs et pratiques. Rev Pratic 12/2: 915-929

Kissel P, Dureux JB, Schmitt J (1963) La sclérose tubéreuse de Bourneville. EMC 17: 165, C10

Koenen J (1932) Eine familare hereditare Form von tuberoser Sklerose. Acta Psychiatr Scand 1: 813-821

Konwaler BE, Keasbey LE, Kaplan L (1955) Subcutaneous pseudosarcomatous fibromatosis (fasciitis). Report of 8 cases. Am J Clin Pathol 25: 241-252

Lakhanpal VP, Yalau SS, Sastry RK, Krishnamurthy CS (1978) Recurring digital fibromas of infancy a case report. Acta Orthop Scand 49: 147-150

Larregue M, Poitou C, Bressieux JP, De Giacomoni P, Vant E (1977) Fibromatose congénitale généralisée. Ann Dermatol Vénéreol 104: 349-354

Lattes R (1964) Tumor seminar case 8. Tex Med 60: 430-433

Lecene P, Delamare G (1908) Remarques sur les fibromes de la paroi abdominale. Bull Med Paris 22/3: 25-27

Lereboullet J (1970) La sclérose tubéreuse de Bourneville. Rev Pratic 20/28: 4425-4436

Lichtenstein L, Goldman RL (1964a) The cartilage analogue of fibromatosis; a reinterpretation of the condition called «juvenile aponeurotic fibroma». Cancer 17: 810-816

Lichtenstein L, Goldman RL (1964b) Cartilage tumors in soft tissues, particulary in the hand and foot. Cancer 17: 1203-1208

Lobuono P, Jothikumart R, Kornbleel O (1979) Acquired digital fibrokeratoma. Cutis 24: 50-51

Mackensie DH (1964) Fibroma: A dangerous diagnosis. A review of 205 cases of fibrosarcoma of soft tissues. Br J Surg 51/8: 607-612
Macfarlane J (1832) Clinical reports of the surgical practice of the Glasgow infirmary. Robertson, Glasgow, pp 63-66
McKenzie AW, Innes FLP, Rack JM, Breathnach AS, Gross M (1970) Digital fibrous swellings in children. Br J Dermatol 83: 446-458
Mande M, Loubry P, Clou P, Marie S (1965) Fibromatose congénitale diffuse du nouveau-né en évolution régressive. Ann Pédiatr 12: 692-701
Manelle JC, Salle S, Daudet M, Rosenber D, Brunat M, Monnet P (1969) Fibrosarcome et fibromatose congénitaux. Sem Hôp Paris 45, 16/4: 1031, 255
Martin B (1981) Les tumeurs de Koenen. Thèse, Faculté de Médecine Broussais, Hôtel Dieu
Masson JB (1930) Desmoid tumors. Ann Surg 92: 444-453
Mauclaire P, Legry H (1915) Fibromatose généralisée de nature congénitale. Bull Mem Soc Chir Paris 41: 1156-1157
Meary R, Danan JP, Tomeno B, Forest M, Cirotteau Y (1978) Fibromes desmoïdes des membres. Rev Chir Orthop 64/3: 195-204
Mehregan AH (1966) Nodular fasciitis. Arch Dermatol 93: 204-210
Mehregan AH, Nabai H, Matthewa JE (1972) Recurring digital fibrous tumor of childhood. Arch Dermatol 106: 375-378
Mercadier M, Clot JP, Camplez PH, Darcy M (1967) Les sarcomes des tissus mous et des coulées conjonctives, à propos d'1 observation. Ann Chir 21/11 et 12: 686-690
Merklen FP, Cottenot F, Melki GR (1970) Histiocytofibromes multiples des doigts. Bull Soc Fr Dermatol Syphiligr 77: 890
Millesi H (1972) Evolution clinique et morphologique de la maladie de Dupuytren. In: Tubiana R, Hueston JT (eds) Maladie de Dupuytren, 2ème edn. L'expansion, Paris, p 67
Moncorps C (1931) Cornu cutaneum. In: Jadassohn J (Hrsg) Handbuch der Haut- und Geschlechtskrankheiten, Bd 8/2. Springer, Berlin, S 436
Nezeloff C, Tubiana R (1958) La maladie de Dupuytren: Étude histologique. Sem Hop Paris 34: 1102
Nichols RW (1923) Desmoid tumors. A report of thirty-one cases. Arch Surg 7: 227-235
O'Gorman DJ, Fairbum EA (1974) Infantile digital fibromatosis. Proc R Soc Med 67: 44
Pack GT, Ehrlich ME (1944) Neoplasm of the anterior abdominal wall with special consideration of desmoid tumors. Surg Gynecol Obstet 79/3: 177-198
Phelan JT, Grace JT Jr, Pickren JW (1968) Juvenile fibromatosis. Surgery 64: 492-497
Pickren JW, Smith AG, Stevenson TW, Stout AP (1951) Fibromatosis of the plantor fascia. Cancer 4: 846
Pinkus H (1968) Discussion de Bart et coll
Pinol-Aguade J, Mascaro JM, Galy Mascaro C, Capdevila J (1971) Etude clinique, histologique et ultra-structurale de la fibromatose juvénile bénigne des doigts (F.J.B.D.) Arch Méd 3: 179-183
Pohjampelto P, Ahlquist J, Hurme K (1967) Recurring digital fibrous tumour of childhood: II. Isolation of a cell transforming agent. Acta Pathol Microbiol Immunol Scand [A] 70: 297-299
Poirier J, Rancurel G (1968) La sclérose tubéreuse de Bourneville. Concours Méd 14/12: 90-50, 8611-8616
Ponsot G, Lyon G (1977) Sclérose tubéreuse de Bourneville, étude clinique et génétique des 59 observations chez l'enfant. Arch Fr Pédiatr 34: 9-22
Poppen NK, Niebauer JJ (1977) Recurring digital fibrous tumor of childhood. J Hand Surg 2: 253-255
Price EB, Silliphant WM, Shuman R (1961) Nodular fasciitis: A clinicopathologic analysis of 65 cases. Am J Clin Pathol 35/2: 122-136
Prior JT, Sisson BJ (1954) Dermal and fascial fibromatosis. Ann Surg 139: 453-467
Rabhan WN, Rosai J (1968) Desmoplastic fibroma. J Bone Joint Surg [Am] 50/3: 487-502
Ramos E, Silva J (1956) Coussinets des phalanges: «pulvillus digiti». Ann Dermatol 83: 22-33
Ramsey RH (1955) The pathology, diagnosis and treatment of extra-abdominal desmoid tumors. J Bone Joint Surg [Am] 37/5: 1012-1018
Reed RJ, Elner LC (1971) Mutiple acral fibrokeratomas (a variant of prurigo nodularis) discussion of classification of acral fibrous nodules and of histogénesis of acral fibro-keratomas. Arch Dermatol 103: 286-297
Reresse G (1969) Les manifestations cutanées des phakomatoses chez l'enfant. Progr Med 97: 315-322
Reye RDK (1956) A consideration of certain subnormal «fibromatous tumors» of infancy. J Pathol 72: 149-154
Reye RDK (1965) Recurring digital fibrous tumour of childhood. Arch Pathol 80: 228-231
Richardson WR, Dewar JP (1964) Problems in managing fibrous tissue tumors in infants and children. Surgery 56: 426
Rios-Dalenz JL, Kim JS, McDowell FW (1965) The so-called «juvenile aponeurotic fibroma». Am J Clin Pathol 44/6: 632-635
Ritter MA, Marshall JL, Straub LR (1969) Extra-abdominal desmoid of the hand. J Bone Joint Surg [Am] 51/8: 1641-1644
Rubisz, Brzezinska J et al. (1973) Acquired digital fibrokeratoma. Przegl Dermatol 60: 347-352
Sakamoto K et al. (1970) Deux cas de fibrokératome digital acquis. Acta Dermatol (Kyoto) 65: 305-307
Samman PD (1965) The nails in disease. Heinemann, London, p 121
Santa Cruz DJ, Reine CB (1978) Recurrent digital fibroma of childhood. J Cutan Pathol 5: 339-346
Saylant T, Marks R, Jones W (1971) Reactive nodular hyperplasia of hands and oral mucasa. Dermatologica 143: 368-375
Schaffzin EA, Chung SMK, Kaye R (1972) Fibromatose généralisée congénitale avec rétrocession complète spontanée. J Bone Joint Surg 54/3: 657-662

Schenkar DL, Kleinert HE (1977) Desmoplastic fibroma of the hand. Plast Reconstr Surg 59/1: 128-133
Schnitzler XL, Verret JL, Varlet PH, Ginies JL, Armand B (1978) Fibromatose congénitale généralisée. Méd Hyg 36/1285: 1600-1605
Schreiber MM, Shapiro S, Sampsel J (1965) Pseudosarcomatous fibromatosis. Arch Dermatol 92: 661-665
Seel DJ, Booher RJ, Joel R (1964) Fibrous musculo-aponeurotic tumors. Surgery 56/3: 497-504
Selmanowitz VJ (1971) Fibro-kératome affectant les extrémités. Int J Dermatol 10: 166-169
Shapiro L (1969) Infantile digital fibromatosis and aponeurotic fibroma. Case reports of two rare pseudo-sarcomas and a review of the literature. Arch Dermatol 99: 37-42
Shnitka JT, Horner RH (1958) Congenital generalized fibromatosis. Cancer 11: 627
Shuman R (1966) Mesenchymal tumors. In: Anderson WAD (ed) Pathology, 5th edn. Mosby, St. Louis
Soule EH (1962) Proliferative (nodular) fasciitis. Arch Pathol 73: 437-444
Specht EE, Konkin LA (1975) Juvenile aponeurotic fibroma; the cartilage analogue of fibromatosis. JAMA 234: 626-628
Specht EE, Staheli LT (1977) Juvenile aponeurotic fibroma. J Hand Surg 2/4: 256-257
Stack G (1960) Tumors of the hand. Br Med J I: 517, 919-922
Steel HH (1965) Garlic clove fibroma. JAMA 191: 104-105
Stevanovic D (1977) Fibromatose multiple continue et progressive. Ann Dermatol Vénéreol 104: 141-146
Stewart FW, Foot GW (1951) Pseudosarcomatous fasciitis. Proc New York State Assoc of Public Health Cen
Stout AP (1948) Fibrosarcoma - The malignant tumor of fibroblasts. Cancer 1: 30
Stout AP (1953) Tumors of the soft tissues. AFIP atlas of tumor pathology, sec 2, fas 5. Armed Forces Institute of Pathology, Washington, pp 1-104
Stout AP (1954) Juvenile fibromatosis. Cancer 7: 953-978
Stout AP (1960) Fibrous tumors of soft tissues. Minn Med 43: 455-459
Stout AP (1961) Pseudosarcomatous fasciitis in children. Cancer 14: 1216-1222
Stout AP (1967) Tumors of the soft tissues. In: Atlas of tumor pathology, serie 2, fasc 1. Armed Forces Institute of Pathology, Washington
Sugiura I (1976) Desmoplastic fibroma. J Bone Joint Surg [Am] 58/1: 126-131
Tapas K, Gupta D, Brasfield RD, O'Hara J (1969) Extraabdominal desmoids: A clinicopathological study. Ann Surg 170/1: 109-119
Thomas WC (1954) Juvenile aponeurotic fibroma. Am J Pathol 30: 625
Thomine E, Nesmann C, Nezelof C (1968) Les fibromes envahissants de l'enfant (à propos de 9 cas). Discussion nosologique et diagnostique. Ann Pathol 13/3: 303-324
Touraine A (1942) Coussinets des phalanges. Bull Soc Fr Dermatol Syphiligr 49: 481
Touraine A, Horowitz A, Lambergeon S, Chertier M (1969) Epiloia familial. Bull Soc Fr Dermatol Syphiligr 5: 68
Touraine A, Scemama P (1936) Adénomes sébacés symétriques et sclérosés tubéreux de Bourneville. Bull Soc Fr Dermatol Syphiligr 43: 1276-1280
Touraine A, Ruel H (1945) La polyfibromatose héréditaire. Ann Dermatol Venereol 5: 1-5
Tubiana R, Hueston JT (1972) Maladie de Dupuytren, 2ème edn. L'expansion, Paris, p 185
Tubiana R, Michon J (1961) Evaluation chiffrée précise de la déformation dans la maladie de Dupuytren. Sa valeur pronostique. Mem Acad Chir 87: 886
Unna PG (1896) The histopathology of the diseases of the skin. MacMillian, New York
Veralo VVM (1968) Acquired digital fibrokeratomas. Br J Dermatol 80: 730-736
Voglino A (1968) Su un caso difibrocheratoma digitale acquisito. Cronache 3: 196
Vouillon G (1971) Fibrokératome digital sus-ungueal. Bull Soc Fr Dermatol Syphiligr 78: 177
Wallace PF, Fitzmorris CS Jr (1977) Juvenile aponeurotic fibroma: A case report. J Hand Surg 2/4: 258-260
Wara EM, Phillips TL, Hill DR et al. (1977) Desmoid tumors. Treatment and prognosis. Radiology 124: 255-256
Weschler J (1972) Contribution à l'étude anatomoclinique des proliférations tumorales fibroblastiques chez l'enfant. Recherche des critères histologiques et pronostiques sur 65 observations. Thèse Med Paris, Créteil
Winkelmann RK, Sams WM (1969) Elastofibroma. Report of a case with special histo-chemical and electron-microscopic studies. Cancer 23: 406
Wise E (1944) Knuckle pads. Arch Dermatol 49: 144
Wyatt EH, Cowan MA (1970) Digital fibrous tumours of childhood. Trans John Hosp Dermatol Soc 56: 162-164
Zeide MS, Wiesel S, Terry RL (1978) Juvenile aponeurotic fibroma. Case report. Plast Reconstr Surg 61/6: 922-923

Fibrosarcoma and Fibrous Histiocytosarcoma

Recent progress in the histological and cytological examination of tumours has made for more critical analysis and raises questions about changing long-accepted terminology. The term fibrosarcoma in particular needs to be qualified. Originally, it grouped together a heterogeneous collection of tumours which included synovial sarcomata, leiomyosarcomata, liposarcomata, histiocytosarcomata and certain types of rhabdomyosarcomata. Since then, each of these tumours has been defined as an entity in itself, and examination of slides of former patients has led to the rectification of many diagnoses.

Fibrosarcomata, once considered relatively common, now amount in some statistics to no more than 0.5% of all malignant tumours (van der Werf-Messing and van Unnik 1965). However, the nosological problems are far from being completely elucidated, and new entities are proposed by some authors based on various classifications with different subgroups linked to separate histological interpretations. It is of particular importance in a malignant tumour to distinguish the cytological element responsible for the proliferation. The structure of normal dense connective tissue does not allow divisions according to the relations existing between different constituents. The fibroblast, certainly the most important, almost exclusive, cell component of connective tissue, is a very elongated and active cell. It has a cytoplasm rich in organelles and a very developed, corrugated, endoplasmic reticulum, with a prominent Golgi apparatus; it is responsible for the synthesis of enzymes and protein structures.

The histiocyte, while less important quantitatively, also forms part of the dense connective tissue. Aside from its role in phagocytosis, it is additionally an essential component of the reticuloendothelial system and may form the point of origin of a tumoral proliferation (Ozzello et al. 1963). The remaining cells making up connective tissue are mast cells, lymphocytes, plasmocytes and eosinophil polymorphs. These cellular elements live in the aqueous environment of the ground substance, accompanied by fibrillary structures - collagen and reticulin - that are aggregates of polypeptide macromolecules synthesised by the fibroblasts. The problem becomes complicated in the light of the knowledge that fibroblasts and histiocytes may also derive from the functional transformation of another cell, which thus acquires new properties in different circumstances. These are then labelled "fibroblast-like" and "histiocyte-like", terms themselves not definitive but indicative of the uncertainty about the responsible cell.

Cell cultures (Ozzello et al. 1963) have made it clear that the fibroblast-like cells are derived from histiocytes, which gain the capacity to secrete collagen and reticulin. Other cells are also capable of such a transformation, e.g. monocytes, Schwann cells, mesenchymal cells and reticular cells (Stout 1953). Cell cultures have also made it possible to group together fibrous histiocytosarcoma, xanthofibrosarcoma and dermatofibrosarcoma, the different aspects of which are linked with the polymorphism of one and the same cell, the histiocyte. Thus, many authors now regard the histiocyte as responsible for numerous tumours hitherto classed globally as fibrosarcomata. The more prudent term of histiocyte-like cell used by Kahn (cited by Stout 1960) allows for the reticence of certain authors for whom this tumoral cell, capable of phagocytosis, containing intracytoplasmic lipids and provided with a rugose endoplasmic reticulum, may itself be only a stage of functional transformation of another cell.

Hence, in the absence of decisive arguments and as reported by Fu et al. (1975) three hypotheses have been proposed to explain the presence in certain tumours of these two predominant types of cell population:

1. The existence of two different cell lines
2. A histiocytic origin, supported by the work of Ozzello et al. (1963) who show that in culture these facultative fibroblasts regain the characteristics of their cell of origin (the histiocyte), and also by the electron microscope studies of Merkow et al. (1971)
3. Finally, as proposed by Weiss and Enzinger (1978), a common mesenchymal origin for both cell lines, this pluripotential cell being capable of transformation into fibroblasts and histiocytes under the influence of the neoplastic process.

The authors support this by noting the presence in such tumours of undifferentiated cells and of cells which, though rare, have the characteristics of both histiocytes and fibroblasts.

We shall study separately fibrosarcoma and fibrous histiocytosarcoma, tumours that are grouped together by Stout (1953) under the name of fusiform-cell sarcoma.

Fibrosarcoma

Fibrosarcomata are malignant tumours arising by degeneration in connective tissue and in preexisting benign connective tissue tumours. Different authors assess their incidence very diversely. Thorbjarnarson (1964) considers that they form 47% of all malignant tumours of the soft tissues, an incidence falling to 44% according to Shieber and Graham (1962) and Hare and Cerny (1963), 30% for Stout (1948), 5.4% for Pack and Ariel (1952) and 5% for Haber et al. (1965). However this may be, the tumour appears especially frequently in the head and neck region and at the base of the limbs; it is very rare in the hand, where it accounts for 3.5% of cases of fibrosarcoma (Pritchard and Soule 1974, 7 of 199 cases). Haber et al. (1965) report 6 fibrosarcomata from 2321 tumours of the hand, i.e. 5% of malignant tumours of this region and Posch (1956) found 3 cases in 679 tumours. Localisation may be superficial or deep; any age may be affected, with a maximal incidence around 40-50 years. The lesions reported in the hand may be situated on the fingers, the palm or the dorsum. Though these patients frequently give a history of antecedent trauma, single or repeated, most authors regard this as coincidental.

Clinical Features

Fibrosarcoma presents as a hard mass of variable size, 2-3 cm on average, with little or no pain, often deep-seated, fixed and adherent to the superficial and underlying planes and possibly ulcerating through the overlying skin (Fig. 65). The course is usually slow and progressive, spread over several months, but sometimes a rapid and painful exacerbation may occur over a few days. At this stage the clinical features of this tumour are disturbing (Fig. 66), but initially a fibrosarcoma, while still indolent and mobile, may completely simulate a benign tumour. Occasionally, a regional adenopathy or even a remote metastasis may be the first sign of the disease. Radiographs of the hand are usually normal, aside from the swelling of the soft parts; rarely, intratumoral calcification is detected.

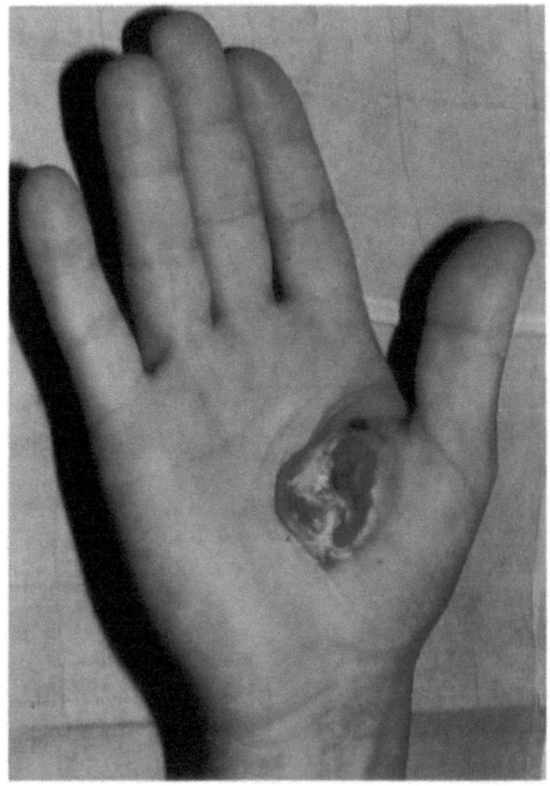

Fig. 65. Fibrosarcoma: ulcerating form

Histology

Histological examination of the lesion is the only means of confirming the diagnosis but often difficulties are encountered as to the nature of the tumour and its malignant potential.

Grossly, the lesion is whitish, firm, rounded or lobulated, infiltrating, destructive and endowed with a pseudo-capsule formed by compression of adjacent tissues.

Classically, fibrosarcoma presents as a fibrocellular proliferation poor in collagen and with an increased cell population composed of fibroblasts with some degree of mitoses and a number of variable monstrosities. The relative proportion of these two elements varies from one tumour to another and sometimes even within the same tumour. Even the characters of the cells and their arrangement are variable, sometimes grouped in regular bundles,

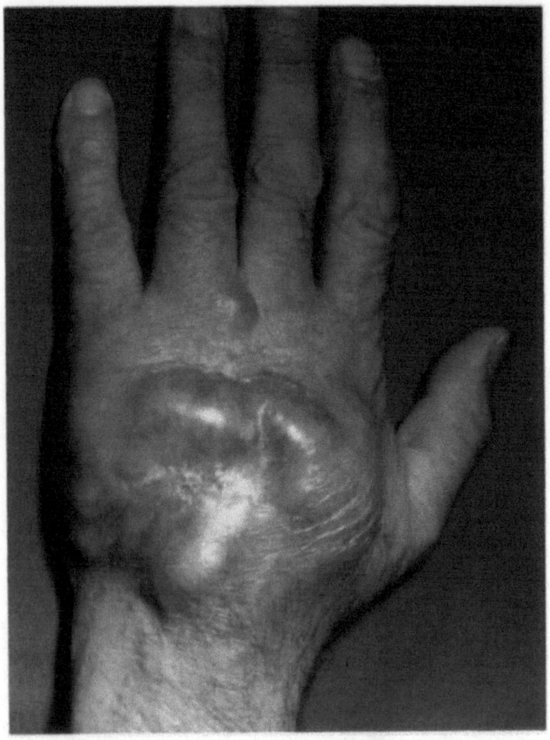

Fig. 66. Advanced fibrosarcoma. (Courtesy of Professor Génin)

sometimes in a polymorphous structure. Vascularisation is considerable, with numerous vascular lacunae that promote bloodstream dissemination. Signs of necrosis and haemorrhagic streaks also exist. Thus, different histological types of fibrosarcoma are postulated according to the degree of differentiation (fasciculation, fibrinogenesis) and the signs of cell malignancy (cell population, mitoses, monstrosities). Stout (1953) distinguishes the differentiated fibrosarcoma - well-fasciculated, more or less rich in mitoses and without monstrosities - from the undifferentiated sarcoma in which the cells are less marked in appearance, with variations in shape and size, numerous mitoses and very evident monstrosities.

The classification of Broders et al. (1939) is more qualified and takes into account the existence of intermediate forms. It provides for four grades, from grade I in which the fibrocellular tumour is markedly fasciculated, without monstrosities, and the arrangement is that of a "fern-shoot", uniform and relatively constant, to Grade IV in which the tumour is anarchic with evidence of numerous giant cells. These different classifications have been proposed in order to simplify the diagnostic and prognostic problems raised by fibrosarcomata. For, at the diagnostic level, certain types of fibrosarcoma may be confused with synovial sarcoma, fibrous histiocytosarcoma or giant-cell sarcoma, this last in any case having long been considered as an undifferentiated fibrosarcoma. Again, distinction from a desmoid tumour is often difficult. Classically, the desmoid tumour is defined as a proliferation of fibroblasts and collagen fibres arranged irregularly in interlacing bundles and with a marked fibrous predominance. Thus, it is quite separate from the differentiated fibrosarcoma, in which the cell density is more marked and particularly more uniform, the mitoses more numerous with atypical cells and the reticulin framework more evident (Stout 1953).

However, the intermediate forms are such that it is admittedly often awkward to make distinctions, and the criteria of malignancy depend on the convergence of a collection of arguments, none of which is sufficient in itself. These criteria of malignancy are as follows (Weschler 1972): cell density, a well-developed collagen framework (in favour of a benign proliferation), atypical mitoses, the morphology of the vessels, necrotic and haemorrhagic changes (common in fibrosarcomata), inflammatory infiltrate (more suggestive of fibromatosis, limitation of the tumour, and a structure that is monomorphous and suggestive of fibrosarcoma or polymorphous and more suggestive of a fibromatosis.

Differential Diagnosis

This may be troublesome in the case of an atypical fibroxanthoma, a tumour described by Halpert and Hackney in 1949 (cited by Helwig 1963) and so denominated by Helwig (1963) in view of the xanthomatous aspect of some of its cells. This lesion, which is clinically benign but whose histological appearance simulates a malignant tumour, was considered as a paradoxical fibrosarcoma of the skin by Bourne (1963) and as a pseudosarcomatous dermatofibroma by Levan et al. (1963). Its chief sites are the neck and face. Kempson and McGavran (1964) found 1 case at the wrist in 21 cases, and Fretzin and Helwig (1973) described digital and palmar lesions. Clinically, the lesion presents in the form of a solitary nodule, firm or ulcerated, rarely painful, with an average size of 1-2 cm and usually occurring in the elderly. This tumour has a slowly progressive course; it may recur after inadequate excision but it never gives rise to metastases. Histologically, there is abundant cell proliferation, essentially fibroblastic, without any particular arrangement. The fibrous element is defective. There are also histiocytic cells and mono- or multinucleated giant cells with numerous mitoses that may simulate a malignant tumour. There is a rich capillary network, Simple excision suffices for the treatment of this tumour; however, it may be confused with a dermatofibrosarcoma protuberans, a fibroxanthosarcoma, a liposarcoma or a basal-cell or squamous-cell epithelioma.

Prognosis

Evidently, histological study is not always decisive in the diagnosis of a fibrosarcoma. Some authors have also suggested criteria for evaluating the prognosis of this disorder. Pritchard and Soule (1974) stress four features: the degree of cellularity, the degree of anaplasia, the amount of fibrous tissue and the mitotic index. Van der Werf-Messing and van Unnik (1965) consider the degree of fibre formation, the degree of cell polymorphism, peritumoral infiltration and the mitotic index. However, these different approaches are sometimes confounded by the frequent coexistence within the same tumour of zones of different structures.

The prognosis is always grave, with a significant level of recurrences and metastatic spread. As the affection is so rare, statistical conclusions can only be problematic. However, most authors report a recurrence rate of between 60% and 70% (Stout 1948; Pritchard and Soule 1974; Van der Werf-Messing and Van Unnik 1965); recurrences aggravate the prognosis and increase the risk of spread. Metastases occur mainly via the bloodstream (40%, chiefly pulmonary) but also by the lymphatic route (10%), to the epitrochlear and axillary nodes for tumours of the hand.

Treatment

This is controversial, but it seems clear that the histological type is more important than the surgical procedure or the size or site of the tumour. Thus, differentiated fibrosarcomata have a mainly local malignancy, a recurrence rate below 50% and give rise to few or no metastases (Stout 1948). On the other hand, the undifferentiated forms have a much gloomier prognosis with fatality in over 50% of cases (Stout 1948). Although simple excision always proves inadequate and leads to recurrence, there is no generally accepted option between radical excision and amputation. Though some (Howard, cited in Haber et al. 1965) favour routine amputation, it is best reserved for recurrent forms (Haber et al. 1965) or undifferentiated types. Many authors consider that radical excision offers the same prognosis (Pritchard and Soule 1974; Hare and Cerny 1963), while others restrict this to the differentiated forms. Van der Werf-Messing and van Unnik (1965) suggest the use of radiotherapy as a supplementary measure or by itself when surgery is not possible. In spite of the well-known radioresistance of fibrosarcomata, Hare and Cerny (1963) report a marked reduction of the tumour mass with doses exceeding 3500 rads.

Clinical Types

Among the clinical types of fibrosarcoma, one in particular deserves special mention, congenital fibrosarcoma. This very rare tumour is present from birth. It is usually solitary, often localised to the limbs and characterised by its tendency to local recurrence and nodal invasion. Various cases have been reported in the hand, situated on the thenar eminence (Stout 1962; Balsaver and Butler 1967), the palm (Goslee et al. 1964) or a finger. This swelling is detectable from birth and grows rapidly in size within several weeks, invading the soft parts and destroying the adjacent bony structures. Grossly, it is a circumscribed but non-encapsulated lesion, firm and whitish, with areas of necrotic and haemorrhagic destruction. The histological appearance is that of a major cellular reaction composed of fibroblasts arranged in sinuous, interlacing bundles with numerous mitoses. There also exists a collagenous fibrous reaction that sometimes raises insoluble problems of differential diagnosis, in particular from a fibromatosis. The prognosis of this disorder is dominated by the risk of local recurrence and metastatic spread. Surgical treatment of this tumour, consisting of wide resection or more often amputation, has occasionally led to reports of cure and long-term survival.

Fibrous Histiocytosarcoma

In 1964 O'Brien and Stout reported 15 cases of patients suffering from tumours that they were convinced were malignant forms of fibrous xanthoma with obvious metastases. The term of fibrous xanthosarcoma was modified by Stout in 1953 in favour of fibrous histiocytosarcoma. Numerous series since published (Kempson and Kyriakos 1972; Soule and Enriquez 1972; Wasserman and Stuard 1974) have confirmed the existence of this lesional

entity. This group of tumours, whose origin remains the subject of discussion (whether histiocytic or mesenchymal), includes the xanthosarcomata, the xanthofibrosarcomata and giant cell sarcomata.
Fibrous histiocytosarcoma occurs more often between the ages of 50 and 60 years and more usually in men (60%) than women. Weiss and Enzinger (1978) have reported on 200 cases and consider that this tumour, though rare, is the most common malignant lesion of the soft parts in adults. They found 19% of their cases in the upper limb, though Soule and Enriquez (1972) give a figure of 57%. In the hand, however, this tumour is extremely rare and has been found twice by Weiss and Enzinger (1978) (1% of cases) and only once by McDowell and Henceroth (1977) and by Fu et al. (1975).

Clinical Features

This superficial, subcutaneous or deep-seated tumour is usually asymptomatic. It appears as a solitary, irregular, painless swelling, progressively increasing in size.

Histology

Grossly, the tumour is grey or whitish, firm, polycyclic, and with zones of softening corresponding to necrotic foci. It is more often deep-seated, with an average diameter of 5 cm and a pseudo-capsule. Histologically, two cellular elements predominate: fusiform fibroblast-like cells and round histiocyte-like cells arranged in the classic cart wheel or "roller-blind" pattern characteristic of histiocytic origin (Stout 1962; Lattes 1964); mitoses are frequent in both types of cell. Also present are xanthomatous cells, giant cells, undifferentiated cells and cells of inflammatory type. The extracellular setting consists of collagen fibres. The appearance produced is often polymorphous and less organised with the cells arranged at random, sometimes even with zones of osteoid or chondroid metaplasia. More rarely, the appearance is monomorphous with a predominance of either fibroblasts simulating a fibrosarcoma or of histiocytes simulating a histiocytosarcoma.
Weiss and Enzinger (1978) stress the frequency of a subgroup poor in cells but with abundant myxoid stroma. In many cases optic microscopy suffices to establish the diagnosis of fibrous histiocytosarcoma, but it is sometimes difficult to differentiate this from a liposarcoma or a polymorphous rhabdomyosarcoma; in such cases the electron microscope is useful in identifying the cellular components. The fibroblast appears as a cell with a large, elongated nucleus containing one or two nucleoli and surrounded by an irregular nuclear membrane; its cytoplasm contains a granular endoplasmic reticulum and a well-developed Golgi apparatus. The histiocyte has an oval or kidney-shaped nucleus, a large Golgi apparatus, phagocytic vacuoles and intracytoplasmic inclusions. The nuclei of both these fibroblasts and histiocytes exhibit inclusions or nuclear bodies in the form of radially disposed microfilaments. Their cytoplasm also contains tubular structures. Finally, systems of intercellular junctions of macula adherens type exist between the fibroblasts and histiocytes. The undifferentiated cells are small and ovoid and have a round nucleus. Their cytoplasm contains some ribosomes and mitochondria.

Prognosis

This must always be cautious and is marked by the development of recurrences and of metastases spread by the bloodstream and lymphatic pathways. Soule and Enriquez (1972) report 65% survival at 5 years and 38% at 10 years. A small superficial tumour of predominantly myxoid structure appears to have a relatively good prognosis (Taxy and Battifora 1977), whereas an anaplastic form has a much gloomier outlook (Hubbard and Burton 1977). Weiss and Enzinger (1978) report 32% of results stable enough to be considered good, 26% with recurrences and 42% with metastases (lungs 89%, lymph nodes 32%), some appearing after an interval of 20 years.

Treatment

The therapeutic guidelines are as controversial as for the fibrosarcomata, and many authors consider that amputation offers no more security than wide resection. Adjuvant treatments (chemotherapy with methotrexate and actinomycin D and radiotherapy), though producing reduction in size of the tumour, are not followed by lasting effects (Hubbard and Burton 1977).

Clinical Types

Among the clinical types of fibrous histiocytosarcoma, one deserves special mention: dermatofibrosarcoma protuberans. This was described in 1924 by Darier and Ferrand under the name of progressive recurrent dermatofibroma or fibrosarcoma of the skin, but it had been reported in 1890 by Taylor and owes its English label of dermatofibrosarcoma protuberans to Hoffmann (1925). It is an uncommon fibroblastic skin tumour, preferentially situated on the trunk and proximal ends of the limbs, only exceptionally affecting the distal extremities. It usually occurs in young subjects, more often in males, but may be seen at any age and even from birth. In the hand, very rare cases have been reported on the fingers and dorsum by Wirman and Sherman (1981) and a congenital form by Schvarcz (1977). Clinically, the lesion presents initially in the form of a small nodule embedded in the dermis, adherent to the skin, mobile on the deeper layers, slightly pink, firm, and stable over several years (Fig. 67). Other nodules may appear progressively and coalesce, producing a projecting tumour with atrophy of the overlying skin. Extension in depth is more limited and retarded by the presence of aponeuroses and fasciae. Radiography does not show any bony invasion.

Histologically, this is a fibroblastic tumour originating from the middle dermis. It is never encapsulated and extends laterally and in depth. It is usually characterised by a monomorphous fibroblastic proliferation, fasciculated in herring bone pattern, and by a variable cell population and the presence of collagen fibres. Nucleocytoplasmic anomalies are frequent, but mitoses are usually regular rather than atypical. The epidermis is thinned or ulcerated but never acanthotic, and the hypodermis is always invaded. Taylor and Helwig (1962) analysed 115 cases of this lesion and demonstrated a very characteristic appearance, i.e. the "cart wheel" arrangement of bundles of fibroblasts with a radial architecture centered on capillaries, collagen fibres or a clump of cells. This appearance has been noted by many authors (Bednar described it as "roller-blind" (1957); it is of great diagnostic value but may also be found in histiocytofibrosarcomata, neurofibrosarcomata and histiocytofibromata. Certain forms have been described with myxoid or hyaline transformation.

Owing to the great variability of the histological appearances, the diagnosis of this disorder rests on a collation of the anatomicoclinical features, their relation to those of other fibrous tumours making for ease of identification.

The surgical treatment of dermatofibrosarcoma protuberans has radically changed the prognosis of this tumour, in which recurrence was hitherto considered inevitable. It consists of wide excision well away from the tumour margin, both peripherally and in depth, with removal of fasciae and aponeuroses and closure by skin grafting. An inadequate excision carries the risk of rebound phenomena with fulminating exacerbations. The question of metastases from this lesion has not yet been clarified. Though reported rarely by some authors, no case of remote spread has been noted in large series (Stout 1960; Pack and Tabah 1951; Taylor and Helwig 1962). This tumour is regarded as a sarcoma possessing local malignancy; it seems difficult to accept unreservedly that dermatofibrosarcoma can give rise to metastases.

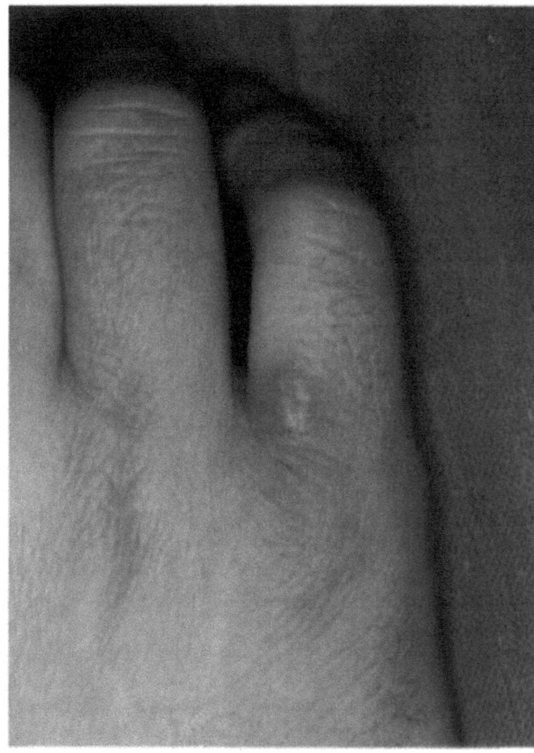

Fig. 67. Darier-Ferrand dermatofibrosarcoma

Bibliography

Balsaver AM, Butler JJ (1967) Congenital fibrosarcoma. Cancer 20: 1607-1615

Battfora H (1976) Spindle cell carcinoma. Ultrastructural evidence of squamous origin and collagen production by the tumor cells. Cancer 37: 2275-2282

Bednar B (1957) Storiform neurofibromas of skin. Pigmented and non-pigmented. Cancer 10: 368-376

Bourne RG (1963) Paradoxical fibrosarcoma of skin. Med J Aust 1: 504-510

Broders AC, Hargraves RL, Meyerding HW (1939) Pathological features of soft tissue fibrosarcoma with special

reference to the grading of its malignancy. Surg Gynecol Obstet 69: 267
Bridges MW (1975) Multifocal malignant histiocytoma. J Clin Pathol 28: 825
Brindley HH, Philips C et al (1955) Fibrosarcoma of the extremities. J Bone Joint Surg [Am] 37: 602
Chomette G (1963) Fibrosarcomes de l'enfant. Arch Anat Pathol 11/2: 115
Churg AM, Kahn LB (1977) Myofibroblasts and related cells in malignant fibrous and fibrohistiocytic tumors. Hum Pathol 8: 205-218
Darier J, Ferrand M (1924) Dermatofibromes progressifs et récidivants ou fibrosarcomes de la peau. Ann Dermatol Venereol 5: 545-562
Del Regato JA et al (1963) Radiotherapy of soft tissue sarcomas. JAMA 185/3: 216-218
Dugois MMP, Colomb L (1959) Fibrosarcomes vrais pigmentés plurinodulaires d'emblée. Réunion de Lyon. Soc Dermatol Syphiligr 15: 140
Eckert C (1963) Surgical treatment of soft tissue sarcomas. JAMA 184/1: 65
Feldman F, Norman D (1972) Intra- and extraosseous malignant histiocytoma (malignant fibrous xanthoma). Radiology 104: 497-508
Fretzin DF, Helwig EB (1973) A typical fibroxanthoma of the skin. Cancer 31: 1541-1551
Fu YS, Gabbiani G, Kaye GI et al (1975) Malignant soft tissue tumors of probable histiocyte origin (malignant fibrous histiocytomas). General considerations and electron microscopic and tissue culture studies. Cancer 35: 176-198
Gentele H (1951) Malignant fibroblastic tumors of skin, clinical and pathological anatomical studies of 129 cases of malignant fibroblastic tumors from cutaneous and subcutaneous loyers observed at radiumhemmet during period 1927-1947. Acta Dermatol Venereol [Suppl] 27: 1-180
Goslee L, Clermon V, Bernstein J et al (1964) Superficial connective tissue tumors in early infancy. J Pediatr 65: 377-387
Guccion JG, Enzinger FM (1972) Malignant giant cell tumor of soft parts. An analysis of 32 cases. Cancer 29: 1518-1529
Haber MH, Alter AH, Wheelock MC (1965) Tumors of the hand. Surg Gynecol Obstet 121: 1073-1080
Hare HF, Cerny MJ (1963) Soft tissue sarcoma. Cancer 16/10: 1332-1335
Helwig EB (1963) Atypical fibroxanthoma. In tumor seminar. Proceedings of 18th annual tumor Seminar of San Antonio Society of Pathologists, 1961. Tex Med 59: 664-667
Hitchens EM, Platt DS (1972) Fibrosarcoma. Cancer 29/5: 1369-1375
Hoffmann E (1925) Über das knollentreibende Fibrosarkom der Haut. Dermatol Z 43: 1
Hubbard LF, Burton RI (1977) Malignant fibrous histiocytoma of the forearm; report of a case and review of the literature. J Hand Surg 2: 292-296
Hughes JP, Mills NL, Lynch RC et al (1975) Malignant fibrous histiocytoma. South Med J 68: 1219-1222
Ivins JC, Dockerty MB et al (1950) Fibrosarcoma of soft tissues of extremities; review of 78 cases. Surgery 28: 495-508
Karev A (1979) Malignant histiocytoma of the arm in a four year old boy. Hand 11/1: 106-107
Kauffman SL, Stout AP (1961) Histiocytic tumors (fibrous xanthoma and histiocytoma) in children. Cancer 14: 469-482
Kempson RL, Kyriakos M (1972) Fibroxanthosarcoma of the soft tissues. A type of malignant fibrous histiocytoma. Cancer 29: 961-976
Kempson RL, McGavran MH (1964) Atypical fibroxanthomas of the skin. Cancer 17: 1463-1471
Krementz ET et al (1963) The treatment of soft tissue sarcoma by chemotherapy. JAMA 184/2: 149
Kroe DJ, Pitcock JA (1969) Atypical fibroxanthoma of the skin: Report of ten cases. Am J Clin Pathol 51: 487-492
Kyriakos M, Kempson RL (1976) Inflammatory fibrous histiocytoma: An aggressive and lethal lesion. Cancer 37: 1584-1606
Lattes R (1964) Tumor seminar case 8. Tex Med 60: 430-433
Levan NE, Hirsch P, Kwong MQ (1963) Pseudosarcomatous dermatofibroma. Arch Dermatol 88: 276-280
Mamelle JC, Salle B (1969) Fibrosarcome et fibromatose congénitaux. Ann Pediatr 16: 255-261
Mandalenakis N, Venne L et al (1980) Dermatofibrosarcome protubérant évoluant depuis 25 ans avec 27 récidives. Ann Pathol 25/2: 139-150
McDowell CL, Henceroth WD (1977) Malignant fibrous histiocytoma of the hand; a case report. J Hand Surg 2/4: 297-298
Merkow LP, Frich LP et al (1971) Ultrastructure of a fibroxanthosarcoma (malignant fibroxanthoma). Cancer 28: 372-383
Miara JM, Bullough PG, Marcove RC (1974) Malignant fibrous histiocytoma and osteosarcoma in association with bone infarcts. J Bone Joint Surg [Am] 56: 932
Michael RH, Dorfman HD (1976) Malignant fibrous histiocytoma associated with bone infarcts. Clin Orthop 118: 182
Mopper C (1953) Primary fibrosarcoma of skin. JAMA 152: 570-572
O'Brien JE, Stout AP (1964) Malignant fibrous xanthomas. Cancer 17: 1445-1455
Ozzello L, Stout AP, Murray MR (1963) Cultural characteristics of malignant histiocytomas and fibrous xanthomas. Cancer 16: 331-344
Pack GT, Ariel IM (1952) Fibrosarcoma of soft somatic tissues: Clinical and pathologic study. Surgery 31: 443-478
Pack GT, Ariel IM (1954) Sarcomas of the soft somatic tissues in infants and children. Surg Gynecol Obstet 98: 675
Pack GT, Tabah EJ (1951) Dermatofibrosarcoma protuberans. Arch Surg 62: 391

Perry H, Chu FCH (1962) Radiation therapy in palliative management of soft tissue sarcomas. Cancer 15: 179-183

Petit VD, Chamness ST et al (1954) Fibromatosis and fibrosarcoma following irradiation therapy. Cancer 7: 149-158

Posch JL (1956) Tumors of the hand. J Bone Joint Surg [Am] 38: 517-540

Pritchard DJ, Soule EH (1974) Fibrosarcoma. A clinicopathologic and statistical study of 199 tumors of the soft tissues of the extremities and trunk. Cancer 33/3: 888-895

Schvarcz LW (1977) Congenital dermatofibrosarcoma protuberans of the hand. Hand 9/2: 182-186

Shieber W, Graham P (1962) An experience with sarcomas of the soft tissues in adults. Surgery 52/2: 295-298

Soule EH, Enriquez P (1972) Atypical fibrous histiocytoma. Malignant fibrous histiocytoma. Malignant histiocytoma and epithelioid sarcoma. A comparative study of 65 tumors. Cancer 30: 128-143

Stevanovic DV (1976) Fibrosarcome disséminé de la peau. Réunion de Paris. Soc Dermatol Syphiligr 83: 273

Stout AP (1948) Fibrosarcoma: Malignant tumor of fibroblast. Cancer 1: 30-63

Stout AP (1953) Tumors of the soft tissues. In: Atlas of tumor pathology. Armed Forces Institute of Pathology, Washington

Stout AP (1960) Tumeurs d'origine mésenchymateuse chez l'enfant. Ann Pathol 5/4: 433-440

Stout AP (1962) Fibrosarcoma in infants and children. Cancer 15/5: 1028-1039

Taxy J, Battifora H (1977) Malignant fibrous histiocytoma: A clinicopathologic and ultrastructural study. Cancer 40: 254-267

Taylor HB, Helwig EB (1962) Dermatofibrosarcoma protuberans. A study of 115 cases. Cancer 15: 717-725

Taylor RW (1890) Sarcomatous tumors ressembling in some respects keloid. J Cutan Genito Urin Dis 8: 384-387

Thorbjarnarson B (1964) Desmoid tumors in treatment of cancer and allied diseases. In: Hoebner P (ed) Tumors of the soft somatic tissues and bone. Hoebner, New York

Van der Werf-Messing B, van Unnik JA (1965) Fibrosarcoma of the soft tissues, a clinicopathologic study. Cancer 18: 1113-1123

Warren S, Sommer GNJ (1936) Fibrosarcoma of soft parts with special reference to recurrence and metastasis. Arch Surg 33: 425-450

Warszawer L (1977) Congenital dermatofibrosarcoma protuberans of the hand. Hand 9/2: 182-187

Wasserman TH, Stuard ID (1974) Malignant fibrous histiocytoma with widespread metastases. Autopsy study. Cancer 33: 141-146

Weedon D, Kerr JFR (1975) Atypical fibroxanthoma of skin: An electron microscopic study. Pathology 7: 173-177

Weiss SW, Enzinger FM (1977) Myxoïd variant of malignant fibrous histiocytoma. Cancer 39: 1672-1685

Weiss SW, Enzinger FM (1978) Malignant fibrous histiocytoma. Cancer 41: 2250-2266

Weschler J (1972) Contribution à l'étude anatomoclinique des proliférations tumorales fibroblastiques chez l'enfant. Recherche des critères histologiques et pronostiques sur 65 observations. Thèse Med, Paris Créteil

Windeyer B, Dische S (1966) The place of radiotherapy in the management of fibrosarcoma of the soft tissues. Clin Radiol 17: 32-40

Wirman JA, Sherman S (1981) Dermatofibrosarcoma protuberans arising of the hand. Hand 13/2: 187

Chondroma and Chondrosarcoma

Chondroma

Chondromata of the soft parts are rare, unlike the cartilaginous tumours found within the bony structures. They may be situated in the subcutaneous areolar tissue in the vicinity of a joint (para-articular chondroma), within a tendon sheath (cartilaginous metaplasia), or they may arise from the synovial membrane (synovial chondromatosis) or from the posterior capsule of a joint of the hand (bilateral, dorsal, digital chondromatosis of Heiple and Elmer 1972). This condition may recur but never gives rise to metastases; however, it is difficult to be dogmatic about its benign nature, even on histological, grounds, since the cellular appearance often raises fears of malignancy.

Cartilaginous tumours may occur at any age and are equally common in both sexes. In a series of 70 chondromata of the limbs, Dahlin (1956) found 20 cases in the foot and 50 in the hand, of which 7 were at the wrist, 32 in the fingers (including 4 in the thumb, though classically this is spared), 7 over the metacarpals and 4 opposite the metacarpophalangeal joints. Lichtenstein (1955) reported 7 cases of cartilaginous tumours, of which 3 were palmar and 4 digital (2 in a nail bed).

Clinical Features

The lesion appears as an irregular, multilobular swelling, often present for many years and gradually increasing in size. Of varying sensitivity it is mobile on the superficial and deep planes and, if situated nearby a neural structure, does not produce any sensori motor deficit. Often there is a moderate increase in size of the finger, of an oedematous nature.

Radiography reveals a tumour of the soft tissues; sometimes a standard film will show disseminated calcification. The related bony structures are normal in all cases apart from superficial erosion of the bone in contact with the tumour.

Pathology

Gross examination during operation may suggest the diagnosis, the tumour usually being multilobular, irregular and exhibiting obvious cartilaginous composition and calcified deposits on section.

Histologically, the lesion is circumscribed and consists of an amorphous eosinophilic stroma and round cells of moderate size often surrounded by a halo. Sometimes there is a major fibrous component leading to the diagnosis of fibrochondroma. The microscopic appearances may be deceptive; they may simulate a xanthoma (giant-cell tumour), and in particular they may exhibit cellular and nuclear anomalies very suggestive of malignancy. This is a very real histological problem and has been encountered by all authors. Nevertheless, and despite this fear, this is a benign tumour not undergoing malignant transformation or metastasis (Stout and Verner 1953).

Recurrences, however, are frequent (Dahlin 1956 gives 18% of cases) and may appear 1-10 years after excision of the tumour; Dahlin even quotes the case of a patient having ten recurrences. In view of the fact that recurrences mainly arise from the histologically less-differentiated forms, Lichtenstein (1955) suggests a classification in terms of the degree of maturation as an aid to prognosis. Although his study was limited to ten cases, it is interesting to note that, in the four cases with well-differentiated hyaline cartilage, the tumour (well-circumscribed) only once raised the suspicion of possible malignancy. The course was routinely satisfactory, and there was no recurrence. In the six other cases the less-differentiated chondroid tissue corresponded to pseudolobular forms with a strong suspicion of malignancy, and despite subsequent excision there were three cases of recurrence.

Aetiology

The origin of these cartilaginous tumours remains unexplained. Stout and Verner (1953) postulate a metaplasia. Despite the proximity of the synovium, Lichtenstein (1955) does not regard this as responsible for the tumour's genesis. Murphy et al. (1962), however, consider that there is a cartilaginous transformation initially situated on the superficial cell layer of the synovium. The metaplastic foci are at first sessile and become pedunculated, at which time they become detached from the membrane to give rise to solitary cartilaginous or osteocartilaginous bodies. Repetitive joint trauma has been noted in some sufferers from this condition. There is no familial incidence.

Clinical Types

Among the various possible presentations of chondromata, one clinical type deserves special mention. Described by Heiple and Elmer (1972) and further investigated by Hensinger et al. (1974) and named "multiple chondromatous hamartoma", it is characterised by multiple, bilateral, cartilaginous lesions of extra-osseous origin situated on the dorsal aspect of the fingers. The reported cases occurred in children and began as a thickening of the dorsal skin of a finger, with progressive impairment of joint function until the finger became deformed and useless. The lesion can affect several fingers and both hands. There is a fibrocartilaginous infiltration adherent to the dorsal aspect of the proximal and distal interphalangeal joints, consisting of a cartilaginous and myxoid matrix and dispersed collagen fibres. In advanced forms the functional handicap becomes very severe, and the authors advise early intervention to preserve mobility and reestablish the appearance of the finger.

Treatment

The therapeutic approach to chondromata is not unequivocal. The tumour is routinely excised in all cases under general or local/regional anaesthesia. Depending on the size of the lesion and the quality of the overlying skin, excision of the superficial layers may be indicated, sometimes requiring a full-thickness skin graft. Lichtenstein and Goldman (1964) advise peroperative histological examination so that treatment may be guided by the degree of maturation, with limited excision if there is hyaline cartilage or rather wider excision for less-differentiated forms to prevent recurrences. Murphy et al. (1962) propose synovectomy when the synovial membrane exhibits foci of cartilaginous metaplasia.

Chondrosarcoma

Chondrosarcomata are exceptionally primitive lesions (Stout and Verner 1953; Ream et al. 1973; Vichard et al. 1980a). They are readily diagnosed when they occur in the amputation scar of a known skeletal chondrosarcoma (Jacobson and Spjut 1960; Patel et al. 1977; Trias et al. 1978). They have a slow course. Exploration often shows them to be adherent to the tendinous apparatus, so that Dahlin and Salvador (1974) suggest a synovial origin. Histology confirms their extraskeletal nature. Their prognosis seems sensibly better than that of other chondrosarcomata, provided there is very wide excision (Vichard et al. 1980).

Bibliography

Dahlin DC (1956) Chondromyxoïd fibroma of bone, with emphasis on its morphological relationship to benign chondroblastoma. Cancer 9: 195-203

Dahlin DC, Salvador AH (1974) Cartilaginous tumors of the soft tissues of the hands and feet. Mayo Clin Proc 48: 721-726

Freund E (1937) Chondromatosis of the joints. Arch Surg 34: 670-686

Heiple KG, Elmer RM (1972) Chondromatous hamartomas arising from the volar digital plates. A case report. J Bone Joint Surg [Am] 54/2: 393-398

Hensinger RN, Wilmington A, Delaware P, Rhyne E (1974) Multiple chondromatous hamartomas. J Bone Joint Surg [Am] 56/5: 1068-1070

Holmes HS, Bovenmeyer DA (1976) Cutaneous cartilaginous tumor. Arch Dermatol 112: 839-840

Jacobson E, Spjut HJ (1960) Chondrosarcoma of the bones of the hand. Acta Radiol 54: 426

Jeffreys TE (1967) Synovial chondromatosis. J Bone Joint Surg [Br] 49: 530-534

Kettelkamp DB, Dolan J (1966) Synovial chondromatosis of an interphalangeal joint of a finger. J Bone Joint Surg [Am] 48/2: 329-332

Lewis MM, Marshall JL (1974) Synovial chondromatosis of the thumb: a case report and review of the literature. J Bone Joint Surg [Am] 56/1: 136-144

Lichtenstein L (1955) Tumors of synovial joints, bursae, and tendon sheaths. Cancer 8: 816-830

Lichtenstein L, Bernstein D (1959) Unusual benign and malignant chondroid tumors of bone, survey of some mesenchymal cartilage tumors and malignant chondroblastic tumors, including a few multicentric one, as well as many atypical benign chondroblastomas and chondromyxoid fibromas. Cancer 12: 1142-1157

Lichtenstein L, Goldman E (1964) Cartilage tumors in soft tissues, particularly in the hand and foot. Cancer 17: 1203-1208

Linden O (1934) Case of chondromatosis of a metacarpophalangeal joint. Acta Chir Scand 75: 181-184

Mosher JF, Kettelkamp DB, Campbelle CJ (1966) Intrascapular or para-articular chondroma. A report of three cases. J Bone Joint Surg [Am] 48: 1561-1569

Murphy FP, Dahlin CD, Sullivan E (1962) Articular synovial chondromatosis. J Bone Joint Surg [Am] 44: 77-86

Mussey RD, Henderson MS (1949) Osteochondromatosis. J Bone Joint Surg [Am] 31: 619-627

Patel M, Pearlman HS et al (1977) Chondrosarcoma of the proximal phalanx of the finger. J Bone Joint Surg [Am] 59: 401-403

Purser DW (1956) Extraskeletal Osteochondromata. J Bone Joint Surg [Br] 38/4: 871-873

Ream JR, Corson JM et al (1973) Chondrosarcoma of the extra-squeletal soft tissue of the finger. Clin Orthop 97: 148

Rosenfeld N, Kurzer A (1980) Soft tissue «Chondroma» of the hand. Hand 12/2: 189-192

Stout AP, Verner EW (1953) Chondrosarcoma of the extraskeletal soft tissues. Cancer 6: 581-590

Trias A, Basora J et al (1978) Chondrosarcoma of the hand. Clin Orthop 134: 297

Vichard P, Pinon P et al (1980a) Le chondrosarcome juxta-osseux. Deux observations. Rev Chir Orthop 66: 111-114

Vichard P, Tropet Y et al (1980b) Chondrosarcome extrasquelettique de la main. Ann Chir 34: 743-746

Wilmoth S (1941) Osteochondromatosis. J Bone Joint Surg 23: 367-374

Leiomyoma

Leiomyomata are benign tumours arising from smooth-muscle fibres. The first cases reported in the literature are attributed to Virchow (1854, cited by Stout 1953) and Forster (1858 cited by Howard 1948), and the most complete studies are those of Stout 1937, Magner and Mill (1961), and Neviaser and Newman (1977). Lever (1967) distinguishes two forms of this lesion, multiple cutaneous leiomyomata and solitary angioleiomyoma, depending on the origin of the tumour. A tumour may arise either from the erector muscles of the hairs and sweat glands or from the vascular (especially the venous) walls. In any case, these lesions seem relatively common, especially in the upper limb. In a series of 84 cases, Neviaser and Newman (1977) reported 24 situated below the elbow (30%), of which 10 were in the forearm, 2 at the wrist and 12 in the hand. The dorsum of the hand is more often involved (Stout 1953), but other sites have been described, such as the thumb (Matton and Defloor 1974) and the first interosseous space (Hauswald et al. 1975). Leiomyomata may occur at any age, particularly around 50 years old, affecting both sexes and both sides with equal frequency.

Clinical Features

This nodular lesion may appear in solitary or multiple form. It is pink or yellowish, usually a few millimeters in size, of firm consistence and often painful, sometimes with paroxysmal episodes. Many authors have remarked on two clinical features suggestive of the diagnosis: sensitivity to pressure and painful hardening on exposure to cold and palpation. Although the number of tumours may increase, the lesion always remains benign, never recurs after excision and never undergoes malignant degeneration. Some authors consider that there are clinical variants related to the origin of these lesions. Thus, the multiple forms are located in the skin and embedded in the dermis, very painful and often pruriginous, while the solitary forms of vascular origin differ by being more deeply situated and are often painless (Neviaser and Newman 1977).

Histology

The appearance produced is that of a well-encapsulated tumour, composed of intertwined smooth-muscle fibres. Rarely, nervous structures have been found under the electron microscope. In angioleiomyomata the venous axis is at the centre of the proliferating muscle fibres, and there is a variable thickening of the vein walls. Although these lesions may sometimes simulate clinically a lipoma or giant-cell tumour, they pose no therapeutic problems and are definitively eliminated by simple surgical excision under local anaesthesis.
Finally, very rare cases of leiomyosarcoma have been reported (Stout 1953), but no lesion of the hand has been described. This is a rapidly growing tumour, eventually greater than 3 cm in size, with marked mitotic activity seen upon histological examination, and metastatic spread by both blood and lymphatic pathways.

Bibliography

Hauswald KR, Kasdan ML, Weiss DL (1975) Vascular leiomyoma of the hand. Plast Reconstr Surg 55/1: 89-91

Howard LD Jr (1948) Tumors of the hand. In: Bunnell (ed) Surgery of the hand, 2nd edn. Lippincott, Philadelphia, p 909

Lever WF (1967) Histopathology of the skin. Lippincott, Philadelphia, pp 675-677

Magner D, Mill DP (1961) Encapsulated angiomyoma of the skin. 2nd subcutaneous tissues. Am J Clin Pathol 35: 137

Matton G, Defloor E (1974) Benign tumors of the hand. Acta Orthop Belg 40/2: 184-185

Neviaser RJ, Newman W (1977) Dermal angiomyoma of the upper extremity. J Hand Surg 2/4: 271-274

Stout AP (1937) Solitary cutaneous and subcutaneous leiomyomas. Am J Cancer 29: 435

Stout AP (1953) Tumors of soft tissues. Armed forces institute of pathology, Washington

Rhabdomyosarcoma

In 1946 Stout defined rhabdomyosarcomata of skeletal muscle as malignant tumours of mesenchymal origin arising from striated muscles or in their close proximity and exhibiting, at some point in the tumour, cell analogous to muscle fibres. Since then, the definition of these lesions has become more precise. In particular, different classifications have been established distinguishing rhabdomyosarcomata on the basis of histological and cytological criteria.

Thus, Masson (1956) distinguishes two different aspects of this tumour: the rhabdopoietic and the rhabdolytic rhabdomyosarcoma. In 1956, Riopelle and Theriault demonstrated that certain "round-celled sarcomata" were in fact a particular variety: alveolar rhabdomyosarcoma.

This hypothesis was confirmed in 1958 by Horn and Enterline, who distinguish polymorphous, embryonal, alveolar and polypoid forms, the last of these being characterised by its urogenital localisation.

The cases reported in the world literature are relatively few (Stout 1946, 121 cases; Pack and Eberhart 1952, 100 cases, and lesions of the hand are rare. The statistics relating to histological type are more interesting. Albores-Saavedra et al. in 1963 collected 35 rhabdomyosarcomata distributed as follows: 9 polymorphous forms, 19 embryonal forms and 7 alveolar forms, with only 1 localisation in the hand (hypothenar eminence) corresponding to this last histological type. Linscheid et al. (1965) reported 87 cases of polymorphous rhabdomyosarcomata of the limbs and limb girdles, 5 of which were in the hand and 4 at the wrist. This type may occur at any age, with a maximum incidence around 55 years, whereas alveolar rhabdomyosarcoma appears at a mean age of 20 years and has a more marked male predominance. Of 110 cases collected by Enzinger and Shiraki (1969), 24 were found in the hand and forearm (flexor muscles). Mutz and Curl (1977) reported the case of a lesion located in the first dorsal interosseous space in a child 11 years of age. Finally, in the 12 cases of rhabdomyosarcomata of the hand reported by Potenza and Winslow (1961), 9 alveolar types predominated as against 2 polymorphous and 1 embryonal, with an equal incidence in the thenar and hypothenar regions.

Clinical Features

The clinical aspect exhibits no specificity. It is a solitary tumour, usually painless, without inflammatory signs or systemic effects. The lesion is often linked to a previous trauma, causes little trouble and does not usually lead to consultation until after some months. The *radiological* aspect is that of a tumour of the soft parts, showing density without calcification; in more advanced forms the adjacent bone may be invaded with irregular cortical erosion.

Histology

Histological examination is the essential contribution to diagnosis. Grossly, the tumour is well-circumscribed, firm and non-encapsulated, with foci of necrosis, haemorrhage and cystic degeneration (Fig. 68). The alveolar rhabdomyosarcoma is greyish on section, friable in places, and contains a whitish or purulent fluid (Riopelle and Theriault 1956). Microscopic study, as classified by Horn and Enterline (1958), defines the following types:

1. Polymorphous rhabdomyosarcoma combines cells of all shapes, fusiform and spherical, and all sizes, together with numerous multinucleated cells, some of which contain vacuolar inclusions. The cytoplasm is abundant and eosinophilic, and the matrix is richly vascularised. In Masson's (1956) view, this is a malignant proliferation of degenerative elements incapable of reconstructing double-striated fibrils. Linscheid et al. (1965) and Albores-Saavedra et al. (1963) report the presence of myofibrils or analogous cells showing double striation.
2. Alveolar rhabdomyosarcoma is a poorly differentiated tumour, whose muscular origin is suggested by the presence of giant multinucleated cells, large cells with eosinophilic cytoplasm, and of rhabdomyoblasts, tumour cells with an often typical structure. The alveolar appearance is due to

the presence of numerous septa, thick connective tissue partitions which bind small compartments filled with tumour cells packed together and free in the alveolar lumen.

3. Embryonal rhabdomyosarcoma is composed of interlaced bundles of long cells with a single oval nucleus and granular eosinophilic cytoplasm; there are also round cells and multinucleated cells in the midst of avascular myxoid areas.

Thus, only histological examination is capable of specifying the diagnosis, making it possible to exclude other hypotheses often put forward, such as synovial sarcoma, reticulosarcoma, angiosarcoma or adenocarcinoma.

Prognosis

This must always be very cautious. The tumour concerned is highly malignant and spreads by the blood and lymph systems. The metastatic locations reported (Potenza and Winslow 1961) are diverse: lung, pleura, mediastinum, scalp, dura mater, bone and axillary nodes. These appear very rapidly as a rule, usually within 3 months, and the average length of survival is 9 months, though there are very rare cases of survival of over 2 years without recurrences (Enzinger and Shiraki 1969).

Treatment

The possibilities of treatment are limited when confronted with such a tumour, and only very early diagnosis and treatment can sometimes arrest its course. The statistics of Potenza and Winslow (1961) are extremely pessimistic: whether the treatment consisted of local or radical resection (7 cases), amputation (4) or only radiotherapy (1), no instance of survival was reported. Mutz and Curl (1977) describe an alveolar rhabdomyosarcoma in a girl 11 years old, treated by primary amputation at the wrist and chemotherapy, who was still alive without recurrence 4 years later. Albores-Saavedra et al. (1963) suggest the addition of axillary nodal eradication and make the following therapeutic distinctions:

- polymorphous rhabdomyosarcoma, resistant to radiation and chemotherapy;
- alveolar rhabdomyosarcoma, more radiosensitive;
- embryonal rhabdomyosarcoma, radiosensitive and positively affected by actinomycin D.

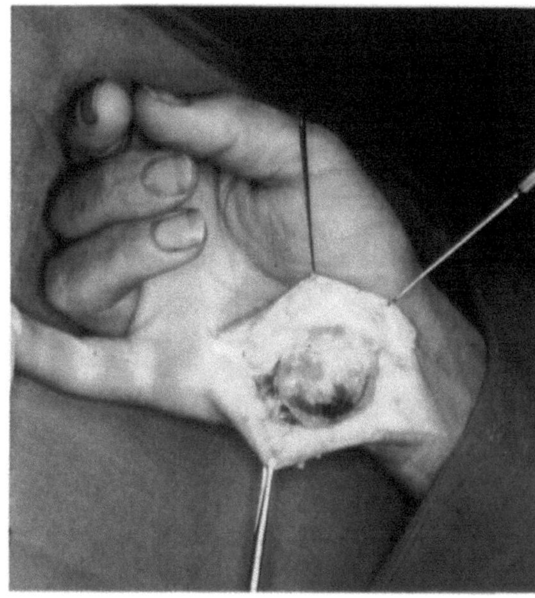

Fig. 68. Alveolar rhabdomyosarcoma: gross appearance. (Courtesy of Professor Génin)

Bibliography

Albores-Saavedra J, Martin RG et al (1963) Rhabdomyosarcoma: A study of 35 cases. Ann Surg 157: 186-197

Caulet T, Adnet JJ et al (1967) Sarcome alvéolaire, à propos d'un cas. Arch Anat Pathol 15: A34-A37

Cureton RJR, Griffiths JD (1955) Rhabdomyosarcoma of the hand following sever trauma. Br J Surg 44: 509-513

Delarue J, Chomette G (1960) Sarcome rhabdopoïétique de structure alvéolaire. Etudes d'une observation anatomique. Ann Pathol 5/2: 291-305

Enterline HT, Horn RC (1958) Alveolar rhabdomyosarcoma: A distinctive tumor type. Am J Clin Pathol 29: 356

Enzinger FM, Shiraki M (1969) Alveolar rhabdomyosarcoma, an analysis of 110 cases. Cancer 24: 18

Fabre J, Fabre MTh, Ette MB (1970) Arguments ultrastructuraux en faveur de l'origine musculaire du sarcome alvéolaire des parties molles. Ann Pathol 15/13: 369-382

Feroldi J, Lesbros F (1971) Deux observations de rhabdomyosarcome alvéolaire du muscle strié. Arch Anat Pathol 19/1: 77-80

Horn RC, Enterline HT (1958) Rhabdomyosarcoma: A clinicopathological study of 39 cases. Cancer 11: 181-199

Linscheid RL, Soule E et al (1965) Pleomorphic rhabdomyosarcoma of the extremities and limb girdles. J Bone Joint Surg [Am] 47: 715-725

Martin JF, Plauchu M, Cabanne F (1959) Une tumeur en quête d'une appellation d'origine «Le sarcome alvéolaire des parties molles». Ann Pathol 4/2: 390-409

Martin JF, Cabanne F, Tommasi M (1961) Le sarcome alvéolaire des parties molles. Arch Anat Pathol 9/4: 263-267

Masson P (1956) Histologie des tumeurs humaines. Maloine, Paris

Mostofi FK, Morse WH (1952) Polypoïd rhabdomyosarcoma (sarcoma botryoïdes) of bladder in children. J Urol 67: 681

Mutz SB, Curl W (1977) Alveolar cell rhabdomyosarcoma of the hand. Case report with four years survival and no evidence of recurrence. J Hand Surg 2/4: 283-284

Pack GT, Eberhart WF (1952) Rhabdomyosarcoma of skeletal muscle. A report of 100 cases. Surgery 32: 1023-1064

Potenza AD, Winslow DJ (1961) Rhabdomyosarcoma of the hand. J Bone Joint Surg [Am] 43: 700

Riopelle JL, Theriault JP (1956) Sur une forme méconnue de sarcome des parties molles: Le rhabdomyosarcome alvéolaire. Ann Pathol 1: 88-111

Stout AP (1946) Rhabdomyosarcoma of skeletal muscles. Ann Surg 123: 447

Tumours of Nerves

Tumours of the peripheral nerves are rare and unfamiliar. The diagnosis is rarely made before operation. Treatment is difficult and risks producing irreversible nerve damage if poorly carried out. Although they are the most common lesions of the peripheral nerves, neuromata are not considered in this chapter since they are not true tumours but rather a proliferation of otherwise normal neural tissue. We also exclude intrinsic tumours that are often responsible for signs of nerve compression.

Depending on the series quoted, nerve tumours constitute 1.2%–4.9% of tumours of the hand. Haber et al. (1965) found 27%, but these included neuromata. Posch (1956) likewise found 3.5%. The series of Woods et al. (1970) is interesting as, of 44 tumours of the hand in children, they found 2 which were neurofibrolipomata.

Classification

Very great confusion exists as regards terminology of neural tumours of the hand and their classification. A dozen appellations have been applied to each of these tumours since the beginning of the century. This confusion is further aggravated by the difficulties of histological diagnosis and by the persistent uncertainty as to the precise origin of these tumours in the nerve tissue. It seems to be agreed by all present-day authors that three main types of peripheral nerve tumours exist, all arising from the Schwann cell and secondarily from the fibroblasts of the endoneurium and perineurium: schwannomata (or neurilemmomata), neurofibromata, and nerve tumours malignant from the outset (neurofibrosarcomata). To these we would add two recently demarcated entities: fibrolipomata and pacinian tumours.

Schwannoma (Neurilemmoma)

Schwannomata are the most common nerve tumours. They are always benign and often asymptomatic and do not always lead to medical consultation. The diagnosis is not always suggested by the clinical findings, but rather by the gross appearance of the exposed tumour – regular, encapsulated, compressing but not invading the nerve bundles. Surgical excision, respecting the nerve bundles, is usually easy, and recurrence is exceptional; whether malignant change ever occurs is very debatable.

It has always been difficult to make a distinction between schwannomata and neurofibromata. For some authors the two terms were equivalent, others described the schwannomata in the context of von Recklinghausen's disease, while yet others ascribed to them different histological origins. We owe the definitive differentiation of these two tumours to the numerous studies of Stout (1949), but certain aetiopathogenetic problems are still unresolved.

Clinical Features

A schwannoma usually presents as a solitary swelling, rounded, regular, firm to palpation and rarely very large. It is mobile transversely in relation to the superficial and underlying planes, but not in the longitudinal direction. The overlying skin is normal (Fig. 69). The tumour is generally situated on the front of the forearm and the palmar aspect of the hand and fingers. The swelling is usually asymptomatic but can sometimes be associated with features of nerve irritation such as distal dysaesthesia or paraesthesia. Percussion of the tumour may elicit a positive Tinel's sign.

Pathology

Macroscopically, a schwannoma presents as a well-encapsulated tumour, easily enucleated from the adjacent layers. Its colour is usually greyish-pink, but it may also be beige, whitish or even brownish. The nerve bundles are crowded together, either laterally or in a fan shape over the tumour (Fig. 70), but they are never invaded. The tumour itself is often eccentric. When sectioned, the specimen may appear homogeneous, or it may be cystic, with alternating cellular and necrotic zones.

Its histology has been familiar since the start of the century (Verocay 1908; Antoni 1920). The schwan-

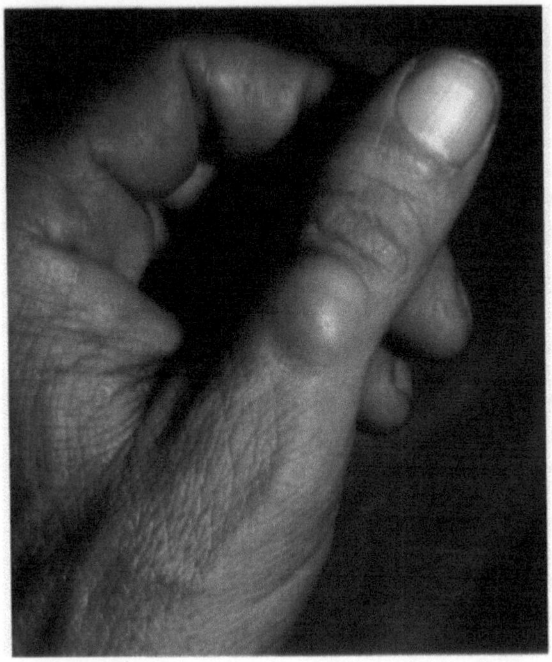

Fig. 69. Schwannoma of a collateral nerve of the thumb

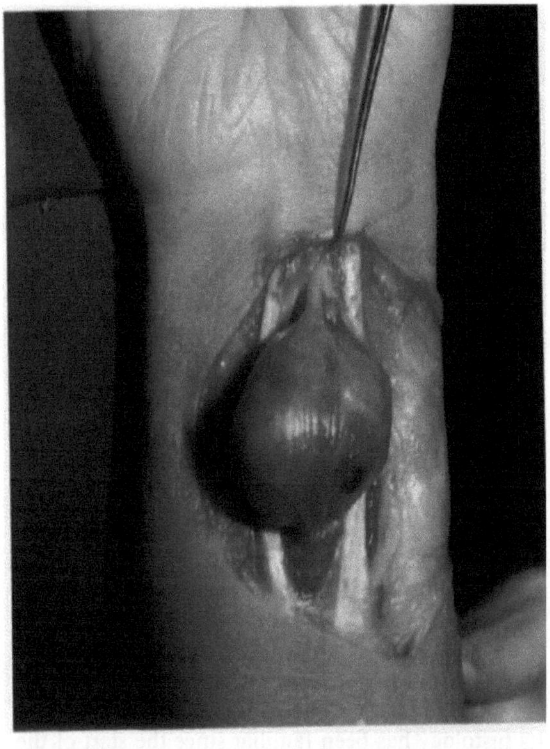

Fig. 70. Large schwannoma of the median nerve: note the nerve bundles fanned out over the periphery of the tumour

noma consists of three components: elongated cells, numerous intercellular collagen fibres and a mucinous hyaline material. Two main histological types of schwannoma can be distinguished:

1. The Antoni type A. This is compact and solid, and the cells are arranged in very orderly fashion in bands, loops, spirals and particularly in palisades. This last is the most characteristic feature, forming the so-called Verocay bodies.
2. The Antoni type B. This has a much looser structure, often oedematous, and the cells are plumper and placed at random without any particular arrangement. There are intercellular microcysts.

Both histological types may be found within one and the same schwannoma. Some authors (Masson 1932) consider that type B is only an aged appearance of type A, the microcysts corresponding to zones of cell necrosis. The tumour's fibrous capsule is a lamellar envelope derived from both the perineurium and the epineurium. The essential feature is that no axons are ever found within a schwannoma; some axons may be present in the immediate vicinity of the capsule but never within the tumour itself, and this is one of the important distinctions from neurofibroma. The vessels often exhibit a characteristic thickening of their walls.

Aetiopathogenesis

Like all tumours of the peripheral nerves, the schwannoma arises from extra-axonal nerve tissue. Since the beginning of the century, there have been two main conflicting schools of thought as regards their precise origin within the nerve tissue. Masson (1932) and Stout (1935) see the schwannoma as derived from the Schwann cells, and this theory is supported by the cell cultures of Murray and Stout (1940) and the histological studies of a large number of cases by Willis and Winston-Evans (1948). Conversely, Mallory (1914) and Penfield (1927) consider that the tumour originates in the fibroblastic tissues of the perineurium and endoneurium. The confusion is further increased by the fact that, even in a normal nerve, it is very difficult to distinguish between the cells of the endoneurium and those of the Schwann sheath. Currently, all authors seem agreed in attributing the origin to the Schwann cells, with or without the participation of the fibroblasts of the perineurium and endoneurium. It is, therefore, a neuroectodermal tumour.

Development

The course of schwannoma is always slow, with a progressive increase in size, and it is extremely rare to find tumours exceeding 3 or 4 cm in diameter. The appearance of signs of nerve compression is late and not always related to the size of the tumour. The course is benign; though the majority of authors raise the very rare possibility of malignant change, this has never been encountered and must be altogether exceptional. Some instances of recurrence have been described, most of them after incomplete excision.

Clinical Types

A few rare cases of multiple schwannomata have been described (Fig. 71; White et al. 1962), and in particular along the same nerve trunk (Phalen 1976). Conley and Miller (1942) and Wirth and Bray (1977) have described intraosseous neurilemmomata, and these are sometimes difficult to remove.

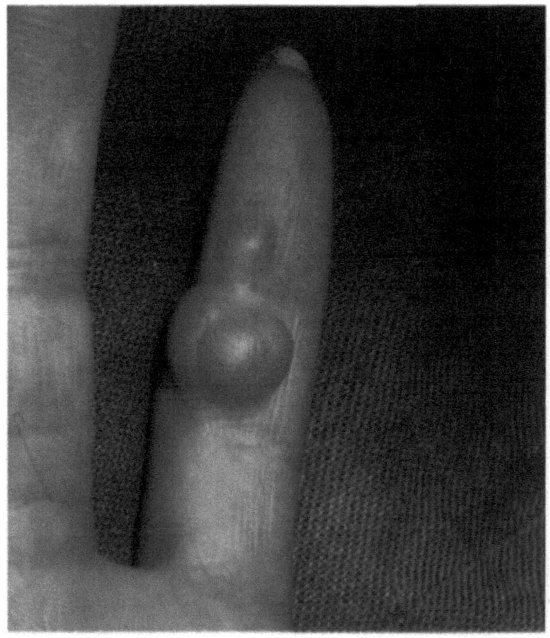

Fig. 71. Multiple schwannomata of lateral collateral nerve of 5th finger

Treatment

Operative measures are not essential, given the frequent absence of symptoms, the slow course and the lack of malignancy; but operation is entirely justified, not only with a view to diagnosis but because of the usual ease of the surgical procedure. An incision over the swelling exposes the schwannoma, which is usually eccentric in relation to the nerve trunk and sometimes elevates several fibres, which may be spread out in a fan shape over the outside of the tumour. Enucleation is easy; any adherence of the tumour to adjacent tissues should arouse doubt as to the diagnosis. It is usually possible to safeguard the integrity of the nerve fibres; some authors advise the use of an operating microscope. The continuity of the nerve must always be respected, and it is better to leave a schwannoma in place if the approach is difficult than to sacrifice a nerve trunk.

Neurofibroma

Neurofibromata are rarely solitary and are usually associated with von Recklinghausen's disease. They may present in two ways, either as a skin tumour (molluscum fibrosum), which is a tumour of neural origin, or as a neurofibroma properly so-called.
The context of a von Recklinghausen's neurofibromatosis arises when these tumours are associated with the following: café-au-lait spots (four of which suffice to establish the diagnosis of neurofibromatosis); multiple nerve tumours (apart from the limbs, these may affect the cranial and particularly the orbital nerves); skeletal anomalies (inequality of limb length, scoliosis; Fig. 72); visceral involvement; and meningeal changes (meningioma). The cardinal difference between neurofibroma and schwannoma is the risk of malignant degeneration, which is considerable for the former (10%-15%) and virtually nil for the latter: hence the importance of diagnostic certitude with solitary lesions.
The neurofibromata were well categorised following the description of von Recklinghausen in 1882. In 1921, van der Hove (cited by Kissel et al. 1970) introduced the term "phakomatosis" to indicate a pathological grouping including neurofibromatosis, Bourneville's tuberous sclerosis, Sturge-Weber's angiomatosis and von Hippel-Lindau's angiomatosis. However, it is Stout (1946, 1949) who deserves the credit for having clearly individualised the neurofibromata, distinguishing them from schwannomata in particular, even though there still persist numerous uncertainties as to the pathogenesis (see the section on schwannomata).

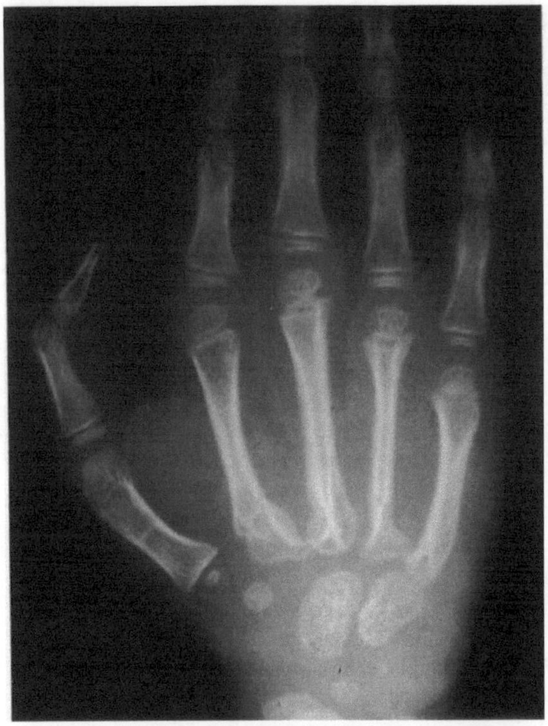

◁ *Fig. 72.* Neurofibromatosis: involvement of the skeleton

Clinical Features

Skin tumours (molluscum fibrosum). Although these are not nerve tumours properly speaking, they are of neural origin since they are situated at the termination of cutaneous branches. They are usually small, soft, ill-defined swellings, tending to be situated on the dorsal aspect of the hand and fingers and bulging a pigmented epidermis outwards. Though generally limited to the skin and subcutaneous tissues, they can invade the subjacent layers, particularly the muscles. They are sometimes solitary but occur much more often multiply and may give the appearance of being "seeded" (Fig. 73 a).

Neurofibromata. These can be solitary, more often multiple, but rarely occur outside the framework of von Recklinghausen's disease. Their size varies; if left to develop, they may become enormous (She-

Fig. 73. ***a*** Cutaneous neurofibromata: molluscum fibrosum. ***b*** Two neurofibromata on the same nerve tract ▽

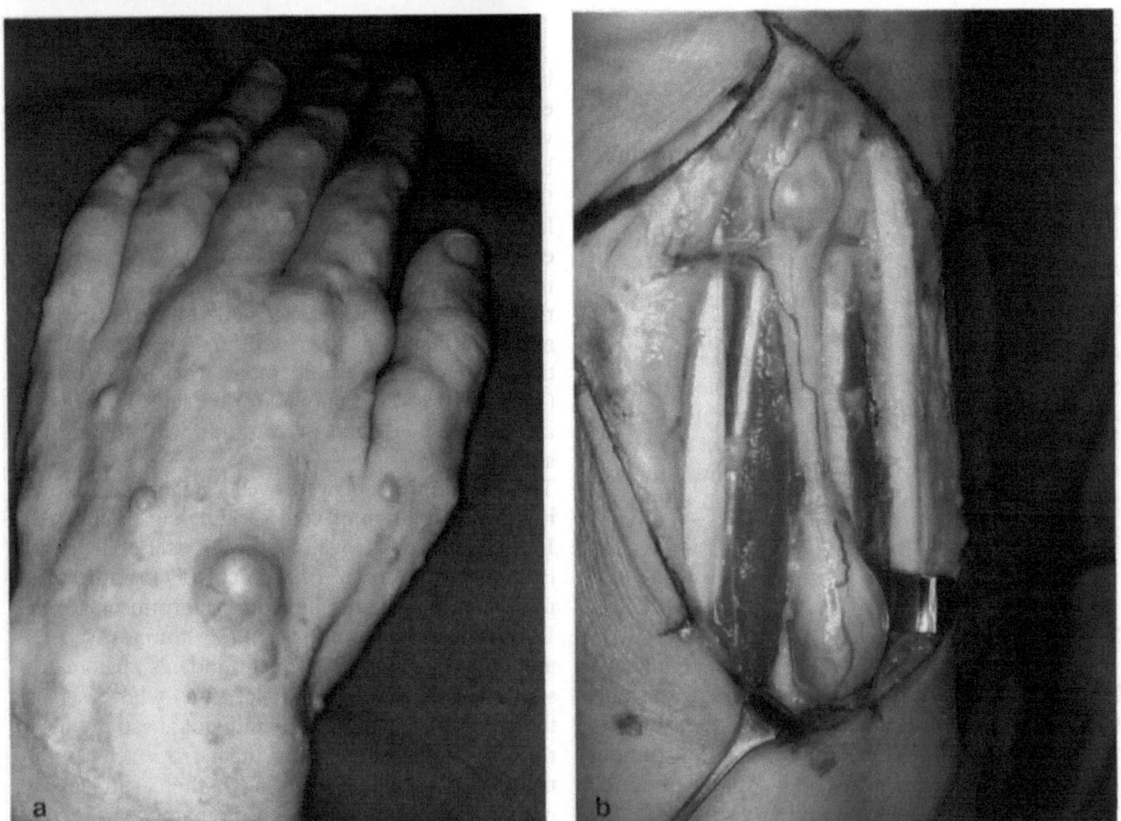

reff et al. 1980). There may be a solitary rounded swelling or more often, several swellings like a string of beads along the same nerve (Fig. 73b); there can be a segmental overall increase in the size of the nerve, giving the cordlike appearance of a parcel tied up with string (plexiform neuroma). Classically situated near flexion creases, these swellings may be situated on any nerve path.
Functional symptomatology varies and is usually proportional to the size of the tumour. Sometimes symptoms are absent; in other cases there are pain and distal features of the paraesthesic or neurological deficit type. Owing to their size, large tumours may limit function of the hand. On palpation, neurofibromata are firm and non-uniform, sometimes giving an encephaloid felling. Percussion may elicit a Tinel's sign. The overlying skin is often thinned, sometimes frayed.

Pathology

Macroscopically, the neurofibroma is often located centrally on the nerve trunk (unlike a schwannoma, which is often eccentric). It sometimes has a pseudo-capsule produced by stretching of the epineurium and perineural tissue. Occasionally, on the other hand, it is very poorly defined and invades the adjacent tissue. It is whitish in colour, often with an oedematous glistening appearance, not uniform in texture, multilobular and sometimes encephaloid. Two features often make it possible to differentiate neurofibroma from schwannoma macroscopically.

1. The tumour frequently encloses nerve bundles, which can be clearly seen entering the tumour and emerging from the other side; a schwannoma stretches the nerve fibres and compresses them but never invades them.
2. Sectioning the specimen never reveals cystic or necrotic zones as in schwannoma.

Microscopically, the striking feature is the absence of any cellular organisation. The neurofibroma consists of a loosely textured stroma traversed by randomly disposed nerve fibres, between which there are small fusiform spindle cells, sometimes interlaced, and collagen fibres disposed in irregular bundles. However, other and much less characteristic aspects also exist: a palisaded nuclear pattern, a cellular organisation reminiscent of the Verocay bodies typical of schwannoma; neuroectodermal elements suggestive of regeneration of nerve fibres and evocative of neuroma; pseudo-corpuscles of Meissner or Pacini; and melanotic cells like those of a blue naevus. There may be aberrant structure, such as striated muscle, bone, cartilage or fatty tissue. It is evident that, without a context of von Recklinghausen's neurofibromatosis, the histological diagnosis is sometimes difficult.

Aetiopathogenesis

In the chapter on schwannomata we mentioned the persisting uncertainties regarding the origin of these two tumours. Just as for schwannomata, most authors currently agree that the neurofibroma is of neuroectodermal origin, involving the Schwann's cells and secondarily the fibroblasts of the endoneurium and perineurium.

Malignant Degeneration

Depending on the authority, this affects between 13% and 22% of neurofibromata (Hosoi 1931, 13%; Stout 1953, 13%; Heard 1962, 17%; Geschickter 1935, 22.5%). It takes the form of a neurofibrosarcoma, still referred to by some as a malignant schwannoma, neurogenic sarcoma or even perineural fibroblastoma. A malignant course is suspected if there is a recent and rapid increase in size of the tumour, the development of pain, indications of nervous system involvement or especially a rapid recurrence after excision. Though sometimes slow, the course is usually swiftly fatal, with bloodstream metastasis. The mortality rate is 90%.
The histological diagnosis depends on the abundant cellularity of the tumour, with cell polymorphism. The nuclei of the spindle cells become large and hyperchromatic, and large, polygonal, multinucleated cells appear with numerous mitoses and zones of necrosis. Treatment consists of radical excision as for any malignant lesion, including sometimes amputation. The lesions are radiorestistant. It was long believed (Jenkins 1952) that surgical procedures in themselves exacerbated the risk of malignant degeneration, and the greatest caution was advised in deciding to excise neurofibromata, which should be left alone unless they were incapacitating. It now seems that there is no proof of this hypothesis, and no series has ever confirmed it.

Clinical Types

Forms with limb hypertrophy: Numerous cases have been reported of segmental hypertrophy of an upper limb but very few cases of global hypertrophy (Anzinger 1931; Holt and Wright 1948; McCarroll 1950; Shereff et al. 1980). The hypertrophy is diffuse and affects all the structures, leading

to an increase in the length and diameter of the upper limb. The skin is generally very soft and marked with thick folds. Not uncommonly, several subjacent nerve tumours can be palpated.
Associated schwannoma and neurofibroma: Several cases have been reported of an association of a schwannoma with a neurofibroma or of a schwannoma developing in the context of a neurofibromatosis. All these cases are quite dated, and it is possible that they may be linked to some confusion in terminology.

Treatment

Ideally, this would consist of complete excision of the tumour without nerve injury. However, neurofibromata are known to be often adherent to the surrounding tissues, are not encapsulated and frequently enclose a certain number of nerve fibres which are difficult to separate from the rest of the tumour. Indeed, some authors have even stated that the only treatment possible is resection of the nerve (Cutler and Gross 1936; Jenkins 1952) and that secondary nerve repair would usually be impossible (Seddon 1972). It does seem possible to perform excision with a minimum amount of nerve damage, and recently several workers have advised the use of an operating microscope (Strickland and Steichen 1977; Kleinert et al. 1973). However this may be, the question of operation must be carefully weighed; it is naturally imperative to remove any neurofibroma suspected of malignancy. On the other hand, manifestly benign lesions should be operated upon only if troublesome. It may be necessary to renounce treatment if the lesion is situated on a large nerve trunk and cannot be removed without serious nerve damage.

Fibrolipomatous Infiltration of the Median Nerve, Fibrolipoma

The clinical, histological and therapeutic features of fibrolipoma of the median nerve are such as to make it an entity quite distinct from classical lipomata. Rather than a true tumour, it is a fibro-adipose infiltration of the nerve. It is benign, and it separates the bundles without invading them. The reported cases are few, since the lesion was not recognised as such before 1930. In his classification of tumours of the peripheral nerves, Stout (1953) mentions the possibility of an encapsulated lipoma of the median nerve. In their series of 2782 tumours of the peripheral nerves, Adair and MacClean (cited by Posch 1956) report no case of fibrolipoma of the median nerve. Similarly, fibro-adipose infiltration of the nerve is not mentioned in any of the large series of tumours of the soft parts of the hand: Straus (1931), Cutler and Gross (1936), Mason (1937), Jenkins (1952), Posch (1956), Haber et al. (1965), Woods et al. (1970). So far 30 cases have been reported in the literature, and thus it is possible to assess the clinical features and histology of the lesion as well as the therapeutic approach (Table 4).
Fibrolipoma of the median nerve occurs equally often in men (14 cases) and women (16 cases) and affects young persons. The average age of operated cases is 18.5 years, but usually the tumour has existed from early infancy, sometimes even from birth (5 cases). It may be situated in any part of the hand (wrist, 12; palm, 24; thumb, 4; fingers, 4), and although the affected side is not always given by the authors, there seems to be an equal distribution between right (12 cases) and left (14 cases).

Clinical Features

Fibrolipoma presents as a usually asymptomatic swelling, always latent for a long period, of soft lipomatous consistency, mobile in relation to the superficial and deep planes, solitary and without regional lymphadenopathy. It lies on the front of the hand, increases very slowly in size, and its bulk may be the cause of dysfunction (Fig. 74). Often the swelling is much reduced in size in the carpal tunnel and reappears in the palm, and then there may be pseudo-fluctuation between the two poles of the lesion. Neurological examination to establish the sensory and motor state and trophic condition is an important step in diagnosis, deciding treatment and prognosis. Pain (11 cases) is observed mainly in lesions of long standing or those of rapid development. This also applies to the sensory disturbance (paraesthesia of the 1st, 2nd and 3rd fingers) seen in 12 cases and the motor deficit (paresis, wasting) seen in 9 cases.
The clinical diagnosis of this lesion is difficult, mainly because of its rarity. However, it ought to be suggested by the occurrence of a carpal tunnel syndrome in a child or young adult, especially if a swelling of the wrist, palm or fingers has been present from birth or early infancy. Increase in size of the tumour may take place from the wrist towards the fingers but can also occur in a retrograde direction, as shown in the second case of Friedlander et al. (1969).

Table 4. Fibrolipoma of the median nerve: a review of published cases to date

Reference	Sex	Age (onset)	Site and side	Clincal features	Neurological signs		Diagnosis/treatment	Course
Brunelli (1964)	F	29 (childhood)	Forearm, wrist, palm R	Soft, painless swelling	Sensory Motor	+ +	Fibrolipoma Decompression of carpal tunnel, nerve biopsy	Neurological signs gone in 3 months
Mikhail (1964)	F	16 (6)	Wrist, thenar, thumb, index R	Pain, dysfunction, increase in size	Sensory Motor	+ +	Lipomatous macrodystrophy. Partial excision osteotomy of index	Residual anaesthesia (thumb and index)
	F	18 (8)	Wrist, thenar L	Painless tumour, dysfunction, increase in size	Sensory Motor	+ +	Fibrolipoma Decompression, nerve biopsy	Sensorimotor sequelae
Pulvertaft (1964)	M	4 (birth)	Palm L	Solitary, painless, asymptomatic mass	Sensory Motor	– –	Fibrolipoma Decompression, nerve biopsy	No sequelae
	F	19 (9)	Forearm, palm R	Solitary, painless mass	Sensory Motor	– –	Fibrolipoma Decompression, nerve biopsy	No deficit, stable
Yeoman (1964)	F	11 (birth)	Palm, ring finger L	Painful mass, growth in size	Sensory Motor	+ +	Fibrolipoma Decompression, nerve biopsy	Stable
	F	16 (13)	Palm, wrist L	Mass and motor power normal	Sensory Motor	+ –	Fibrolipoma Decompression. nerve biopsy	
	F	21 childhood)	Base of index L	Solitary swelling	Sensory Motor	– –	Fibrolipoma Decompression	
Emmett (1965)	F	6 (3)	Palm L	Painless mass, growth in size, dysfunction	Sensory Motor	– –	Fibrolipoma Decompression	Apparent regression
Rowland (1967)	M	4 (3)	Palm L	Painful mass, growth in size	Sensory Motor	– –	Fibrolipoma Decompression, nerve biopsy	Loss of pain, slight thenar atrophy
Callison et al. (1968)	M	28 (childhood)	Forearm, palm L	Painful swelling	Sensory Motor	+ +	Fibrolipoma Decompression, nerve biopsy	Neurological signs unchanged at 2 years
	F	24 (8)	Palm R	Painful palmar mass	Sensory Motor	+ +	Fibrolipoma Carpal tunnel decompression	Decrease of mass, regression of sensory signs, persistent atrophy
	M	30 (7)	Wrist, palm L	Pain, palmar swelling, palmar fullness	Sensory Motor	+ –	Fibrolipoma Carpal decompression, nerve biopsy	Persistent mass, regression of sensory signs
	F	3½ (birth)	Forearm, wrist, palm L	Painful swelling	Sensory Motor	– –	Fibrolipoma Carpal decompression, nerve biopsy	Persistent mass, appearance of sensory motor signs after 12 years
Mazingarbe (1968)	M	17 (childhood)	Wrist, palm L	Painful swelling	Sensory Motor	+ –	Fibrolipoma Decompression, decortication of nerve	Regression of symptoms
Johnson and Bonfiglio (1969)	F	23 (childhood)	Forearm, wrist, palm	Painful swelling	Sensory Motor	+ +	Lipofibromatous hamartoma Total excision	Disappearance of mass, persistent sensorimotor deficit

Table 4 (continued)

Reference	Sex	Age (onset)	Site and side	Clincal features	Neurological signs	Diagnosis/treatment	Course
Friedlander et al. (1969)	M	24 (11)	Forearm, hand L	Painful swelling, poor grip	Sensory – Motor –	Fibrolipoma Carpal decompression nerve biopsy	Decrease of pain, no neurological deficit
	F	46	Index, palm, wrist R	Bulky mass deformity of index	Sensory – Motor –	Fibrolipoma Fusion of TIP joint of index, decompression, resection radial collateral of index and medial collateral of thumb	Decrease of pain, sensory loss (thumb and index)
Bergman et al. (1970)	F	9 (birth)	Wrist L	Painful swelling	Sensory – Motor –	Fibrolipoma Complete resection of tumour, excision of nerve	Disappearance of mass, no neurological deficit
Touzard (1970)	M	21	Palm R	Awkward, large swelling	Sensory – Motor –	Fibrolipoma Decompression, section collaterals	Moderate sensory disorder
	F	41 (16)	Wrist, thumb R	Swelling hindering abduction and opposition of thumb	Sensory + Motor +	Fibrolipoma Carpal tunnel decompression	Regression of neurological signs
Paletta and Rybka (1972)	M	8	Palm	Painful swelling	Sensory – Motor –	Fibrolipomatous hamartoma Decompression nerve biopsy	Increased size of mass
	M	12	Wrist L	Solitary swelling	Sensory – Motor –	As above	Stable, symptom-free
	M	2	Palm	Rapidly growing swelling	Sensory – Motor –	As above Decompression resection of mass plus section of digital branches	No neurological deficit
Louis and Dick (1973)	M	13 (8)	Forearm, palm R	Two separate asymptomatic swellings	Sensory – Motor –	Ossifying fibrolipoma Decompression, nerve biopsy	Stable
Gournet (1973)	M	23	Forearm, palm R	Macrodactyly (index and medius)	Sensory + Motor +	Fibrolipoma Decompression, nerve biopsy	Regression of symptoms, good result
Ross and Jaffa (1975)	F	17 (12)	Palm R	Painless mass hindering grip	Sensory – Motor –	Fibrolipomatous hamartoma Decompression, nerve biopsy	Stable
Terzis et al. (1978)	F	16 (birth)	Thumb R	Solitary swelling	Sensory – Motor –	Fibrolipoma Microsurgical excision, nerve biopsy	Shrinkage of size of thumb, no neurological deficit
Patel et al. (1979)	M	24 (8)	Palm R	Solitary swelling	Sensory – Motor –	Fibrolipoma Decompression, nerve biopsy	Shrinkage of tumour, no neurological signs
	M	6	Forearm, wrist	Solitary swelling	Sensory – Motor –	Fibrolipoma Excision of tumour with part of nerve	Normal opposition, moderate sensory deficit, no recurrence

Radiology

The radiograph shows a large opacity of the soft tissues of the palmar aspect of the hand. Louis and Dick (1973) report a case of ossifying fibrolipoma with linear calcification in the palm and forearm and punctate calcification of the thenar eminence.

Pathology

Grossly, the tumour is fusiform, orange-yellow, situated on the hypertrophied trunk of the median nerve and may measure 2-4 cm in diameter. It develops within the epineurium and does not involve the adjacent tissues, which are normal. The infiltration often ceases where the nerve divides into its branches, but the latter, especially the thenar branch, may be involved (Fig. 75). The yellowish-white nerve sheath is easily separated from the adjacent structures. The different constituents of the nerve are increased in size, but their architecture is preserved.
The histological feature characteristic of this lesion is the proliferation of fibro-adipose connective tissue composed of fat cells and fibrous tissue without a capsule, separating the nerve bundles, to which they are closely adherent. The collagen fibres are arranged in concentric perineural lamellae. All the structures of the nerve (sheath, epineurium, perineurium and endoneurium) are involved in this infiltration, which always contains fibrous and lipomatous components but in varying proportion. Blood capillaries and lymphatics are also seen. On the other hand, and unlike the disease of Dejerine and Sottas, myelin degeneration and Schwann cell proliferation remain unobtrusive.

Terminology

The histological features peculiar to this lesion have given rise to a varied terminology. Stout (1953) applied the name of benign mesenchymoma to a benign tumour exhibiting at least three different tissue components (fibrous, adipose and vascular). Mikhail (1964) used the term lipoma of the median nerve. Yeoman (1964), in view of the absence of a

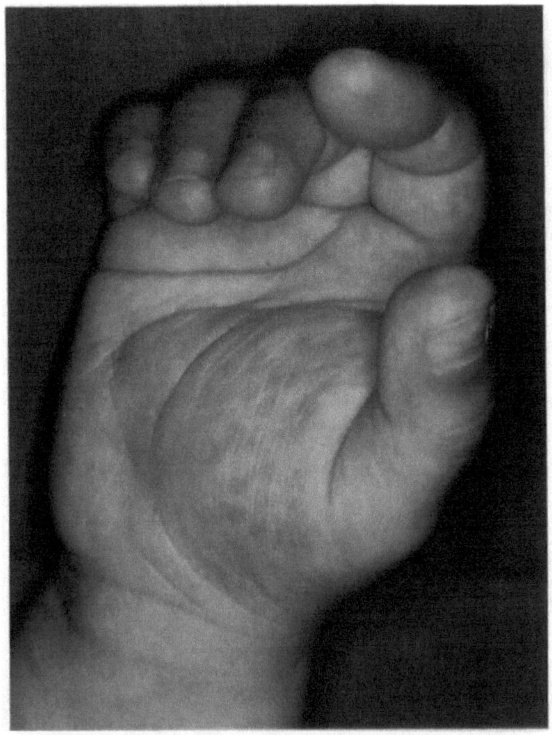

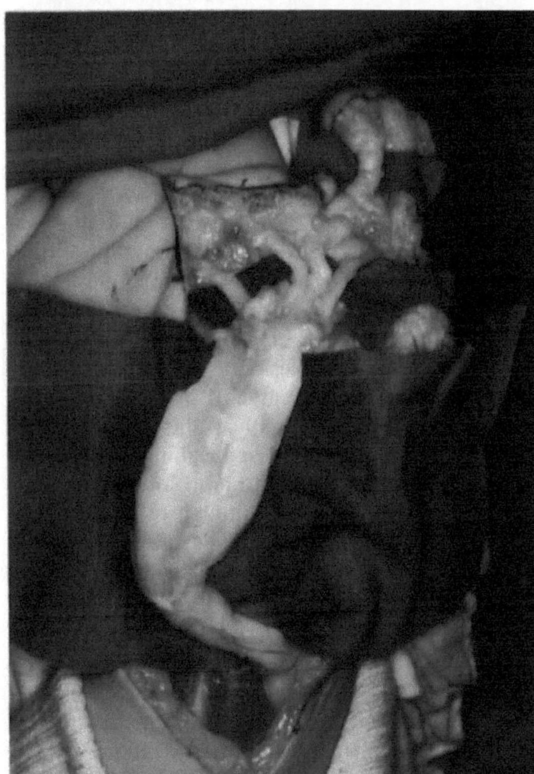

Fig. 74. Fibrolipoma of median nerve: note enlargement of the thumb and index finger *(above)*

Fig. 75. Fibrolipoma of median nerve: gross appearance. Note hypertrophy of collateral nerves of thumb *(right)*

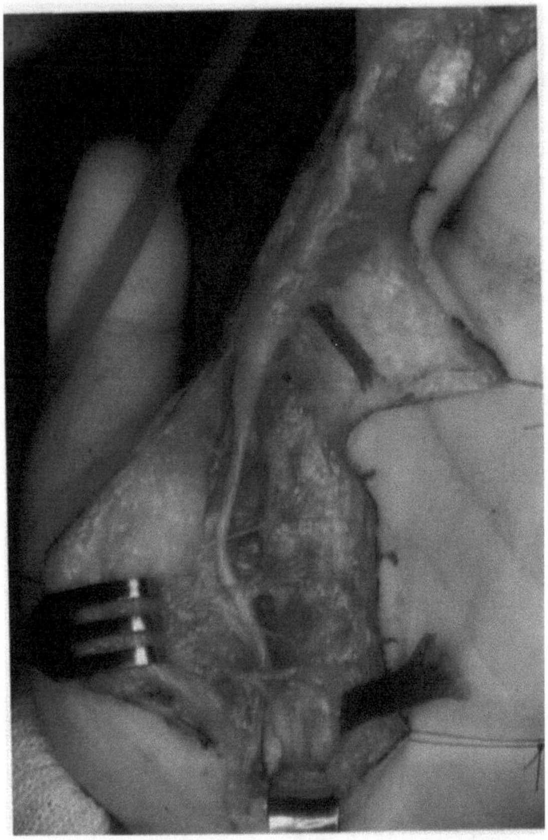

Fig. 76. Solitary fibrolipoma of radial collateral nerve of index finger

clearly marked capsule, considered that this was not a true tumour and spoke of "fibroadipose infiltration of the median nerve". Morley's (1964) term, "intraneural lipoma", seems unjustified since the nerve bundles are separated but not invaded. Emmett (1965) regards the lesion as a lipomatous hamartoma, while Pulvertaft (1964) speaks of a special type of neurofibroma. Finally, Callison et al. (1968) agree with Yeoman (1964) in speaking of fibro-adipose proliferation or infiltration of the median nerve, a description which, though rather vague, is probably the most suitable, given the absence of a true tumour or any degenerative pathology or inflammatory reaction.

Aetiology

The aetiology of fibrolipoma of the median nerve, as of lipomata in general, remains unknown. The youth of the patients and the absence of hereditary, toxic, traumatic or infective antecedents argue in favour of a congential origin (the coexistence of injury, sometimes mentioned by the patient, seems purely coincidental). Sunderland (1972) studied the protective role of the connective tissue on the peripheral nerve and concluded that a reactive proliferation might develop following chronic irritation. Yeoman (1964) suggests that fibrolipoma is due to abnormal compression of the median nerve by the transverse carpal ligament during growth; this hypothesis fits in well with the site of the tumour, the youth of the patients and the stability (or even the regression) observed in the infiltration after simple resection of the transverse carpal ligament. However, certain observations seem to contradict this hypothesis.

Thus, Yeoman (1964) admits that although the proliferation of fibrous tissue is compatible with a chronic irritative aetiology, the abundance of adipose tissue remains an enigma. Again, this infiltration is not seen only in the carpal tunnel but may begin (Friedlander et al. 1969) in, or remain strictly localised to, a finger (Fig. 76) (Patel et al. 1979), in areas removed from any possible zone of compression. In fact, the infiltration may be situated at any part of the nerve. Finally, analogous lesions of the nerve are not observed in the carpal tunnel syndrome of older patients, in which the transverse carpal ligament is incriminated.

Clinical Types

In addition to isolated infiltration of the median nerve, a particular type associated with macrodactyly should be mentioned.

In 1936, Fevre (cited by Gournet 1973) noted the existence of an "abnormal nerve cord which created and formed the center of hypertrophy of the finger". This is considered by many authors as the essential feature and actual cause of the digital hypertrophy. Gornet (1973) studied the role of nerve involvement in 22 cases of macrodactyly and found 20 tumours of the median nerve, one tumour of the ulnar nerve and 1 case with a tumour of both the median and ulnar nerves. The histological lesions he found were not specific: eight fibrolipomata, four neurofibrolipomata, three plexiform neuromata, one neurofibroma, one lipoma and five without precise aetiology. In every case the hypertrophic finger exhibited infiltration of its collateral nerves. However, Gournet (1973) did not admit the primary role of this infiltration in macrodactyly and advanced three other arguments: embryologically, development of the finger is independent of the collateral nerve; from the morbid anatomical aspect the nerve is not really hypertrophied but infiltrated; and, finally, digital hemiatrophy has not been observed after resection of an affected collateral. For this author the relatively frequent association of these two lesions justifies a systematic search for a fibrolipomatous infiltration of the wrist and the existence of a carpal tunnel syndrome if there is macrodactyly in a young subject. There is also the lipomatous macrodystrophy described by Werthemann (1952, cited by Mikhail 1964) and reported by Mikhail (1964, first case) associated with a lesion of the median nerve and characterised by a fibrolipomatous infiltration of all the tissues of the finger, with hypertrophy of the cutaneous, subcutaneous and bony tissues.

Differential Diagnosis

The diagnosis of fibrolipoma of the median nerve can only be based on the gross peroperative findings and the histology. It may resemble a ganglion or lipoma of the wrist. Morley (1964) and also Watson-Jones (1964) have reported cases of true lipomata of the median nerve, benign encapsulated tumours, easily enucleated and composed of typical fatty lobules without hypertrophy or dissociation of the neural elements. True hamartomata have been reported by Lamphier (1954) and by Coventry et al. (cited by Posch 1957) exhibiting adipose, fibrous and cartilaginous tissues. The inherited disorder of Dejerine and Sottas presents more diffuse lesions and a particular type of histological involvement of the nerve. Finally, a neurofibroma is often seen within the framework of von Recklinghausen's disease.

Treatment

This is performed under general or local/regional anaesthesia, with a pneumatic tourniquet. The incision is made over the tumour, often beginning in the lower part of the forearm and continuing towards the thenar eminence in the opposition crease of the thumb. After incision of the aponeurotic layer and the transverse carpal ligament, a bulky, yellow, tumoral mass of lipomatous appearance bulges out, requiring identification of the nerve above the tumour in a healthy area. The swelling is progressively dissected and is never found adherent to the flexor tendons. The nerve is surrounded by a transparent connective tissue sheath and often exhibits an hourglass shape with narrowing opposite the transverse carpal ligament. The tumoral infiltration, which is soft, yellowish and of fatty appearance, dissociates the nerve fibres while always respecting their continuity. This infiltration may involve the thenar branch of the median nerve; it becomes attenuated distally and ceases where the terminal branches are given off. Total excision of the lesion is usually not possible and always carries a risk of sensorimotor disorder, which is very serious for the surgery of a benign tumour. Microsurgical decortication of the bundles is difficult, so much so that virtually all authors prefer conservative measures. In fact, section or resection of the transverse carpal ligament together with nerve biopsy for diagnostic purposes produces a decompression adequate to relieve the symptoms.

Such treatment, carried out in 20 cases (Patel et al. 1979), usually resulted in regression or stabilization of the neurological signs. The tumoral mass became less palpable, not because of any shrinkage of the tumour but because of the decompression of the nerve. In one case, however, there was a postoperative increase in size of the tumour, and in another case progressive sensorimotor disorder developed. However, we should also mention the good result observed after microsurgical decortication (Terzis et al. 1978) and three interesting cases of resection of the median nerve without sensorimotor deficit (Bergman et al. 1970; Paletta and Rybka 1972; Patel et al. 1979). This surprising result was due to the existence of an anastomosis between the median and ulnar nerves, the anastomosis of Martin-Gruber, which is situated between the elbow and wrist and is present, according to Mannerfelt (1966), in 10% of cases. This anatomical variation makes the ulnar nerve essentially responsible for innervation of the thenar muscles but does not explain the observed absence of any sensory loss. Selective study of the nerve trunks showed that the sensory territory of the median nerve was under the control of the radial nerve.

Apart from these observations, decompression of the median nerve by resection of the transverse carpal ligament suffices to give a good result. There has been no occasion of malignant degeneration to regret, but as the perspective of these cases is still limited, a clinical follow-up of affected patients is advised by some. It should be added, finally, that analogous fibrolipomatous infiltration of the ulnar nerve has been described.

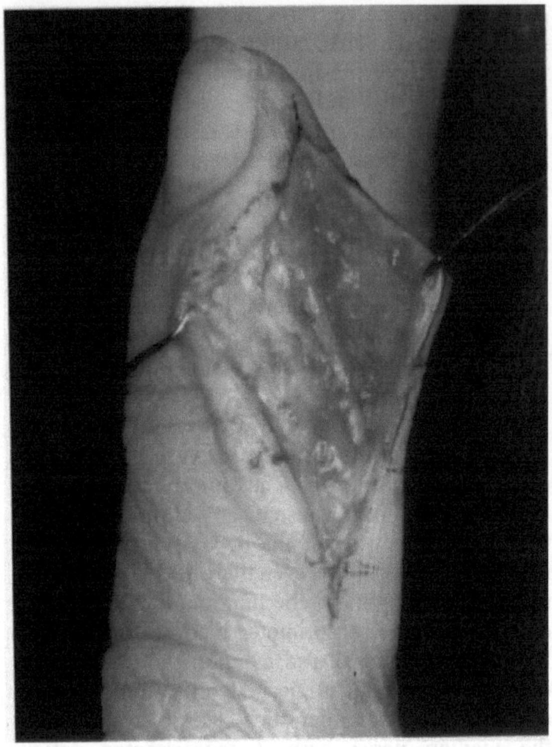

Fig. 77. Pacinian tumour

Pacinian Tumour ("Pacinoma")

Although they are at the nosologic extreme of neural tumours, a rapid reference may be made to the rare pacinian tumour. To our knowledge, the literature to date includes only 8 reports of them in the hand, making 11 cases in all. These tumours generally appear in a post-traumatic context. They are usually revealed by pain on gentle touching or stroking, so that very severe pain prevents the slightest contact. This pain is sometimes absent (Patterson 1956; Zweig and Burn 1968), and then the tumour is noticed by a swelling situated on the palmar aspect of the hand and fingers, more often at the fingertips. Macroscopically, there is a bunch of small pearly-white nodules, 2-8 mm long, attached to the digital nerve, resembling grains of rice (Fig. 77). Microscopically, there is the normal lamellar structure of the pacinian corpuscles, i.e. a concentric arrangement of connective tissue fibres surrounding a central acellular space containing the nerve fibre. There also exist hyperplasia and hypertrophy of the corpuscles with widening of the interlamellar space. These abnormal corpuscles are sometimes situated beneath the epineurium (Zweig and Burn 1968). This is therefore not a tumour in the strict sense of the term but rather a hyperplasia, sometimes associated with a metaplasia.

Surgical excision of these "pacinomata" is usually easy as they are clearly visible macroscopically and simply enucleated from the adjacent planes. If they are situated under the epineurium, the latter must be incised with due care not to damage the nerve bundles. Unlike certain authors who have observed a satisfactory course after simple excision, we have been struck, along with Chavoin et al. (1980) and Hart et al. (1971), by the virtually inevitable recurrence of these lesions after excision. These recurrences, generally very painful, sometimes associated with causalgia in the entire upper limb, have led some to perform radical neurectomy (Hart 1971), while others (Chavoin et al. 1980) have ensheathed the nerve in a silastic tube.

Primary Malignant Tumours

Here again, the nosology remains confused. The term "neurofibrosarcoma" used by the majority of authors is rejected by Winston-Evans (1968) as it suggests a fibroblastic and not a neuroectodermal origin for tumours that are malignant from the outset. The terms malignant schwannoma, malignant neurilemmoma, neurogenic sarcoma and perineural fibroblastoma have also been applied.

Neural tumours malignant from the outset are extremely rare in the hand, but no series has provided precise statistics. Kendall et al. (1969) found no neural tumour in their series of 73 such tumours in the hand. Haber et al. (1965) reported 3.3% of neural tumours among malignant tumours of the hand, but without specifying whether these were primary or secondary. Their series included 4 neurofibrosarcomata of 607 neural tumours of the hand (neuromata included). Byrne (1966) found that 50% of neurofibrosarcomata were secondary to von Recklinghausen's disease, so that it seems that only one of every two of these tumours is primarily malignant from the outset. Although the literature on the subject is scanty, it appears that primary neurofibrosarcomata exhibit the same clinical and evolutionary features as tumours developing on the basis of neurofibromatosis. The development of the first signs is rapid, with early functional symptomatology (pain or paraesthesia, paralysis), and this suggests the diagnosis of malignancy. A rapid increase in the size of the tumour confirms the diagnosis, which generally leads to amputation in view of the site of the lesion. Conservative treatment by local excision is often followed by recurrence. The appearance of pulmonary metastases is late.

Bibliography

Allende BT (1967) Macrodactyly with enlarged median nerve. Plast Reconstr Surg 39: 578-581

Antoni NRE (1920) Über Rückenmarkstumoren und Neurofibrome. Bergmann, München

Anzinger FP (1931) Congenital plexiform neurofibromas and elephantiasis neuromatosa of the right arm and neck. JAMA 96: 1381

Bailey P, Hermann J (1938) The role of the cells of Schwann in the formation of tumors of the peripheral nerves. Am J Pathol 14: 1

Banzet P (1971) Le gigantisme digital. Ann Chir 25/19-20: 965-970

Barber KW, Bianco AJ, Soule EH, MacCarty CS (1962) Benign extraneural soft-tissue tumors of the extremities causing compression of nerves. J Bone Joint Surg [Am] 44: 98-104

Barsky AJ (1967) Macrodactyly. J Bone Joint Surg [Am] 49: 1255-1266

Ben-Bassat M, Casper J, Kaplan I, Laron Z (1966) Congenital macrodactyly. J Bone Joint Surg [Br] 48: 359-364

Bergman FD et al. (1970) Radical excision of a fibrofatty proliferation of median nerve with no neurological loss symptoms. Plast Reconstr Surg 46: 375-380

Bergman FO, Blom SEG, Stenstrom SJ (1970) Radical excision of a fibrofatty proliferation of the median nerve with no neurological loss symptoms. Plast Reconstr Surg 46/4: 375-380

Bird CC, Willis RA (1969) The histogenesis of pigmented neurofibromas. J Pathol 97/6: 631

Biscatti G, Cagini P, Bruschelli M (1973) Dermatoglyphics of the hand in Recklinghausen. Minerva Pediatr 25/8: 329-330

Blair WF (1980) Granular cell schwannoma of the hand. J Hand Surg 5: 51-52

Bogumill GP, Sullivan DJ, Baker GI (1975) Tumors of the hand. Clin Orthop 108: 214-222

Brihaye C (1974) Local gigantism of the hand associated with a plexiform neurofibroma of the ulnar nerve. J Neurosurg Sci 18/4: 271-277

Brown H, Flynn FE (1973) Abdominal pedicle flap for hand neuromas and entrapped nerves. J Bone Joint Surg [Am] 55: 575-579

Brunelli G (1964) Lipoma interfibrillare del nervo mediano con sindrome da compressione nel canale del carpo. Minerva Ortop 15/4: 211-217

Burton RI, Littler JW (1975) Non trauma soft tissues afflictions. Curr Probl Surg: 1-56

Butler ED, Hamill JP (1960) Tumors of the hand: A 10 years survey and report of 437 cases. Am J Surg 100: 293

Byrne JJ, Cahill JM (1961) Tumors of major peripheral nerves. Am J Surg 102: 724

Byrne JJ (1966) Nerve tumors. In: Flynn JE (ed) Hand surgery. Williams & Wilkins, Baltimore

Callison JR, Thommas OJ, White WL (1968) Fibro-fatty proliferation of the median nerve. Plast Reconstr Surg 42/5: 403-413

Campbell CS, Wulf RF (1954) Lipoma producing a lesion of the deep branch of the radial nerve. J Neurosurg 11: 310-311

Chavoin JP, Durrous R, Mansat M et al. (1980) Prolifération tumorale douloureuse des corpuscules de Pacini au niveau de la main. Ann Chir 34: 738

Clifford R, Kelly A (1955) Primary malignant tumors of the hand. Plast Reconstr Surg 15: 227-232

Conley AH, Miller DS (1942) Neurilemmoma of bone. J Bone Joint Surg 24: 684

Coventry MB, Woolner LB, Anderson ME (1949) Proc Mayo Clin 24: 587

Cutler EC, Gross RE (1936) Surgical treatment of tumors of peripheral nerves. Ann Surg 104: 436

Dejerine, Sottas (1893) Sur la névrite intersticielle, hypertrophique et progressive de l'enfance. Compt Rend Soc Biol 5: 63-69

De Palma AF (1959) Macrodactyly. Clin Orthop 15: 154-166

Dupre A, Christol B, Bonafe JL (1974) Névrome cutané à tumeur unique avec importantes altérations des cellules schwanniennes. Ann Dermatol Venerol (Paris) 101/3: 271

Emmet AJ (1965) Lipomatous hamartoma of the median nerve in the palm. Br J Plast Surg 18: 208-213

Epstein MJ (1971) Malignant schwannomas of the hand: A review. Bull Hosp Jt Dis Orthop Inst 32/2: 136-147

Friedlander HL, Rosenberg NJ, Graubard DJ (1969) Intraneural lipoma of the median nerve. J Bone Joint Surg [Am] 51: 352-362

Frykman G, Wood W (1978) Peripheral nerve hamartoma with macrodactyly in the hand. J Hand Surg 3: 307-312

Gaisford JC (1960) Tumors of the hand. Surg Clin North Am 40: 549-566

Gaisford JC (1960) Tumors of the hand. Surg Clin North 40: 549-566

Gama C, Franca LCM (1980) Nerve compression by Pacinian corpuscles. J Hand Surg 5/3: 207-210

Geschickter CF (1934) Lipoïd tumors. Am J Cancer 21: 617-641

Geschickter CF (1935) Tumors of the peripheral nerves. Am J Cancer 25: 377

Gournet C (1973) Macrodactylie et fibrolipomes du médian (phacomatoses). Ann Chir 27/5: 511-517

Greene RG (1964) Soft tissue tumors of the hand and wrist: A 10 years survey. J Med Soc NJ 61: 495

Haber MH, Alter AH, Wheelock MC (1965) Tumors of the hand. Surg Gynecol Obstet 121: 1073-1080

Hallson CH, Marcuse P (1944) Neurofibroma within a lipoma of the neck. Am J Surg 64: 268-270

Hart WR, Thompson NW, Hildreth DH, Abell MR (1971) Hyperplastic pacinian corpuscles: A cause of digital pain. Surgery 70: 730

Hayes JR, Mulholland RC, O'Connor BT (1969) Compression of the deep palmar branch of the ulnar nerve. J Bone Joint Surg [Br] 51: 469-472

Heard HG (1962) Nerve sheath tumors and von Recklinghausen's disease of the central nervous system. Ann R Coll Surg 31: 229

Hecht OA, Hass A (1982) Regional multiplicity of a neurilemmoma. Hand 14: 97

Herbert DC, Degeus J (1975) Post-traumatic lipomas of the abdominal wall. Br J Plast Surg 28: 303-306

Herrick RT, Godsil RD, Widener JH (1980) Lipofibromatous hamartoma of the radial nerve: A case report. J Hand Surg 5/3: 211-213

Holt JF, Wright EM (1948) The radiologie features of neurofibromatosis. Radiology 51: 647

Hosoi K (1931) Multiple neurofibromatosis (von Recklinghausen's disease). Arch Surg 22: 258

Hueston JT, Millroy P (1971) Macrodactyly associated with hamartoma of major peripheral nerves. Aust NZ Surg 37: 394-397

Inglis KL (1950) Local gigantism. Am J Pathol 26: 1059-1083

Jeffery CC (1973) Hamartoma in the hand. A simulator. Hand 5/1: 79-82

Jenkins SA (1952) Solitary tumors of peripheral nerve trunks. J Bone Joint Surg [Br] 34: 401-411

Johnson JR, Bonfiglio M (1969) Lipofibromatous hamartoma of the median nerve. J Bone Joint Surg [Am] 51: 984-990

Johnson RK, Shrewsbury MM (1970) Anatomical course of the thenar branch of the median nerve; usually in a separate tunnel through the transverse carpal ligament. J Bone Joint Surg [Am] 52: 269

Jones KG (1963) Megalodactylism. J Bone Joint Surg [Am] 45: 1704-1708

Kissel P, Dureux JB, Schmitt J (1970) Qu'appelle-t-on phacomatoses? Rev Prat 20: 4353-4364

Kleinert HE, Serafin D, Daniel RK (1973) The place of microsurgery in hand surgery. Orthop Clin North Am 4/4: 929-944

Lassmann H, Gebhart W, Mamoli B, Niebauer G (1977) Nervous lesions in a case of local gigantism. Acta Neuropathol (Berl) 38: 109-115

Leclercq A (1977) Le lipofibrome du nerf médian. Thèse, Paris

Leung PC (1981) Tumours of hand. Hand 13: 169

Lewis RC, Nannini LH, Cocke WM (1981) Multifocal neurilemomas of median and ulnar nerves of same extremities. J Hand Surg 6: 406-408

Louis DS, Dick HM (1973) Ossifying lipofibroma of the median nerve. J Bone Joint Surg [Am] 55: 1082-1084

Mallory FB (1914) The principles of pathologic histology Saunders, Philadelphia London

Mannerfelt L (1966) Studies of the hand in ulnar nerve paralysis. Acta Orthop Scand [Suppl] 87: 86-92

Marmor L (1965) Solitary peripheral nerve tumors. Clin Orthop 43: 183

Mason ML (1937) Tumors of the hand. Surg Gynecol Obstet 64: 129-148

Masson P (1932) Experimental and spontaneous schwannomas. Am J Pathol 8: 367

Matton G, Defloor E (1974) Benign tumors of the hand. Acta Orthop Belg 40/2: 172-198

Mazingarbe A (1968) Lipome du nerf médian. Mem Acad Chir (Paris) 94/12: 389-394

McCarroll HR (1950) Clinical manifestations of congenital neurofibromatosis. J Bone Joint Surg [Am] 32: 601-617

McCarroll HR (1956) Soft tissue neoplasms associated with congenital neurofibromatosis. J Bone Joint Surg [Am] 38: 717-731

McCarthy DM, Door CA, Mackintosh CE (1969) Unilateral localised gigantism of the extremities with lipomatosis, arthropathy and psoriasis. J Bone Joint Surg [Br] 51: 348-353

Mikhail IK (1964) Median nerve lipoma in the hand. J Bone Joint Surg [Br] 46/4: 726-730

Money RA (1950) Tumors of peripheral nerves. Aust NZ J Surg 19: 239

Moon N, Marmor L (1964) Parosteal lipoma of the proximal part of the radius. A clinical entity with frequent radial nerve injury. J Bone Joint Surg [Am] 46: 608-614

Murray MR, Stout AP (1940) Schwann cell versus fibroblast as the origin of the specific nerve sheath tumor. Am J Pathol 16: 41

Paletta FX, Rybka FJ (1972) Treatment of hamartomas of the median nerve. Ann Surg 176/2: 217-222

Patel ME, Silver JW, Lipton DE, Pearlman HS (1979) Lipofibroma of the median nerve in the palm and digits of the hand. J Bone Joint Surg [Am] 61: 393-397

Patterson TJS (1956) Pacinian corpuscle nevroma of the thumb pulp. Br J Plast Surg 9: 230

Penfield W (1927) The encapsulated tumors of the nervous system. Surg Gynecol Obstet 45: 178

Pepin J (1956) Les tumeurs nerveuses des membres. Thèse, Paris

Phalen GS, Kendrick JI, Rodriguez JM (1971) Lipomas of the upper extremity. Am J Surg 121/3: 298-306

Phalen GS (1976) Neurilemmomas of the forearm and hand. Clin Orthop 114: 219-222

Posch JL (1956) Tumors of the hand. J Bone Joint Surg [Am] 38: 517-534

Pulvertaft RG (1964) Unusual tumors of the median nerve. Report of two cases. J Bone Joint Surg [Br] 46: 731-733

Ranawat CS, Arora MM, Singh RG (1968) Macrodystrophia lipomatosa with carpal tunnel syndrome. J Bone Joint Surg [Am] 50: 1242-1244

Rechnagel K (1967) Megalodactylism. Report of 7 cases. Acta Orthop Scand 38: 57-66

Recklinghausen F von (1882) Über die multiplen Fibrome der Haut und ihre Begrenzung zu den multiplen Neuromen. Hirschwald, Berlin

Rhode CM, Jennings WD Jr (1975) Pacinian corpuscule nevroma of digital nerves. South Med J 68/1: 86-89
Richmond DA (1953) Lipoma causing a posterior interosseous nerve lesion. J Bone Joint Surg [Br] 35: 83
Richmond DA (1973) Uncommon causes of nerve compression with hand symptoms. Hand 5/3: 209-213
Ross R, Jaffe S (1975) Painless fibro-fatty hamartoma of the median nerve. Br J Plast Surg 28: 301-302
Rowntreet R (1949) Anomalous innervation of the hand muscles. J Bone Joint Surg [Br] 31: 505
Rowland SA (1967) Lipofibroma of the median nerve in the palm. J Bone Joint Surg [Am] 49: 1309-1313
Rowland SA (1977) Case report: Ten year follow up of lipofibroma of the median nerve in the palm. J Hand Surg 2/4: 316-317
Rudolph R, Jaffe S (1975) Painless fibro-fatty hamartoma of the median nerve. Br J Plast Surg 28: 301-302
Rusko RA, Larsen RD (1981) Intraneural lipoma of the median nerve. J Hand Surg 6: 388-391
Sandzen SC, Baksic RW (1974) Pacinian hyperplasia. Hand 6: 273
Schmitz RL, Keeley JL (1957) Lipomas of the hand. Surgery 42: 696-700
Schuler F, Adamson J (1978) Pacinian neuroma, an unusual case of finger pain. Plast Reconstr Surg 62: 576-579
Seddon H (1972) Surgical disorders of peripheral nerves. Livingstone, Edinburgh London
Shereff MJ, Posner MA, Gordon MH (1980) Upper extremity hypertrophy secondary to neurofibromatosis: Case report. J Hand Surg 5: 355-357
Stern PJ, Nyquist SR (1982) Macrodactyly in ulnar nerve distribution associated with cubital tunnel syndrome. J Hand Surg 7: 569-571
Stout AP (1935) The peripheral manifestations of the specific nerve sheath tumor (neurilemmoma). Am J Cancer 24: 751
Stout AP (1946) Neurofibroma and neurilemmoma. Clin Proc 5: 1
Stout AP (1949) Tumors of peripheral nerves. In: Atlas of tumor pathology, sect 2, fasc 6. Armed Forces Institute of Pathology, Washington
Stout AP (1953) Tumors of the soft tissues. In: Atlas of tumor pathology, sect 3, fasc 5. Armed Forces Institute of Pathology, Washington, pp 29-35
Stowell A, Grice P, Smail D (1963) Tumors of the ulnar and median nerves. Am Surg 29: 442-445
Straus FH (1931) Deep lipomas of the hand. Ann Surg 94: 269
Strickland JW, Steichen JB (1977) Nerve tumors of the hand and forearm. J Hand Surg 2: 285-291
Sunderland S (1972) Nerves and nerve injuries. Livingstone, London, p 751
Terzis JK, Daniel RK, William HB, Spencer PS (1978) Benign fatty tumors of the peripheral nerves. Ann Plast Surg 1/2: 193-216
Touzard RC (1970) 2 cas de lipome du nerf médian à la main. J Chir (Paris) 99/5: 549-550
Tsuge K, Ikuta Y (1973) Macrodactyly and fibro-fatty proliferation of the median nerve. Hiroshima J Med Sci 22: 83-101
Vaunois P (1972) Contribution à l'étude des lipomes du nerf médian à la main. Thèse, Paris (Pitié)
Verocay J (1908) Multiple Geschwülste als Systemerkrankung am nervösen Apparate. Festschrift Hans Chiari. Braunmüller, Vienna Leipzig
Wardle EN (1957) Nerve-sheath tumors. Br J Surg 45: 58-61
Watson-Jones R (1964) Encapsulated lipoma of the median nerve at the Wrist. J Bone Joint Surg [Br] 46: 736
Wertheimer P, Feroldi C, Gilbertas RC (1950) Les tumeurs isolées et primitives des gros troncs nerveux périphériques. J Chir (Paris) 66: 746
White WL, Hanna DC (1962) Troublesome lipomata of the upper extremity. J Bone Joint Surg [Am] 44: 1353-1359
White NB (1967) Neurilemmomas of the extremities. J Bone Joint Surg [Am] 49: 1605-1610
Willis RA, Winston-Evans R (eds) (1948) Tumors of nerve sheaths. In: Pathology of tumors. Butterworth, London, p 846
Winston-Evans R (1968) Tumours of the nerve sheath. In: Histogical appearance of tumours. Livingstone, Edinburgh London, pp 359-381
Wirth W, Bray C (1977) Intra-osseous neurilemmoma. J Bone Joint Surg [Am] 59: 252-255
Woods JE, Murray JE, Vawter GF (1970) Hand tumors in children. Plast Reconstr Surg 46/2: 130-139
Yeoman PM (1964) Fatty infiltration of the median nerve. J Bone Joint Surg [Br] 46: 737-739
Zweig J, Burn H (1968) Compression of digital nerves by Pacinian corpuscles. J Bone Joint Surg [Am] 50: 999-1001

Glomus Tumour

Glomus tumour, or glomangioma, is a rare benign lesion that takes the form of a hypertrophy of the neuro-myo-arterial glomus. It is therefore more of a hamartoma than a true tumour. It is preferentially located in the hand, especially in the subungual region, but is also found in the foot and anywhere the glomus exists under normal conditions (i.e. essentially in the dermis). The glomus tumour is characterised by its hyperalgic symptomatology and is routinely cured by surgical excision. Unfortunately, diagnosis is often delayed until after the failure of sometimes mutilating treatments.

There is a very abundant and long-standing literature on this subject. To the best of our knowledge, a glomus tumour was first reported by Wood in 1812 under the name of "painful subcutaneous nodule". It was subsequently mentioned by many authors (Sucquet 1862; Kolaczek 1878). Masson (1924) became interested in the anatomy of the glomus, to which he gave the name because of its resemblance to the vascular cocygeal gland of Luschka. Popoff (1934) then described its role in peripheral heat regulation. Murray and Stout (1942) based their approach on the histology of the lesion and identified the so-called glomic cells as being pericytes, but more recent ultrastructural studies have compared them to smooth muscle cells.

Incidence

The hand is by far the most common site of occurrence (Carroll and Berman 1972, 30-50% of cases; Iselin 1969, 75%). It remains a relatively rare lesion, stated by various authors as constituting between 1% (Boyes 1970) and 4.5% (Posch 1956) of tumours of the hand (average 1.6%). In decreasing frequency, glomus tumours are found in the foot, all the tegumentary surfaces (lower limbs, trunk, face) and in certain extrategumentary regions (stomach).

Clinical Features

There are two types of glomus tumours which differ notably in their clinical and histological features, the solitary and the multiple. For purposes of description the solitary glomus tumour will be considered here and the multiple form will be dealt with secondarily.

Three symptoms should suggest the diagnosis of glomus tumour. (a) There is the spontaneous lancinating pain, radiating to the forearm and shoulder, with sudden hyperalgic paroxysms. (b) The pain provoked by touch, often even by the gentlest stroking, is very intense and leads the patient to avoid the slightest contact, thus making examination very difficult. However, after establishing confidence, examination with a blunt-tipped instrument often permits very precise localisation of the site. (c) Intolerance to temperature changes is evident, always very marked to cold (e.g. sea bathing), the pain sometimes being relieved by immersion in warm water. The combination of these three features, which are rarely absent, is considered as pathognomonic (Carroll and Berman 1972). On examination, the tumour is sometimes visible, or palpable in superficial forms, as a small rounded mass rarely exceeding 2-3 mm in diameter. More often, a bluish or deep-red discoloration of the dermis (Fig. 78) or overlying nail (Fig. 79) is noticed. With subungual lesions there may also be a groove in the nail.

Other and much rarer signs are deceptive (Cooke 1971): excessive sweating, increased local warmth, atrophy of the finger pulp (probably due to failure to use the finger) or even the Claude Bernard-Horner syndrome (Ottley 1942). However, their association with the three classic symptoms should suggest the diagnosis. In the very great majority of cases it is striking how long the symptoms have continued before the correct diagnosis is made; in the published series the average course length is 7 years (Carroll and Berman 1972; Rettig and Strickland 1977), with extremes reaching to as long as 20 and even 40 years. Several conclusions may be drawn from this: that the lesion is unrecognised by the majority of physicians and surgeons, that it is totally resistant to any treatment other than surgical excision and, finally, that it has an extremely slow and local course.

Among the various localisations in the hand, the subungual lesion is the most common (Carroll and Berman 1972, 65%), followed by the fingertip and then the thenar and hypothenar eminences. All the fingers seem equally affected, although Carroll and Berman (1972) note a preference for the thumb and index finger. Adult life is the most prevalent time of glomus tumour manifestation (average age 40 years); it is rare before 20 and after 60 years of age. Kohout and Stout (1961) particularly studied the childhood forms, assessing them as 6% of cases, some of them congenital. They stress the greater frequency of multiple forms in children (26% as against 10% in adults) and of extracapsular spread (12% as against 0.3% in adults). Localisation to the hand, particularly the subungual region, is most frequently found in women (Mullis et al. 1972, 12 cases; Shugart et al. 1963, 75% of cases), whereas lesions of the forearm are rather the attribute of men. Mathy (1979) finds a correlation between age and sex, women being affected between 30 and 35 years of age and men between 50 and 60.

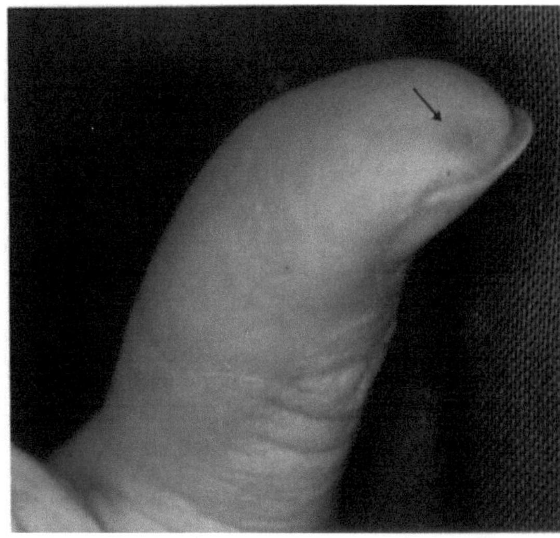

Fig. 78. Glomus tumour of the thumb-tip: note the red coloration of the dermis

Multiple Glomus Tumours

As stressed by Goodman and Abele (1971), one must distinguish between multiple regional glomus tumours and multiple disseminated glomus tumours or glomangiomatosis. The former involve only the juxtaposition in the same place of several glomus tumours (Plewes 1941; Curtis 1966), with clinical and histological characteristics strictly identical to those of a solitary glomus tumour. The only difficulty relates to treatment, as all the tumours must be excised to effect a cure (here arteriography may be important, see below). On the other hand, the disseminated tumours represent a separate entity (Conant and Wiesenfeld 1971). Goodman and Abele (1971) collected 45 cases from the literature. They represent around 10% of all glomus tumours, develop earlier than the solitary tumours and may be familial and hereditary (autosomal dominant transmission, Hueston 1961). They may be disseminated over the entire body, but it is rare to find more than a dozen in the same patient. Subungual localisation is unusual. The most marked difference from the solitary tumor is the absence of spontaneous pain, though they are sometimes sensitive to pressure. A bluish colour of the overlying skin is often found, sometimes resembling a haemangioma (Conant and Wiesenfeld 1971).

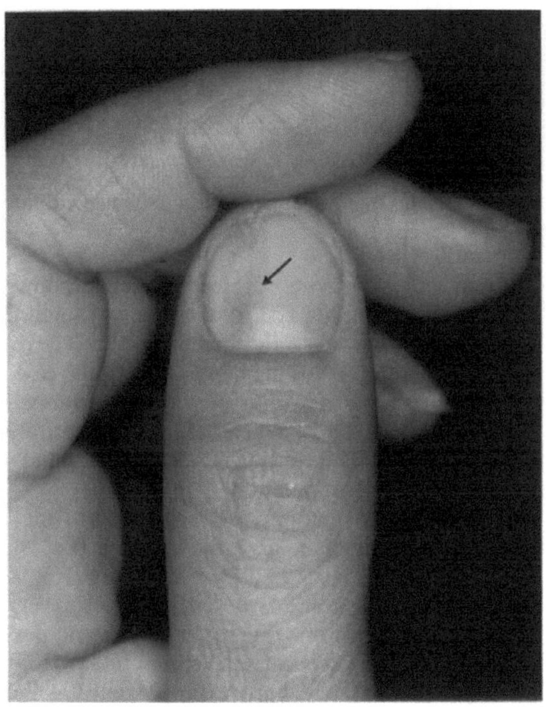

Fig. 79. Subungual glomus tumour: note the coloration of the nail and the distal groove

Special Investigations

Positive *radiographic* findings occur with deep-seated tumours in contact with bone and in advanced tumours. Different researchers assess the incidence at 22%-60% (9 of 15 in the series of Mathis and Schultz 1948). At the very least a dome-shaped de-

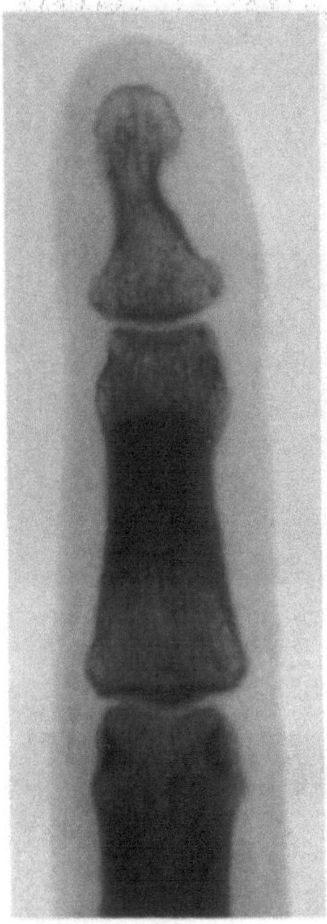

Fig. 80. Glomus tumour: domed defect in phalangeal tuft

pression along the cortex is seen, which is sometimes eroded but always in continuity (Sugiura 1976) (Fig. 80), while at most there is an "open" cavity or bony notch of varying depth. Rarely, the cavity is totally enclosed within the bone (Pietrabissa 1956; Mackensie 1962). When the lesion is entirely distal, on the phalangeal tuft, it may create the appearance of distal amputation.

Arteriography has been practised repeatedly: Natali et al. (1966) Rascol et al. (1968), Mercier et al. (1970), Germain et al. (1970), Becchi et al. (1976). In the arterial phase it shows a small, rounded, very dense image formed of a fine vascular network, persisting into the venous phase. Venous return from the finger is rapid, with early disappearance of the arterial opacification, indicating increased rate of blood flow in the finger due to lessened peripheral resistance at the site of the tumour. These data have been confirmed by the plethysmographic studies of Mantero et al. (1968). In some cases (Mathy 1979) the tumour is not opacified by the contrast medium. Thus, the routine performance of arteriography in the assessment of glomus tumour does not seem

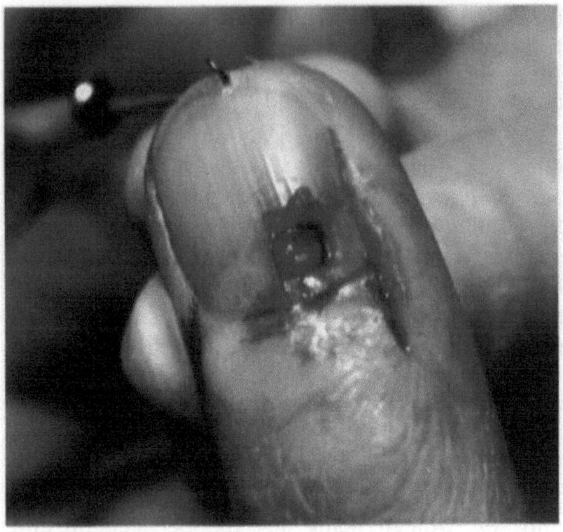

Fig. 81. Glomus tumour: transungual excision

justified. It should be reserved for cases which cannot be localised precisely, either clinically (absence of a precise painful point) or radiographically, and for "recurrences", when it sometimes allows visualisation of several tumours (Mercier et al. 1970).

Bone scintigraphy with technetium or vascular studies with sodium pertechnate do not show zones of accumulation (Mathy 1979).

Pathology

The histological structure of the normal glomus has been studied by Masson (1924) and particularly by Popoff (1934). It consists of an afferent arteriole connected by an anastomotic vessel (the Sucquet-Hoyer channel) to an efferent venule. On the arterial side the vessel wall is surrounded by a "sleeve" of nerve fibres, the whole being embedded in fibrous collagen tissue surrounded by a capsule. The exact role of the glomi has been much debated. They seem to be peripheral vasoregulators engaged, with other structures, in the control of peripheral body temperature.

Grossly, the glomus tumour is a small, round, well-encapsulated mass, easily enucleated from the surrounding dermis and usually measuring 2-10 mm in diameter (Fig. 81). It is purplish or pink in colour. On section, the specimen consists of a tangle of winding tubules twisted on themselves and incorporated in fibrous tissue. Microscopically, the arterial component is disorderly, with abnormal vascular branches ending in multiple venous segments. The boundary between the two sides is indefinite. Histo-

logically, the arterial wall is lined with a single-layered endothelium without a limiting elastic intima, edged by cuboid cells with round nuclei known as glomus cells. It was initially believed (Murray and Stout 1942) that these cells were related to Zimmermann's pericytes, but more recent electron microscopic studies seem to indicate that they derive from smooth muscle cells (Murad et al. 1968; Toker 1969; Venkatachalam and Greally 1969).
Depending on the predominance of the vascular or cellular components, the lesions may be regarded as of vascular (angiomatous) or epithelioid type. Masson (1924) also described a neuromatous type and a degenerative (hyaline) type. The glomus cells are embedded in a loosely textured collagen network which forms the stroma of the tumour. The whole is surrounded by a capsule distinct from the adjacent dermis. In 1%-2% of cases there exists capsular invasion, but never extracapsular invasion. On the other hand, extracapsular spread is not uncommon in the lesions of childhood (see p. 145). In multiple glomus tumours, the lesion, which is of the angiomatous type with a predominance of vascular elements, is not encapsulated. Some authors link these histologically with cavernous angioma (Goodman and Abele 1971).

Differential Diagnosis

Striking delays are sometimes seen between the onset of symptoms and the diagnosis of glomus tumour. Various aetiologies have been suggested: neuroma, schwannoma, gout, arthrosis, subungual melanoma, chronic paronychia or causalgia, and these have sometimes led to increasingly mutilating treatment for persistence of pain (alcohol injection, nerve resection, section of the scalenus anterior, sympathectomy, laminectomy, even amputation of the finger). The radiographic appearance may suggest an epithelial cyst at the distal phalanx.

Aetiopathogenesis

A traumatic aetiology has often been suggested but seems difficult to substantiate; Carroll and Berman (1972) found antecedent trauma in 10% of cases. The same authors put forward the hypothesis that the glomus tumour is only one of the possible tumoral developments of the glomus, the others being haemangiopericytoma (with possible local invasion) and sarcoma (with possible metastases).

Treatment

Surgical excision of the glomus tumour is always curative. "Recurrences", always early, are due to incomplete excision. It is therefore always necessary to identify the exact site of the lesion preoperatively and to bear in mind the possibility of multiple tumours. Iselin (1969) has even described dorsal tumours passing round the edge of the phalanx to gain the palmar aspect.
Subungual tumours pose problems of surgical access. The transungual approach (Fig. 81), either by raising a small nail flap or by stripping off the entire nail, though technically very simple, carries the risk of failure to recognise a multiple lesion and of subsequent dystrophy of the nail. It cannot be used either for lesions of the matrix. The lateral approach, whether latero-ungual (Johnson 1971) or latero-digital (Iselin 1969; Achten and Lejour 1972), permits elevation of the entire nail bed and matrix, gives better exposure of the lesion, monitors any localisations in the pulp and seems less damaging to the nail apparatus.
Symptoms disappear immediately after operation. However, if the excision has been incomplete, they return after some days or weeks with the same intensity as before and call for further excision. Thus the glomus tumour is a benign lesion, the surgical treatment of which is relatively simple and routinely effective; it is unfortunate that too often it is unrecognised.

Bibliography

Achten G, Lejour M (1972) Tumeurs glomiques sous-unguéales; ablation par abord latéro-digital. Bull Soc Fr Dermatol Syphiligr 79: 247-248

Becchi GF, Mantero R, Zanetti PP (1976) Tumeur de Masson des doigts. J Chir (Paris) 111: 201-206

Boyes JH (1970) Tumors. In: Bunnell S (ed) Surgery of the hand, 5th edn. Lippincott, Philadelphia Toronto, pp 666-710

Bureau H et al. (1978) Tumeurs glomiques. In: Pierre M (ed) L'ongle. Monographie du Groupe d'Etude de la Main. Expansion scientifique, Paris, pp 102-106

Carroll RE, Berman AT (1972) Glomus tumors of the hand. J Bone Joint Surg [Am] 54: 691-703

Conant MA, Wiesenfeld SL (1971) Multiple glomus tumors of the skin. Arch Dermatol 103: 481-485

Cooke SAR (1971) Misleading features in the clinical diagnosis of the peripheral glomus tumour. Br J Surg 58: 602-606

Cornell SJ (1981) Multiple glomus tumours in one digit. Hand 13: 301-302

Curtis RM (1966) Tumors of blood and lymphatic vessels. In: Flynn JE (ed) Hand surgery. Williams & Wilkins, Baltimore, pp 1052-1056
Debienne AM (1965) A propos de 6 observations de tumeurs glomiques à localisations inhabituelles. Thèse Doctorat Médecine, Paris
Degos R, Guilaine J, Civatte J (1964) Tumeurs glomiques multiples. Ann Dermatol Venereol 91: 40-41
Duane T, Freier MD (1970) Subcutaneous glomus tumor. Am J Surg 120: 359-362
Francis SJ, Saw HS (1974) Solitary glomus tumors of the fingers. Hand 6: 204-207
Germain H, Mercier R et al. (1970) A propos d'un nouveau cas d'une tumeur glomique sous-unguéale artériographiée. J Chir (Paris) 99: 541-548
Goodman TF, Abele DC (1971) Multiple glomus tumors. A clinical and electron microscopic study. Arch Dermatol 103: 11-23
Gordon B, Heyman AB (1961) Multiple non tender glomus tumors. Arch Dermatol 83: 640-643
Greider JL, Flatt AE (1982) Glomus tumor associated with pacinian hyperplasia. Case report. J Hand Surg 7: 113-117
Hueston JT (1961) Multiple painless glomus tumors. Br Med J I: 1210-1212
Iselin M (1969) Tumeurs vasculo-nerveuses dites tumeurs glomiques des doigts. Bull Acad Natl Med (Paris) 153: 112
Johnson RK (1971) A surgical approach of subungueal glomus tumors. Plast Reconstr Surg 47: 345-346
Kohout E, Stout AP (1961) The glomus tumor in children. Cancer 14: 555-565
Kolaczek L (1878) Über das Angio-sarkom. Dtsch Z Chir 9: 165-227
Leger L, Bitry-Boely C et al. (1960) Tumeurs glomiques multiples ou glomangiomatoses. Presse Med 58: 2261-2264
Mackensie DH (1962) Intraosseous glomus tumors. Report of two cases. J Bone Joint Surg [Br] 44: 648
Mantero R, Furfaro M, Ghigliazza GB (1968) Considérations sur la tumeur de Masson de la main. Riv Chir Mano 6: 133
Masson P (1924) Le glomus neuro-myo-artériel des régions tactiles et ses tumeurs. Lyon Chir 21: 257-280
Mathis WH, Schultz MD (1948) Roentgen diagnosis of glomus tumors. Radiology 51: 71-76
Mathy H (1979) Contribution à l'étude des tumeurs glomiques des doigts. A propos d'une série de 8 cas opérés. Thèse, Paris
Mercier R, Bresson P, Viallet JF, Vanneuville G (1970) Intérêt de l'artériographie dans les tumeurs glomiques sous-unguéales. J Radiol 51/5: 303-304
Mullis WF, Rosato FE et al. (1972) The glomus tumor. Surg Gynecol Obstet 135: 705-707
Murad TM, von Haam E, Murthy MSN (1968) Ultrastructure of a hemangiopericytoma and a glomus tumor. Cancer 22: 1239-1249
Murray MR, Stout AP (1942) The glomus tumor. Investigation of its distribution and behavior and identity of its «Epitheloid cell». Am J Pathol 18: 183-203
Natali J, Ecarlat B et al. (1966) Artériographie d'une tumeur glomique. J Chir (Paris) 92: 481-484
Ottley CM (1942) Glomus tumor. Br J Surg 29: 387-391
Pietrabissa G (1956) Case of glomic, multiple and intraosseal tumor of phalanges of the hand. Minerva Orthop 7: 458-463
Plewes B (1941) Tumeurs glomiques multiples, quatre sur l'extrémité d'un doigt. Can Med Assoc J 44: 361
Popoff NW (1934) The digital vascular system, with reference to the state of glomus in inflammation. Arch Pathol 18: 295-330
Posch JL (1956) Tumors of the hand. J Bone Joint Surg [Am] 38: 517-539
Rascol M, Izard J, Rascol A (1968) Etude ultrastructurale d'une tumeur glomique de Masson. Presse Med 76/35: 1695-1698
Rettig AC, Strickland JW (1977) Glomus tumor of the digits. J Hand Surg 2: 261
Riddel DH, Martin RS (1951) Glomus tumor of unusual size. Ann Surg 133: 401-403
Rypins EL (1945) The roentgenological aspects of the subungueal glomus tumors. AJR 46: 667-672
Shugart RR, Soule EH, Johnson EW (1963) Glomus tumor. Surg Gynecol Obstet 117: 334-340
Sigiura I (1976) Intra-osseous glomus tumor. J Bone Joint Surg [Br] 58: 245
Sluiter JTF, Postma C (1959) Multiple glomus tumors of the skin. Acta Derm Venereol (Stockh) 39: 98-107
Stout AP (1950) News notes: Report on 1346 soft tissue tumors (of the hand). J Bone Joint Surg [Am] 32: 457
Stout AP (1956) Tumors featuring pericytes. Glomus tumor and hemangiopericytoma. Lab Invest 5: 217-223
Sucquet JP (1862) D'une circulation dérivative dans les membres et dans la tête chez l'homme. Delahaye, Paris
Suire P, Pigeaud P (1960) Les tumeurs du glomus neuro-myo-artériel. Ann Chir 14: 179-184
Toker C (1969) Glomangioma; an ultrastructural study. Cancer 23: 487-492
Varian JPW, Cleak DK (1980) Glomus tumors in the hand. Hand 12: 293-299
Venkatachalam MA, Greally JG (1969) Fine structure of glomus tumor: Similarity of glomus cells to smooth muscle. Cancer 23: 1176-1184
Wood W (1812) On painful subcutaneous tubercle. Edinburg Med J 8: 283

Vascular Tumours

Different classifications have been proposed in the attempt to categorise the varied collection of tumours affecting the arteries, veins, capillaries and lymphatics. Considering only the haemangiomatous tumours of the hand, Weisman (1959) proposed four groups:

1. Congenital dysplasias: plane angioma, tuberous angioma, cavernous haemangioma, diffuse haemangiomatosis, congenital arteriovenous aneurysm, glomus tumour, telangiectasis, Rendu-Osler's disease
2. Tumours: haemangioendothelioma, haemangioendotheliosarcoma, haemangioleiomyoma, haemangioleiomyosarcoma, haemangiopericytoma, haemangiopericytosarcoma
3. Traumatic lesions: traumatic arteriovenous aneurysms, true aneurysms, false aneurysms
4. Inflammatory lesions

The classification employed here is based largely on that made by Merland et al. (1982) and includes:

1. Angiodysplasias: haemangioma, lymphangioma
2. Vascular tumours: haemangioendothelioma, haemangiopericytoma, angioleiomyoma, haemangioendotheliosarcoma or angiosarcoma, lymphangiosarcoma

Angiodysplasia

The "angiomata" are congenital or acquired anomalies of the lymphatic or blood vessels; they are therefore angiodysplasias, grouping together a great variety of lesions. We shall consider first the haemangiomata and then the lymphangiomata.

Haemangiomata

The classification proposed by Merland et al. (1982) is based on two criteria:

1. The clinical course:
   a) Immature or phasic angiomata specific to the infant which regress spontaneously or
   b) Mature angiomata without a tendency to regression which form a varied collection of congenital or acquired lesions
2. The angiographic findings, which serve to distinguish:
   a) Capillary angiomata
   b) Capillary-venous malformations
   c) Deep venous dysplasias
   d) Cavernomata of the venous component
   e) Aneurysms
   f) Arteriovenous fistulae
   g) Complex arteriovenous malformations
   h) Regional hypervascularisation

On this basis, Merland et al. (1982) propose the following classification:

1. Immature angiomata: tuberous, subcutaneous, mixed
2. Mature angiomata:
   a) Capillary-plane angioma, telangiectases, angiokeratoma
   b) Capillary-venous
   c) Venous dysplasias of a main trunk, cavernoma
   d) Arterial aneurysms (congenital or acquired), arteriovenous fistulae (congenital or acquired), arteriovenous malformations
   e) Complex disseminated angiomata-neurocutaneous angiomata (Sturge-Weber-Krabbe, Blanc-Bonnet-Dechaume, Cobb, von Hippel-Lindau), osteodystrophic angiomata (Klippel-Trenauney, Parkes-Weber, Rendu-Osler, Bean, Maffucci-Kast, Fabry)

Owing to this great clinical polymorphism and the spontaneous regression of a certain number of vascular tumours, the incidence of these lesions in the hand is difficult to specify. Posch (1976) gives 1% (7 vascular lesions of 679 tumours of the hand), Stack (1960) gives 5% (9 angiomata and 6 traumatic aneurysms in 300 tumours), while Palmari (1983) gives 8% and classes angioma as the fourth most common tumour of the hand after ganglia, giant-cell tumours and mucous cysts. In all cases these are benign tumours which are relatively common in children and may involve all the tissues of the

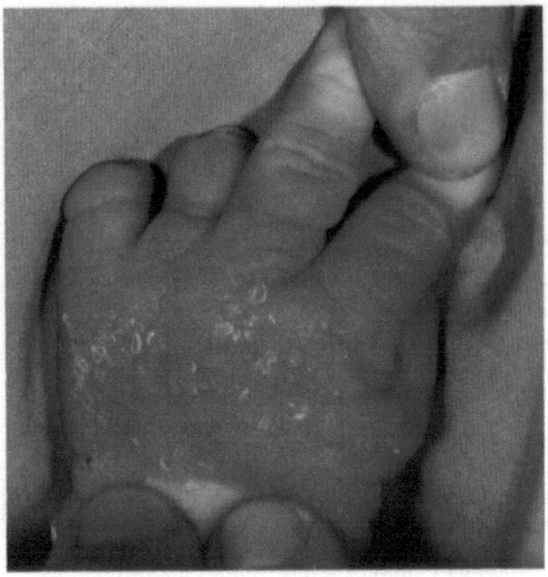

Fig. 82. Angioma: immature tuberous form

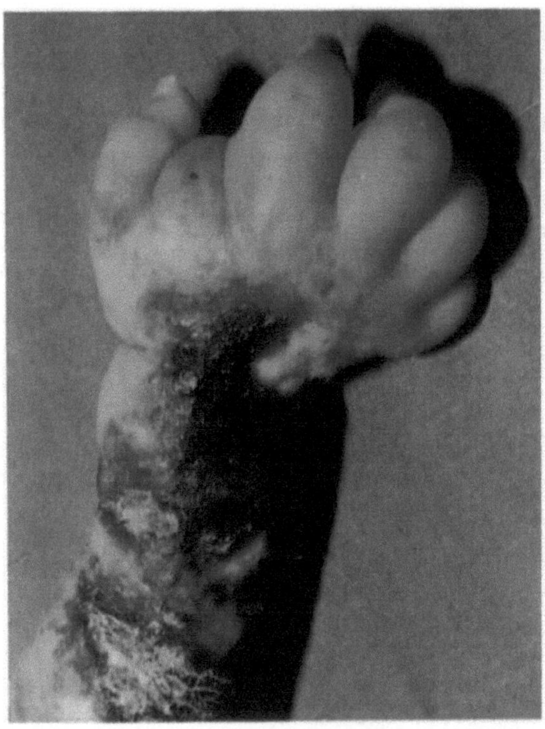

Fig. 83. Immature angioma complicated by extensive skin necrosis

hand: skin, subcutaneous tissue, muscle, tendons and their sheaths, nerves and bone.

Immature Angiomata

Though often minimal or absent at birth, these appear very early, often in the first month of life. They indicate the persistence of embryonal circulatory remnants in contact with the adult vascular system. Their course is characterised by its triphasic nature: a phase of rapid extension with a maximum around the 6th month, a phase of stabilisation up to the 20th month and a phase of slow regression as proved by Lister in 1938 in 92 untreated cases. This regression continues to the age of 6-7 years and is complete in nearly 90% of cases.

Clinical Features

These lesions, which are more common in females, may assume three aspects:

1. Tuberous angioma, the strawberry naevus, a purely dermal, bright red lesion like "a strawberry placed on the skin", composed of young mesenchymal cells arranged in compact fashion (Fig. 82).
2. Subcutaneous angioma, covered with skin that is often normal but may be bluish and telangiectatic, composed of wide vascular cavities and few cells
3. Mixed angioma, combining both components. Tuberous angioma usually regresses rapidly and completely, unlike subcutaneous angioma whose course is slower and often incomplete.

Diagnosis

This is usually clinical; if arteriography is performed, it confirms the pseudotumoral aspect at the capillary phase, without either arteriovenous shunt or precocious venous return. Complications may occur during the phase of extension of these angiomata. There may be superficial necrosis with ulceration, usually not immediately serious (Fig. 83). Rarely, repeated haemorrhage may produce anaemia. Exceptionally, the Kasabach-Merritt syndrome may develop during the extension phase of an immature angioma and must be feared if purpura and ecchymoses appear as evidence of major platelet destruction involving a threat to life. The great majority of cases, however, involve entirely benign lesions, and the only problem is to recognise which forms will regress completely, i.e. therapeutically:

all that is necessary is regular surveillance with serial comparative photography.

Treatment

Surgery has a very limited place with angiomata and is invoked only in types with incomplete regression. As far as medical treatment is concerned, many authors find an indication only in extensive complicated angioma, which may require the use of a corticosteroid.

Mature Angiomata

Capillary Angiodysplasias

Telangiectases

Stellate angiomata, or telangiectases, are small benign lesions often observed on the backs of the hands and characterised by a bright red central point surrounded by a fine arborescence abolished by pressure with a glass slide. Their course is marked by exacerbations, and these angiomata often disappear spontaneously. Treatment for cosmetic purposes may be applied by fine central diathermy. These telangiectases may be viewed in the context of Rendu-Osler's disease, an autosomal dominant hereditary angiomatosis, the prognosis of which is linked to the number and visceral sites of the haemorrhagic lesions as well as the therapeutic possibilities offered by embolisation.
Lastly, there are the stellate angiomata of the backs of the hands, forearm and face occurring in pregnancy and assoicated with diffuse erythema of the hands (Walsh and Becker 1941). These relatively common lesions may recur with each pregnancy and usually disappear in the 6 weeks following delivery.

Angiokeratomata

These are angiectases of the papillary dermis with epidermal hyperkeratosis; they are small lesions of variable numbers and dimensions (Fig. 84). Different types have been described; in the hand there are essentially two distinct disorders:

1. Circumscribed angiokeratoma, present at birth with the appearance of an erythematous macule. This is usually unilateral and increases progressively in size as its colour becomes darker. The definitive appearance arrived at is that of a firm, dark-blue nodule which rather resembles a melanoma. Histological examination confirms the diagnosis, showing the vascular dilatation of the papillary dermis, acanthosis and hyperkeratosis
2. Mibelli's angiokeratoma, an uncommon lesion situated on the dorsal aspect of the fingers and toes, occurring in childhood and adolescence, more often in females. The appearance produced is that of small reddish-violet lesions of lentil size, covered with a verrucous hyperkeratosis. It may be associated with signs of acrocyanosis. Apart from cases in which the lesion disappears spontaneously, treatment by simple excisional surgery or diathermy effects a cure.

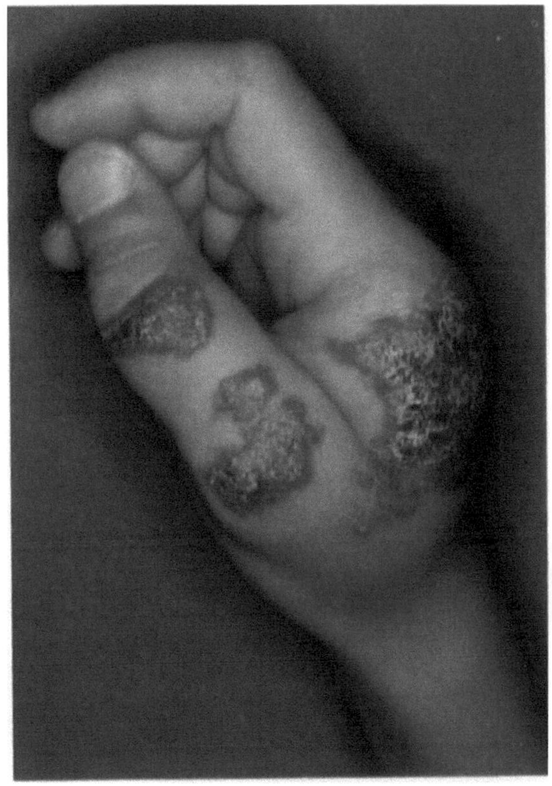

Fig. 84. Angiokeratosis of the dorsum of the hand

Plane Angioma

Commonly known as a "wine-spot", this is an often stable congenital vascular malformation that sometimes develops slowly and never disappears spontaneously (Fig. 85). It is often found on the face and limbs and is obliterated by pressure from a glass slide. The histological appearance is that of vascular dilatations of the superficial dermis, sometimes associated with ectasias but without cell prolifera-

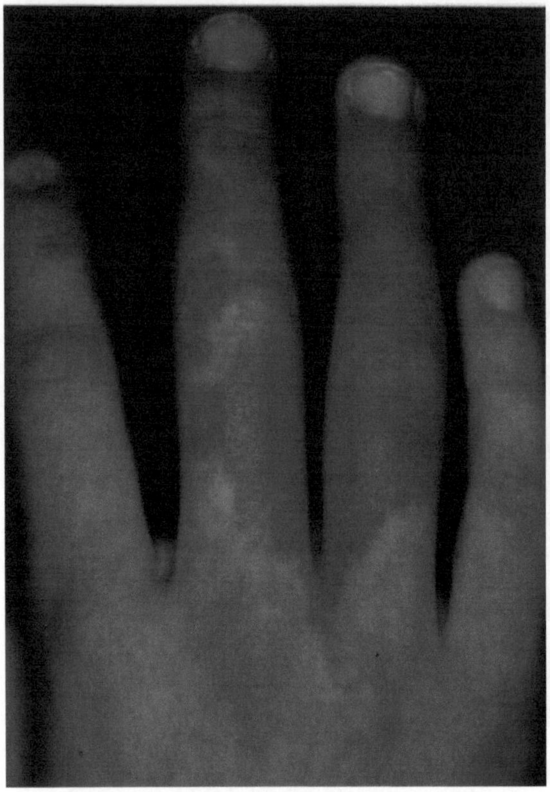

Fig. 85. Plane angioma

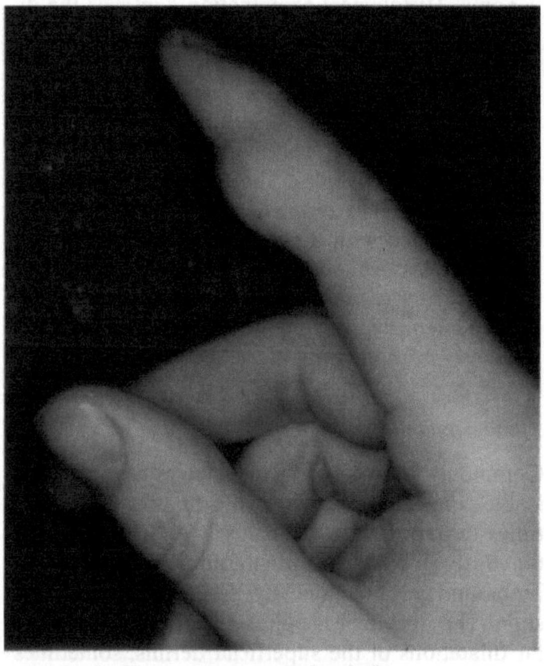

Fig. 86. Capillary-venous angiodysplasia

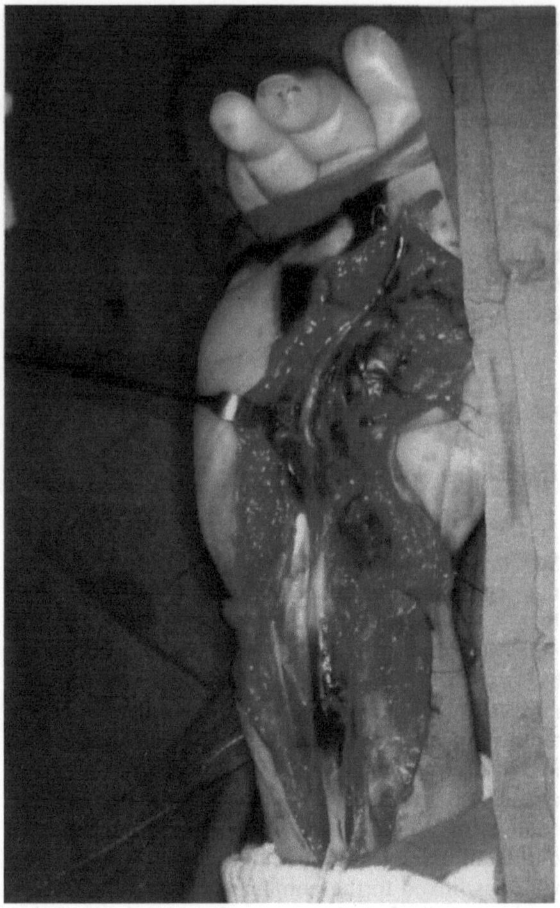

Fig. 87. Extensive capillary-venous angiodysplasia; note particularly the infiltration of the median nerve

tion. Thus, this is a hyperaemic naevus or telangiectatic naevus rather than an angioma. In advanced forms the skin covering may thicken to accompany the appearance of violet keratotic nodules. The treatment of these lesions is very difficult, and only two techniques are used: surgical excision with skin plasty for localised forms or skin graft for more extensive lesions, the cosmetic result being uncertain enough to justify the conservatism observed by many authors, or an argon laser, which carries the risk of pigmentation and keloids. Those using the latter method recommend a preliminary therapeutic trial in a localised area.

In certain cases the plane angioma is not solitary but is accompanied by deep neurocutaneous or osteodystrophic lesions, described in the chapter on complex angiodysplasias.

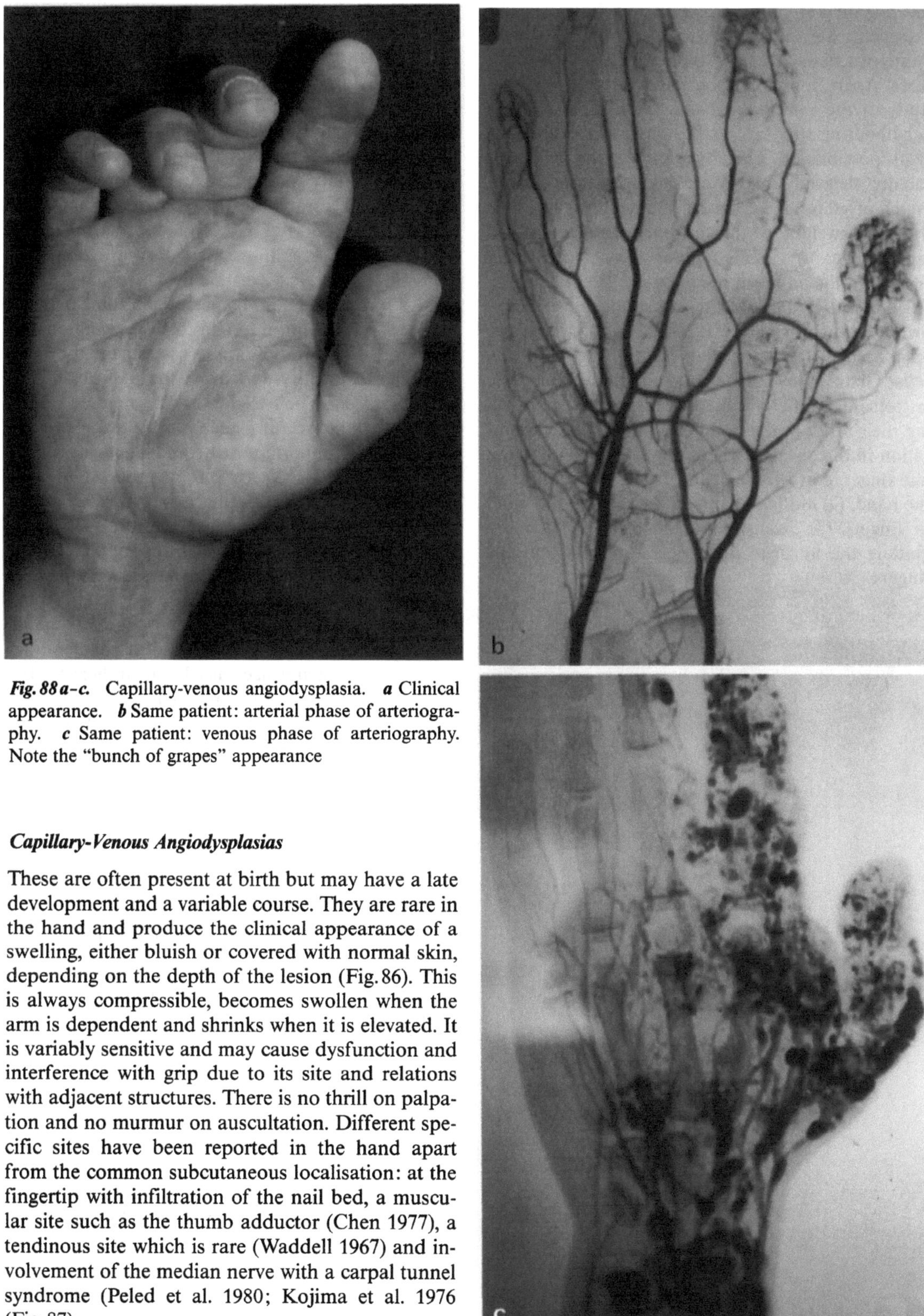

Fig. 88 a–c. Capillary-venous angiodysplasia. ***a*** Clinical appearance. ***b*** Same patient: arterial phase of arteriography. ***c*** Same patient: venous phase of arteriography. Note the "bunch of grapes" appearance

Capillary-Venous Angiodysplasias

These are often present at birth but may have a late development and a variable course. They are rare in the hand and produce the clinical appearance of a swelling, either bluish or covered with normal skin, depending on the depth of the lesion (Fig. 86). This is always compressible, becomes swollen when the arm is dependent and shrinks when it is elevated. It is variably sensitive and may cause dysfunction and interference with grip due to its site and relations with adjacent structures. There is no thrill on palpation and no murmur on auscultation. Different specific sites have been reported in the hand apart from the common subcutaneous localisation: at the fingertip with infiltration of the nail bed, a muscular site such as the thumb adductor (Chen 1977), a tendinous site which is rare (Waddell 1967) and involvement of the median nerve with a carpal tunnel syndrome (Peled et al. 1980; Kojima et al. 1976 (Fig. 87).

Arteriography is an important step in diagnosis and treatment; the technique should be rigorous, with numerous views and early films with multiple exposures. Apart from the topographic interest, it allows study of the extension of the lesion in relation to neighbouring structures and assessment of the surgical possibilities. Opacification of the lesion is usually delayed and gives the appearance of a bunch of grapes (Fig. 88a-c). The preliminary films show the swelling of the soft parts, sometimes also the presence of calcified phleboliths, more rarely erosion of the adjacent bony cortex.
Histological analysis of these angiodysplasias confirms the mature nature of the lesions, which are composed of large vascular cavities and with few cells. Glanz (1969) made a detailed study of these angiomata, formerly called "cavernous", confirming their rarity in the hand. He proposes a classification in three stages: (1) minor angioma, affecting one finger, part of the palm or part of the back of the hand, (2) moderate angioma, affecting the palm or fingers, (3) extensive angioma of the palm including the intrinsic (lumbrical) muscles and the fingers.

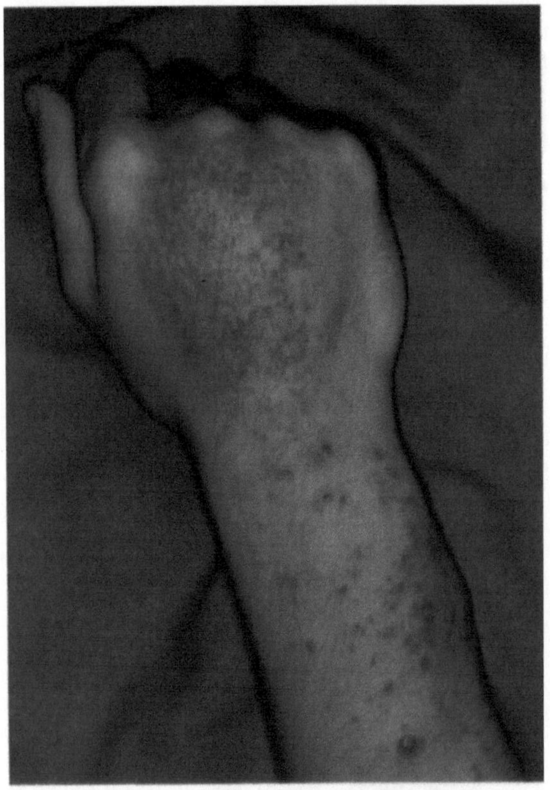

Fig. 89. Extensive capillary-venous angiodysplasia

The treatment of these lesions is essentially surgical. Cryotherapy and sclerosing injections are no longer used, while the argon laser has no effect on subcutaneous angiomata. Embolisation is very debatable in view of the risks of digital necrosis. The existence of such an angiodysplasia in a child is an indication for regular follow-up with serial photographs in order to detect any spontaneous regression. Some authors suggest combining this with moderate compression by an elastic band to induce shrinkage of the lesion by resulting circulatory slowing and thromboses. Surgical treatment is reserved for definitive forms; it should be performed using a pneumatic tourniquet but not on a totally exsanguinated limb and after arteriographic assessment. The objective is to identify and ligate the pedicles of the angiodysplasia. The excision itself is rendered difficult by the absence of a plane of cleavage and the proximity of important structures that are sometimes involved in the dysplastic process. It is in such cases in which resection is incomplete that recurrences appear. In some instances the special nature of the lesional site may call for supplementary measures. Thus, invasion of the related skin may require the use of a skin graft or even a flap (local forearm, pedicled or free). Lesions of the base of the fingers necessitating sacrifice of a collateral artery to avoid recurrence may be an indication for microsurgical vascular reconstruction. Finally, there exist diffuse and extensive forms to which surgery is inapplicable, and conservatism is justified (Fig. 89).

Venous Angiodysplasias

Merland et al. (1982) contrast the dilatation of a large venous trunk with the cavernoma, a diversional ectasia on a vein. These are anomalies of venous structure, very rare in the hand, giving the appearance of a soft bluish swelling, turgid in the dependent position, which may simulate a lipoma or a lymphangioma (Fig. 90a, b). Neither painful nor troublesome, they are often only aesthetically displeasing. Angiography by the venous route or by direct puncture visualises the dilated and deformed venous segment. Surgical treatment is easy for cavernoma (Fig. 91) but is usually much more difficult and disappointing for multiple venous dilatations, in which case several operative stages are sometimes necessary. Some authors have proposed the use of elastic compression.

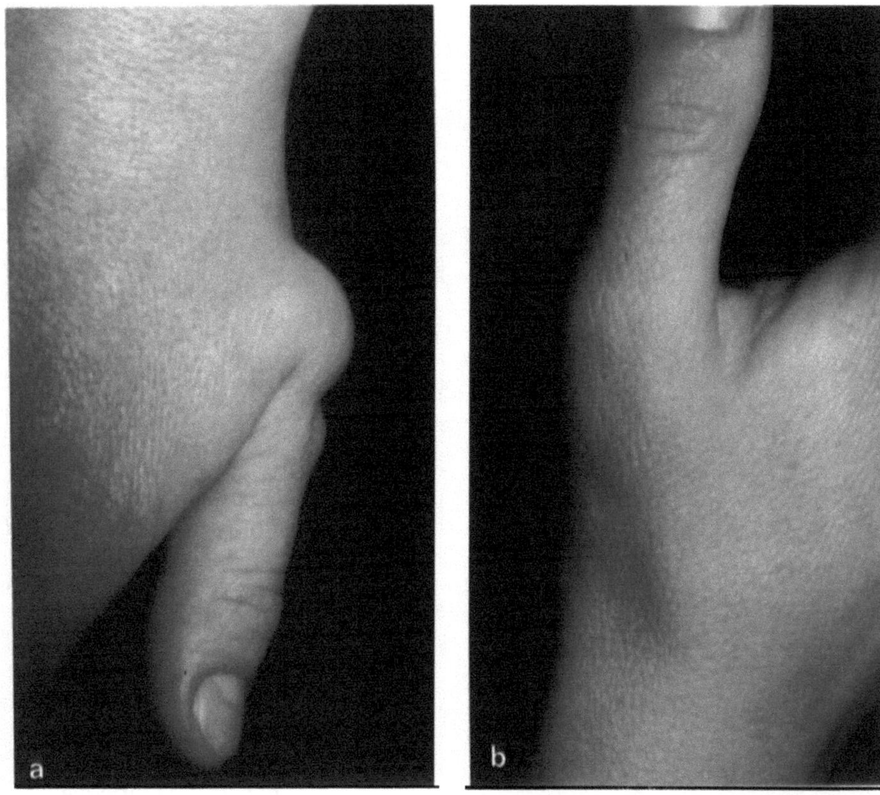

Fig. 90a, b. Cavernoma of the base of the thumb. *a* Engorgement in dependent position, *b* emptying in elevation

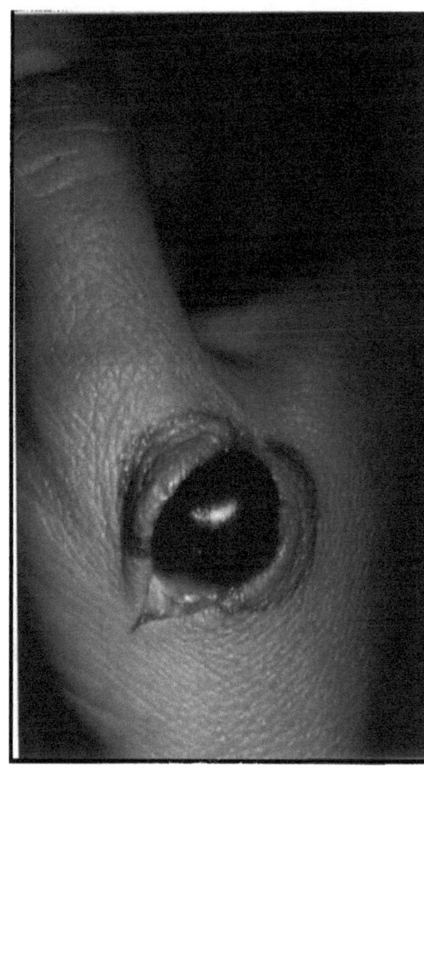

Fig. 91. Cavernoma: operative view *(right)*

Arterial Angiodysplasias

Arterial Aneurysms

Long familiar and studied in numerous publications, aneurysms can be very classically divided into two types, false and true. A false aneurysm originates from a haematoma secondary to an arterial wound. The haematoma is initially colonised by inflammatory cells and then by fibroblasts, which enter into continuity with the arterial lumen. The lesion is saccular and is characterised histologically by the absence of elastic and muscular fibres. A true aneurysm may also be of traumatic origin, and then it is secondary to a non-penetrating injury that weakens the arterial wall, which becomes distended under the effect of pressure. A true aneurysm may also develop within the context of an infection or of arteriosclerosis; it can be an isolated lesion, which is considered idiopathic. It is fusiform and charac-

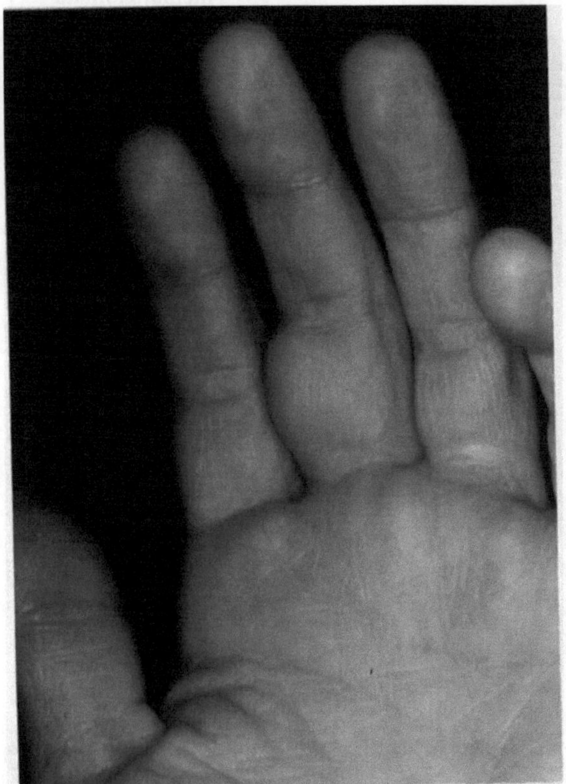

Fig. 92. Arterial aneurysm at a rare site

terised histologically by the presence of elastic and muscular fibres.

In the hand these lesions may be located on any of the vascular axes.

True aneurysms are seen more often; in a series of 35 cases, Smith (1962) reports the following sites: 25 in the hypothenar region (superficial ulnar artery), 5 in the thenar region, 4 at the deep palmar arch and 2 on the dorsum of the hand. False aneurysms are more likely to be found on the superficial palmar arch than on the deep arch, and, paradoxically, digital lesions are extremely rare (Fig. 92) (Hueston 1973; Baruch 1977; Suzuki et al. 1980; Layman et al. 1982).

The *symptomatology* of these lesions is very varied and calls for an exacting enquiry. The traumatic context is clear if the trauma was quite recent, perhaps only a few hours previously, but may escape notice if it happened some years ago. Pain is occasionally absent and sometimes may assume a deceptive pseudo-inflammatory aspect. When the tumour is palpable, it is compressible and rapidly refills, but its pulsatile nature and systolic thrill may be absent or not evident. Associated paraesthesia and tingling will indicate nerve compression. An ulnar compartment syndrome may be the only clinical feature of an aneurysm (Kalishman et al. 1982). Sometimes a hitherto asymptomatic lesion may be revealed by an embolic complication (pain, cyanosis of a finger; May et al. 1982). In all cases the pulse distal to the lesion must be tested, although it may be absent. It is also useful to perform Allen's test, which is very suggestive if it produces shrinkage of the aneurysm and also provides information on the blood supply. It is wise to exclude other possible diagnoses which may arise: ganglion, when the lesion is not pulsatile; abscess when the symptomatology is inflammatory; and arteriovenous aneurysm, which may be pulsatile and is notable for its thrill, bearing in mind that an arteriovenous communication may originate from a false aneurysm.

Adjuvant studies include standard radiographs, which may display parietal calcification. Whether arteriography should be performed is controversial, since there is a risk of spasm and digital necrosis, especially in the presence of Raynaud's disease. It is indicated in some cases, but only when the diagnosis is doubtful, and yields information on the relations of the aneurysm and the quality of the collateral circulation.

Surgical treatment of the aneurysm is done under pneumatic tourniquet and must begin with identification of the adjacent nerves. Usually, it consists of excision between two ligatures (Fig. 93). Microsurgical reconstruction of the arterial axis, as proposed by Kleinert et al. (1973), is not essential if the collateral supply is adequate, and it is difficult to accomplish in the elderly arteriosclerotic patient.

Postoperative monitoring is essential to verify the absence of cyanosis and ischaemia, which are sometimes reversible but may become established and lead to gangrene (hence the importance of preoperative assessment of the quality of the collateral circulation).

Arteriovenous Aneurysms

These fistulae consist of an abnormal communication between an artery and a vein, producing precocious diversion of the blood flow with venous hyperpressure. They are rare in the hand. They may be secondary to a wound (traumatic fistulae) or congenital, the result of persistent embryonic arteriovenous communications. These congenital aneurysms, studied by Curtis (1953), Coursley et al. (1956) and Padovani et al. (1973, cited by Merland et al. 1982), appear in both old and young patients and may be revealed in the aftermath of injury by the opening of latent arteriovenous shunts.

Clinically, the lesions are latent for long periods and often painless. They may then be revealed by the development of varicosities, venous dilatations in the hand or forearm, more marked as a rule in the congenital forms. Clinical examination may detect a thrill and a murmur, and there may be increased local warmth associated with venous hyperpressure. Symptoms may be less obvious and limited to some weakness in the hand and some degree of oedema and cyanosis. Finally, an arteriovenous fistula should be considered when there is isolated hypertrophy of one or more fingers, or when there is a trophic complication such as chronic ulceration.

Arteriography is the key study method, revealing sinuous dilated arteries proximal to the lesion, the distal vessels being barely visible, if at all. The presence of pools of contrast medium indicates the site of the communication, and the display of abnormally early venous return is characteristic (Fig. 94). Congenital fistulae give images very varied from those of traumatic cases as the communications are many and their calibre different. Above all, arteriography specifies the condition of the adjacent structures, which influences the possibilities of treatment. The lesional process can involve the neighbouring muscles and also the bony structures, which gravely complicates the prognosis of these fistulae. This is the case with complex and extensive arteriovenous malformations in which there are numerous shunts, precocious and explosive venous return and the appearance of a nidus in the invaded bone.

The methods of *treatment* differ in each case, and it is important to make a precise assessment of the lesion, best done by arteriography, before deciding on the surgical treatment. Traumatic arteriovenous fistula is usually corrected by ligature of the shunts and excision of the pathologic tissue. It is not usually followed by recurrence or by ischaemic problems in the hand, and it is rare to have to reconstruct the vascular axis. Surgical decisions in congenital arteriovenous aneurysm must be made cautiously, for the operation may provoke the opening-up of hitherto latent and asymptomatic shunts and produce

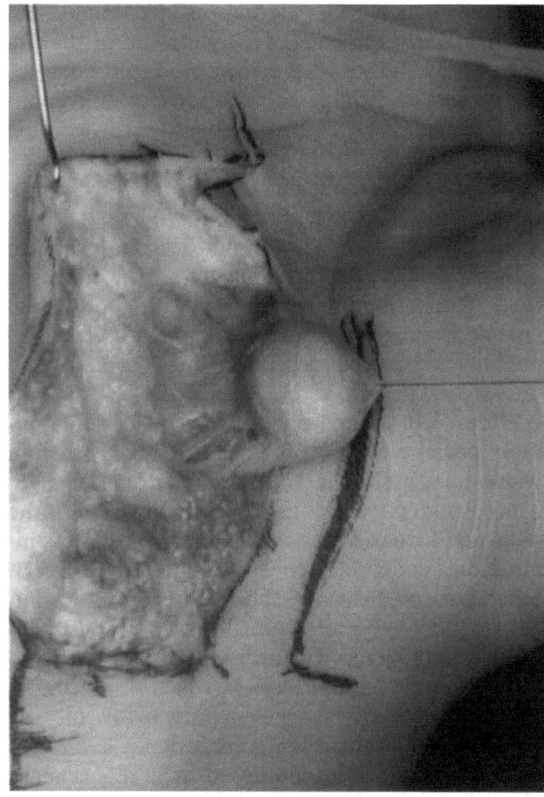

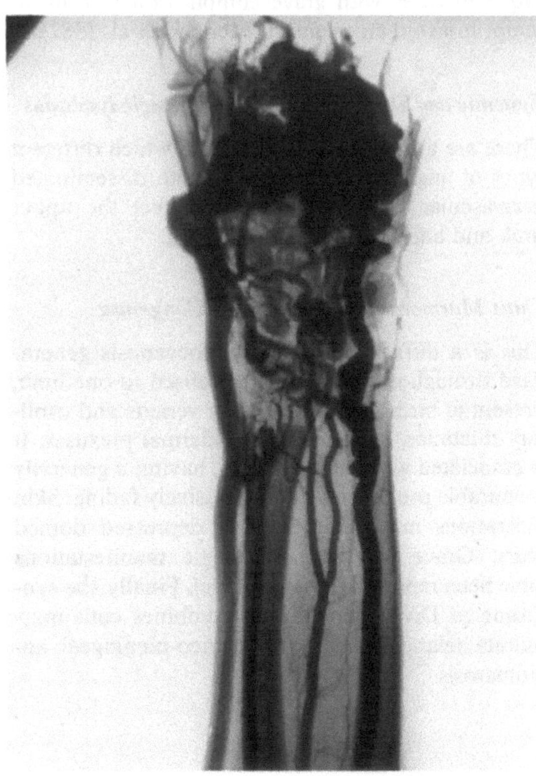

Fig. 93. Aneurysm of ulnar artery: operative appearance. Note the arterial continuity at the ends of the aneurysmal sac *(above)*

Fig. 94. Arteriovenous fistula: arteriography showing early massive venous return *(right)*

an exacerbation with extension of the lesions, loss of function in the hand, even gangrene of the fingers. Many statistics have been produced concerning congenital fistulae and extensive arteriovenous malformations. Of 28 cases submitted for excision of the lesion, Gelberman and Goldner (1978) reported 12 good results but 16 cases in which symptoms persisted or deteriorated, 3 with digital gangrene. Most authors have found less than 50% of positive results over the long term. The localised types, without bony invasion and with good digital vascularisation, can be excised in one or two stages, possibly with an interposed vein graft (Leonard and Vassos 1951).

In cases with limited bony involvement, some authors propose ablation followed by reconstruction with a bone graft, but the risk of recurrence is high. Mulitple arteriovenous fistulae with invasion of the palm and fingers almost always carry a risk of complications and are a contraindication to surgery. Some types with massive bony invasion may compel amputation. Embolisation, as such, is hardly indicated in the hand. Griffin et al. (1978) report the use of isobutyl-2-cyanoacrylate, which produces occlusion of the arterial segments just below the site of injection, but the reported experience is limited. It has, however, been proposed as a rescue technique in cases with grave complications in an attempt to avoid amputation (Merland et al. 1982).

Systemic and Disseminated Complex Angiodysplasias

There are a number of disorders in which different types of angioma are associated with disseminated nonvascular lesions. Those that affect the upper limb and hand include the following.

Cutis Marmorata Telangiectatica Congenita

This is a diffuse reticular erythrocyanosis generalised throughout the body or localised to one limb, present at birth and indicative of venous and capillary dilatation of the dermo-subdermal plexuses. It is associated with telangiectases, having a generally favourable prognosis and progressively fading. Skin ulcerations may occur, leaving depressed domed scars. Grave but rare pathologic manifestations have been reported (renal, cardiac). Finally, the syndrome of Divry Van Bogaert combines cutis marmorata telangiectatica with cortico-meningeal angiomatosis.

Osteodystrophic Angiodysplasias

Klippel-Trenaunay and Parkes-Weber syndromes. These affections combine a plane angioma present at birth, global hypertrophy of a limb and a deep-seated vascular dystrophy that is predominantly venous in Klippel-Trenaunay's syndrome and arteriovenous in the Parkes-Weber syndrome (Fig. 95). This is a declining distinction, as complex associations may be found in either of these syndromes. Two adjuvant investigations are fundamental: phlebography shows the anomalies of the deep venous pathways, while arteriography displays the arteriovenous shunts, sometimes numerous and of different calibres, whose presence determines the gravity of the condition.

Based on clinical and radiological data, Bourde (1974) suggests a classification into five groups, the severest forms having arteriovenous fistulae with a strong flow and very high levels of oxygen saturation in the venous blood. Based on this classification, Bourde sets out a programme of treatment; he advises abstention from surgery in the moderate forms, surgery for the varices plus a bone procedure in the more severe forms, and even amputation in the gravest cases.

Maffucci-Kast Syndrome. This is a congenital angiochondromatosis combining venous capillary dysplasia, mainly in the hands, with a disseminated dyschondroplasia consisting of multiple chondromata of the extremities. The metaphyseal site of the bony lesions is responsible for marked deformities and shortening, causing major functional handicap, exacerbated by multiple fractures and malunion. The therapeutic possibilities are limited, and the prognosis of this disorder is still dominated by its association with visceral and lymphatic lesions and the risk of sarcomatous degeneration of the chondromata.

Rendu-Osler Disease. This is an autosomal dominant hereditary angiomatosis combining alimentary, pulmonary, hepatic and cerebral lesions with telangiectases of the skin and mucosae, mainly on the face and hands. The haemorrhagic visceral lesions determine the prognosis and may sometimes be accessible to embolisation.

Fabry's Disease: Angiokeratoma Corporis Diffusum. This is a hereditary disease with a sex-linked recessive transmission that affects males, females being latent carriers. It comprises: (a) cutaneo-mucous signs, with keratotic telangiectases that may be situated on the pulp of the fingers and toes and (b) visceral lesions - cardiac, alimentary, pulmonary and especially renal - which are responsible for the gravity of the condition. The basis of the disease is a metabolic anomaly producing an excess of glycosphingolipids, seen under optic microscopy as intracellular vacuoles and under the electron microscope as dense, lamellar, filamentous inclusions (striped bodies). Biochemical tests show reduced activity of the α-galactosidase of the white cells and an increase of di- and trihexoside ceramides in the blood and urine. The current treatment of this disorder is symptomatic, but only correction of the enzyme deficiency offers any hope of effective management.

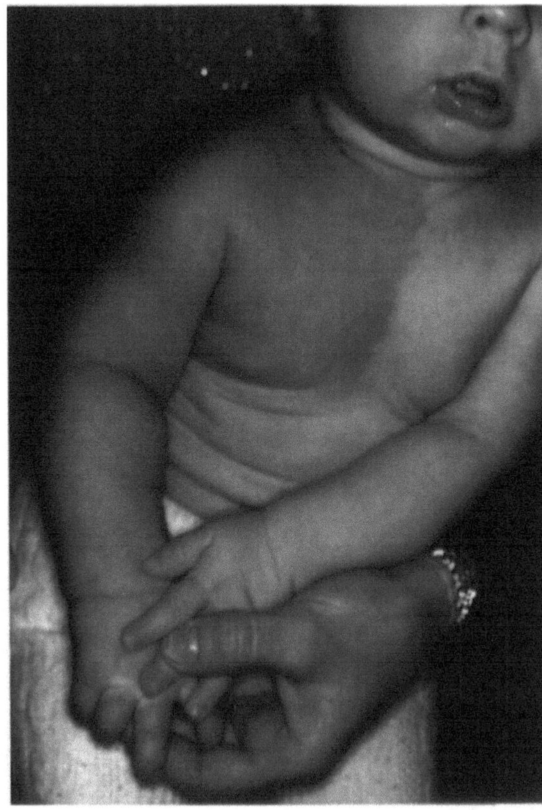

Fig. 95. Osteodystrophic angiodysplasia: note the plane angioma and the overall hypertrophy of the upper limb

Lymphangioma

The lymphangiomata are relatively rare congenital dysplasias of the lymphatic system, still sometimes called lymphatic naevi. They may be situated anywhere in the body but have a clear preference for the trunk and axillary fossa. Landing and Farber (1956) distinguish four groups on the basis of the clinical and histological forms: capillary lymphangiomata, cavernous lymphangiomate, cystic lymphangiomata and haemolymphangiomata. However, most authors seem to be agreed in distinguishing two main types: superficial circumscribed lymphangioma and deep lymphangioma. This lesion appears to be exceptional in the hand: in their series of 177 cases Saijo et al. (1975) report 10% of localisations in the extremities (without further details), while Flanagan and Helwig (1977) describe a personal case of lymphangioma of the 4th and 5th fingers.

Superficial Circumscribed Lymphangioma

This lymphangioma consists of grouped pseudovesicles arranged in an irregular patch, giving the appearance of clear translucent blebs of small size (1-5 cm), tense but yielding and containing lymphatic fluid. The skin between these structures is often normal and sometimes keratotic. They may be associated with red or pink pseudo-vesicles, indicative of a blood-vascular contribution or intralymphocytic haemorrhage, thus constituting a haemolymphangioma. Apart from its displeasing appearance, circumscribed lymphangioma is mainly troublesome because of the chronic persistent lymphatic efflux that follows puncture of a pseudo-vesicle.

These lesions are usually present at birth or appear very early in life; they have a course marked by inflammatory exacerbations and persist indefinitely.

The *clinical diagnosis* of a superficial lymphangioma is usually easy; in particular, the appearance produced is quite different from that of herpes.

Histological examination reveals a cavity consisting of very dilated lymphatic vessels containing coagulated lymph. This cavity is covered by a thin epidermis in most cases but is sometimes thick and hyperkeratotic.

There is no generally accepted hypothesis concerning the physiopathology, and, although there exist forms associated with cavernous lymphangiomata, it seems established in all cases by lymphography that no connexion exists with the deep lymphatic system (Edwards et al. 1972).

The *treatment* is difficult; surgical excision of the various lesions is often preferable to diathermy coagulation, which carries a risk of recurrence.

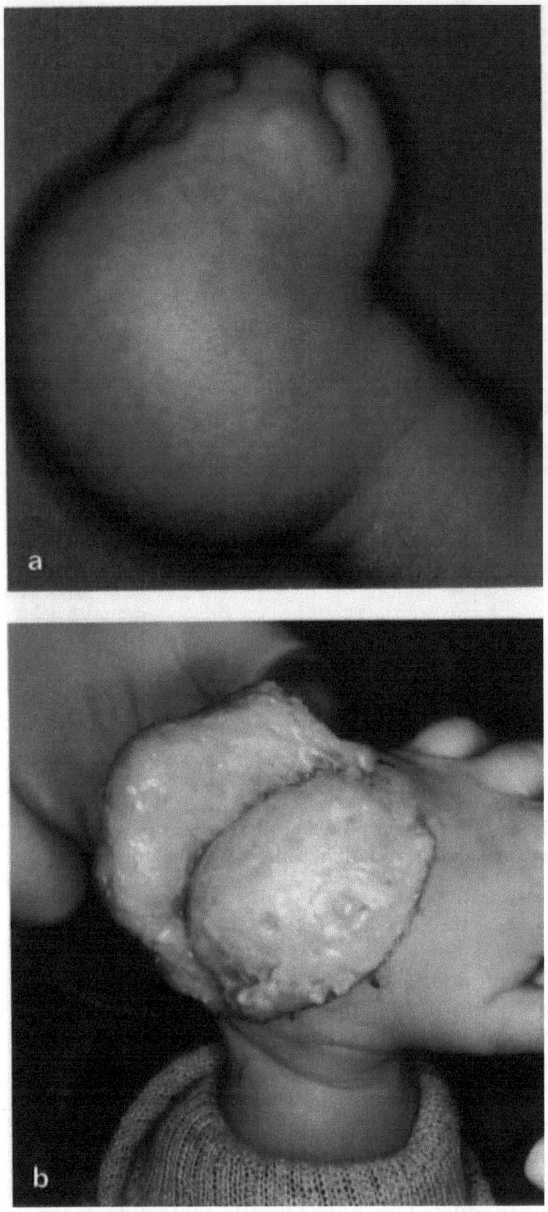

Fig. 96. ***a*** Congenital lymphangioma of dorsum of hand. ***b*** Same patient: operative exposure

Deep Lymphangioma

Deep lymphangioma is most often congenital and simulates the appearance of a cyst or lipoma. It presents in the form of a slightly projecting tumour measuring 1 cm or more, often firm and elastic but sometimes soft and yielding. It is usually painless (Fig. 96). Macroscopically, two forms can be distinguished: cavernous lymphangioma with an ill-defined outline and the better-defined cystic lymphangioma situated electively on the neck (and considered by some as a quite distinct entity). In all cases, deep lymphangiomata may be associated with the superficial circumscribed form, which facilitates the diagnosis, or a blood-vascular dysplasia constituting a haemolymphangioma, as evidenced by the bluish appearance of the swelling. Histological examination shows numerous cavities of variable size, bounded by an endothelium and filled with a fluid containing lymphocytes. The course, also marked by acute exacerbations of inflammatory type, carries no risk of malignant transformation.
As for the *treatment* of such lesions (Fig. 96b), this is a matter for caution as surgical excision quite often proves difficult and incomplete, saddled with the risk of recurrence and with the development of postoperative lymphorrhoea which is sometimes very long in drying up.

Other Vascular Tumours

Haemangioendothelioma

In his original description in 1923, Masson utilised the term of fungating intravascular haemangioendothelioma to describe a benign lesion that was subsequently labelled Masson's pseudo-angiosarcoma because of its histological characteristics. However, Stout in 1943 described under the name of haemangioendothelioma a tumour with a totally different potential course, since this was a malignant lesion of which only the well-differentiated forms occurring in children (9% of cases) carried a good prognosis.
In all cases there is a very rare vascular tumour, which may appear at any age and is preferentially situated on the head and neck or the extremities (feet and hands). The clinical appearance produced is that of a nodule of variable size, often quite large, dark-red or violet and mildly infiltrating but without local nodal involvement. The differential diagnosis is from fibrosarcoma or lipoma.
Histological examination, in the form described by Masson (1923), may lead to confusion since this is a tumour with very abundant cellular elements, characterised by vascular dilatations surrounded by phenomena of thrombosis (whose primary or secondary role is still debatable). Two essential criteria favour the benign nature of this lesion: the absence of atypical nuclei and the strictly intravascular site.

In the form described by Stout (1943) there exist numerous endothelial cells remarkable for their atypical nature, hence the term of malignant haemangioendothelioma used by Patel et al. (1978).
In the hand, Acharya et al. (1980) report three benign forms situated in the right hypothenar region, on the back of the middle phalanx of the right 4th finger and on the right thumb. Julliard and Baviera (1983) describe two tumour sites in the same patient, on the back of the wrist and thumb of the left hand. Moss et al. (1982) report the case of a congenital haemangioendothelioma on the right index finger. All these lesions were considered benign and treated conservatively. Patel et al. (1978) report the case of a malignant haemangioendothelioma of the index finger treated by amputation and free from metastases 4 years later. Thus, the true prognosis of such a tumour seems difficult to ascertain, and it is necessary to trust the histological features. These allow a distinction between the benign, strictly intravascular forms without atypical nuclei, adequately managed conservatively by surveillance, and the malignant forms of haemangioendotheliosarcoma with atypical cells requiring more aggressive treatment.

Haemangiopericytoma

This was described by Stout and Murray in 1942 and is a rare vascular tumour. It extends very slowly, is locally aggressive and infiltrating, and presents in single or multiple nodular form under the skin. It is of firm consistence, elastic or soft, generally well circumscribed, reddish in colour and painless. However, it may ulcerate or be accompanied by satellite nodules. There are two main problems with haemangiopericytoma.

1. It is important to confirm the diagnosis, which is often difficult even on histological criteria. The tumour consists of a capillary proliferation, marked especially by the presence of pericytes, elongated contractile cells of perivascular arrangement, though it is often difficult to differentiate these cells from endothelial cells or fibroblasts.
2. It seems difficult to assess the prognosis of haemangiopericytoma as recurrences have been noted a very long time after excision of the lesion; in particular, malignant change has been reported. Thus, Stout (1943) reports 14 cases with metastases from a series of 258 haemangiopericytomata.

Cases involving the hand have been reported only very rarely in the literature. Kaplan (1949) mentions a palmar lesion without recurrence after 4 years. Converse (1977) describes a haemangiopericytoma of the index finger requiring numerous operations, which recurred after 11 years and led to amputation of the 2nd ray. Sen (1976) reports a case in the right hypothenar region, adherent to the skin, recurring after excision and requiring amputation of the 5th ray. Thus, although in many instances haemangiopericytoma appears to be a benign tumor, malignant forms do seem to exist, with blood stream dissemination. Hence the importance of exact diagnosis, particularly excluding certain lesions of similar appearance such as capillary haemangioma or haemangioendothelioma.

Angioleiomyoma (see Leiomyoma, p. 117)

Angiosarcoma

Arising from the endothelial cells of the vessels, angiosarcoma - or, more precisely, haemangioendotheliosarcoma - is an extensive, primarily malignant tumour with rapid metastatic spread. The diagnosis of this lesion is usually difficult, even given the histological evidence, as the presence of a marked vascular component in a malignant tumour is not necessarily evidence of the vascular origin of the sarcoma. Haemangioendotheliosarcoma is an exceptional tumour. The largest series reported is that of Landells (1958, cited by Merland et al. 1982) with 117 cases. A more recent detailed study (Girard et al. 1970) covers 28 cases. Though it may occur at any age, this lesion appears more often in young children and after 60 years of age. It is preferentially located on the head and neck, though Girard et al. (1970) report three cases in the hand (wrist, hypothenar eminence, index finger).

Clinical Features

The appearance produced is that of an infiltrating erythematous plaque or a raised nodule, dark red, firm, painful and irregular. Initially it resembles an angioma but develops very rapidly, infiltrating the adjacent structures with a superficial fungating proliferation that is multinodular, bossed, purplish and bleeds readily. Nodular lesions appear in the vicinity of the tumour, evidence of local spread, followed by early nodal invasion. Metastatic lesions in the lungs, liver and bones come later, indicative of the terminal stage of the affection.

Histology

Histological examination confirms the diagnosis, though this is difficult in the early stages, which may rather suggest a basal- or squamous-celled epithelioma or a malignant melanoma. There are irregular vascular dilatations of lacunar aspect and numerous atypical cells of large size. Electron microscopic study determines the endothelial origin of these cells.

Treatment

The therapeutic options with such a lesion are limited. Indeed, the prognosis is very disappointing despite the combination of surgery, radiotherapy and chemotherapy. Girard et al. (1970) however, report the case of one haemangioendotheliosarcoma of the wrist requiring amputation at the elbow followed by disarticulation at the shoulder with a survival of 13 years despite the appearance of tumour nodules on the chest wall.

Lymphangiosarcoma

The impossibility of distinguishing angiosarcoma from lymphangiosarcoma has led many writers to reserve the latter term for specific entities: the Stewart-Treves syndrome, or lymphangiosarcoma developing on a post-mastectomy lymphoedema, and lymphangiosarcoma arising on chronic non-tumoral lymphoedema. Although the pathogenesis of these sarcomata is still debatable, the responsibility of lymphatic stases is held to be a promoting factor.

Stewart-Treves Syndrome

This was described in 1948 on the basis of six cases reported by F. W. Stewart and N. Treves, in which lymphangiosarcoma developed in a chronic lymphoedema, itself secondary to axillary clearance for cancer of the breast. However, the incidence of this syndrome is very small, since it is seen in only 0.5% of cases and appears on average 10 years after the nodal eradication. Its occurrence seems to be encouraged by treatment including radiotherapy.

Clinical Features

It is evidenced by the appearance on the lymphoedematous upper limb of purblish-blue nodules rather like those of Kaposi's disease. These nodules multiply rapidly and are associated with increased oedema of the limb. Extensive ulcerations also frequently appear. The course is very rapid, as shown by the appearance of nodal metastases and secondaries in the lungs and liver. The prognosis of this lesion is disastrous, since death occurs on average within 6 months; despite the most aggressive treatment, including surgical excision and chemotherapy, less than 6% of survivals have been reported.

Diagnosis

The diagnosis can be made on the clinical aspects and is confirmed by histological examination. This shows dermo-subdermal and muscular nodules consisting of multiple vascular spaces and clefts separated by cell islands containing lymphocytes.

Aetiopathogenesis

Here, there are two opposing theories. One is that this is a metastatic proliferation related to the breast adenocarcinoma. The other is that this is a tumour developing independently of the mammary lesion, whose occurrence is promoted by the lymphoedematous nature of the limb.

Lymphangiosarcomata Arising on Chronic Non-Tumoral Lymphoedema

These are much rarer tumours, the first case being reported by Lowenstein in 1906. The chronic lymphoedema may be of congenital or acquired origin, though it is possible that apparently causal infections or injuries may simply reveal an underlying congenital malformation.

Clinically, the lesion presents in the form of multiple, reddish-purple, adherent nodules, suggestive of a malignant tumour from the outset. These nodules are easily haemorrhagic and bleed on contact. They may resemble Kaposi's disease, fibrosarcoma or malignant melanoma. The course of these lymphangiosarcomata is equally menacing, causing death from metastatic spread in a few months, despite aggressive treatment combining surgical exicision such as amputation or disarticulation with chemotherapy.

Bibliography

Acharya S, Merritt WH, Teogarag SD (1980) Hemangioendotheliomas of the hand: Case report. J Hand Surg 5: 181-182

André JM, Picard L, Kissel P (1974) Les angiodysplasies systématisées. J Fr Neuroradiol 1: 3-45

Baruch A (1977) False aneurysm of the digital artery. Hand 9: 195-197

Bean WB (1942) A note on the development of cutaneous arterial "spiders" and palmar erythema in persons with liver disease and their development following the administration of estrogens. Am J Med Sci 204: 251

Bean WB (1943) Acquired palmar erythema and cutaneous vascular "spiders." Am Heart J 25: 463

Bean WD, Cogswell R, Dexter M, Embick JF (1949) Vascular changes of the skin during pregnancy. Surg Gynecol Obstet 88: 739-752

Ben-Menachem Y, Epstein MJ (1974) Post-traumatic capillary hemangiomas of the hand. J Bone Joint Surg [Am] 56: 1741-1743

Bingham HG, Lichti E (1970) Use of ultrasound transducer (Doppler) to localize peripheral and arteriovenous fistulae. Plast Reconstr Surg 46: 151-154

Bogumill GP (1977) Clinico-pathological correlation in a case of congenital arterio-venous fistula. Hand 9: 60-64

Booher RJ (1961) Tumors arising from blood vessels in the hands and the feet. Clin Orthop 19: 71-96

Bourde G (1953) Les fistules artério-veineuses congénitales des membres. J Chir (Paris) 69: 728-748

Bourde C (1974) Classification des syndromes de Klippel-Trénaunay et de Parkes-Weber d'après les données angiographiques. Ann Radiol 17/2: 153-160

Bourde C, Bourdoncle EY, Jouve A (1955) Documents sur les fistules artério-veineuses des membres. Essai de classification. Arch Mal Cœur 8: 775-801

Campanacci M, Pizzoferrato A (1971) Osteosarcoma emorragico. Chir Organi Mov 60: 409-421

Carneiro RS (1974) Aneurysm of the wrist. Plast Reconstr Surg 54/4: 483-488

Chen VT (1977) Haemangioma of adductor pollicis. Hand 9/2: 187-194

Civatte J, Harter P (1967) Un cas d'hémangio-endothéliome végétant intra-vasculaire de Masson. Bull Soc Fr Dermatol 74: 520-521

Coursley G, Ivins JC, Barker NW (1956) Congenital arteriovenous fistulas in extremities: Analysis of 69 cases. Angiology 7: 201-217

Cross FS, Glover DM, Simeone FA et al. (1956) Congenital arteriovenous aneurysms. Ann Surg 148: 649-665

Curtis RN (1953) Congential arteriovenous fistulae of the hand. J Bone Joint Surg [Am] 35: 917-928

Curtis RM (1976) Tumors of the blood and lymphatic vessels. In: Flynn JE (ed) Hand surgery, 2nd edn. William & Wilkins, Baltimore, pp 690-691

Dolph JL, Demuth RJ, Miller SH (1981) Angiokeratoma circumscriptum of the index finger in a child. Plast Reconstr Surg 67/2: 221-223

Dubin HV, Creehan P, Headington JT, Arbor A (1974) Lymphangiosarcoma and congenital lymphoedema of the extremity. Arch Dermatol 110: 608-614

Edwards JM, Peachey RDG, Kinmonth T (1972) Lymphangiography and surgery in lymphangioma of the skin. Br J Surg 59/1: 36-41

Engelman RM, Clemens JM, Herrmann JB (1915) Stabwound and traumatic false aneurysms in the extremities. J Trauma 9: 77-87

Enjolras O (1981) Malformations vasculaires, angiomes et angiodysplasies cutanéo-muqueuses. Dermatol Prat 10: 11-41

Fabry J (1898) Ein Beitrag zur Kenntnis der Purpura haemorrhagica nodularis (Purpura papulosa haemorrhagica Hebrae). Arch Dermatol 43: 187

Fabry J (1915) Über ein Fall von Angiokeratoma circumscriptum am linken Oberschenkel. Dermatol Z 22: 1

Flanagan BP, Helwig EB (1977) Cutaneous lymphangioma. Arch Dermatol 113: 24

Flynn JE (1966) Hand surgery. Williams & Wilkins, Baltimore, p 1956

Geiser JH, Eversmann WW (1978) Closed system venography in the evaluation of upper extremity hemangiomas. J Hand Surg 3/2: 173-178

Gelberman RH, Goldner L (1978) Congenital arteriovenous fistulas of the hand. J Hand Surg 3/5: 451-454

Girard C, Johnson WC, Graham JH (1970) Cutaneous angiosarcoma. Cancer 26/4: 868-884

Glanz S (1969) The surgical treatment of cavernous hemangiomas of the hand. Br J Plast Surg 22: 293-301

Godsick PA, Lang EN (1976) Hemangioma of the extensor policis longus. Bull Hosp Joint Dis 37: 172

Green DP (1973) True and false traumatic aneurysms in the hand. Report of two cases and review of the literature. J Bone Joint Surg [Am] 55: 120-128

Griffin JM, Vascones LO, Schatten WE (1978) Congenital arterioveinous malformations of the upper extremity. Plast Reconstr Surg 62/1: 49-57

Griffiths J (1897) A case of spontaneous aneurysm of the ulnar artery in the palm: Excision of the aneurysm: Recovery. Br M J II: 646-647

Guyon F (1861) Note sur une disposition anatomique propre a la face anterieure de la region du poignet et non encore decrite. Bull Soc Anat Paris 6: 184

Hahn MJ, Dawson R, Esterly JA, Joseph DJ (1973) Hemangiopericytoma. Cancer 31: 255-261

Hansson CJ (1940) Kaposi sarcoma, clinical and radiotherapeutic studies on 23 patients. Acta Radiol 21: 457

Hashimoto K, Lieberman P (1976) Angiokeratoma corporis diffusum (Fabry disease). Arch Dermatol 112: 1416

Hentz V, Jackson I, Fogarty D (1978) Case report: False aneurysm of the hand secondary to digital amputation. J Hand Surg 3: 199-200

Herndon WA, Hershey SL, Lambdin CS (1975) Thrombosis of the ulnar artery in the hand. J Bone Joint Surg [Am] 57: 993-995

Horton BT, Ghormley RK (1935) Congenital arteriovenous fistulas of extremities visualized by arteriography. Surg Gynecol Obstet 60: 978-983

Horton BT (1931) Hemihypertrophy of extremities associated with congenital arteriovenous fistulae. Proc Mayo Clin 21: 316-319

Hueston JT (1973) Traumatic aneurysm of the digital artery. A complication of fasciectomy. Hand 5: 332-334

Jaffe HL (1943) Hereditary multiple exostosis. Arch Pathol 36: 355

Julliard FA, Baviera EE (1983) Lésions bénignes de la main simulant un angiosarcoma: L'hémangioendotheliome végétant intra-vasculaire de Masson. Ann Chir 2/1: 59-61

Kalishman M, Laborde K, Wolff T (1982) Ulnar nerve compression secondary to ulnar artery false aneurysm at the Guyon's canal. J Hand Surg 7/2: 137-139

Kaplan EB (1949) Hemangiopericytoma of the hand. Bull Hosp Joint Disease 10: 49

Kaplan EB, Zeide MS (1972) Aneurysm of the minor artery. A Case Report. Bull Hosp Jt Dis 33: 197-199

Kleinert HE, Hayes JE (1971) The ulnar tunnel syndrome. Plast Reconstr Surg 47: 21

Kleinert HE, Valianitis GJ (1965) Thrombosis of the palmar arterial arch and its tributaries: Etiology and newer concepts in treatment. J Trauma 5: 447

Kleinert HE, Burget GC, Morgan JA, Kutz JE, Atasoy E (1973) Aneurysms of the hand. Arch Surg 106: 554-557

Klippel M, Trénaunay P (1900) Du nævus variqueux et ostéo-hypertrophique. Arch Gén Méd 3: 641-672

Kojima T, Ide Y, Marumo E, Ishikawa E, Yamashita H (1976) Haemangioma of the median nerve causing carpal tunnel syndrome. Hand 8: 62

Koman AL, Urbaniak JR (1979) Thrombosis of ulnar artery at the wrist. In: A.A.O.S. Symposium on Microsurgery. Mosby, St Louis, 119-132

Landing BH, Farber S (1956) Tumors of the cardiovascular system. In: Atlas of tumor pathology, sec 3. Armed Forces Institute of Pathology, Washington

Layman CD, Ogden LL, Lister GD (1982) True aneurysm of digital artery. J Hand Surg 7/6: 617-618

Laugier P, Hunziker N (1971) Angiomes stellaires, érythème palmaire, néoformations angiomateuses et grossesse. Rev Fr Gynécol 66/10: 599-602

Leonard FC, Vassos GA Jr (1951) Congenital arteriovenous fistulation of the lower limb. N Engl J Med 245: 885

Lewis DW (1930) Congenital arteriovenous fistulae. Lancet II: 621

Lewis RJ, Ketcham AS (1973) Maffucci's syndrome - Functional and neoplastic significance: Case report and review of the literature. J Bone Joint Surg [Am] 55: 1465-1479

Lewis TR Jr (1957) The treatment of hemangiomas. Plast Reconstr Surg 19: 201-212

Lister WA (1938) The natural history of the strawberry nevi. Lancet I: 1429

Löwenstein SI (1906) Der atiologische Zusammenhang zwischen akutem ein-maligem Trauma und Sarkom. Beitr Z Klin Chir 48: 780

Lowrey CW, Chadwick RO, Waltman EN (1976) Digital vessel trauma from repetitive impact in baseball catchers. J Hand Surg 3: 236-238

Maffucci AD (1881) Di un caso de encondroma ed angioma multiplo. Mov Med Chir Nap 3: 399-412, 565-575

Mangus DJ (1972) Continuous compression treatment of hemangiomata. Plast Reconstr Surg 49: 490

Martin E, Vilain R (1960) Discussion anatomique sur un cas de syndrome de Stewart-Trèves. Arch Anat Pathol (Paris) 8: 246-249

Martorell F (1951) Tumorigenic lymphedema. Angiology 2: 386

Masson P (1923) Hemangioendothelioma vegetant intravasculaire. Bull Soc Anat Paris 93: 517-523

Matsuno T, Unni KK, McLeod RA, Dahlin DC (1976) Telangiectatic osteogenic sarcoma. Cancer 38: 2538-2547

May JW, Grossman JA, Costas B (1982) Cyanotic painful index and long fingers associated with an asymptomatic ulnar artery aneurysm. J Hand Surg 7/6: 622-625

Mays ET (1970) Traumatic aneurysm of the hand. Am Surg 36: 552-557

McCarthy WD, Pack GI (1950) Malignant blood vessel tumors. A report of 56 cases of angiosarcoma and karosis sarcoma. Surg Gynecol Obstet 91: 465

Merland JJ, Riche MC, Enjolras O (1980) Aspects artériographiques et place de l'embolisation dans les angiomes de l'enfant. Cutis 14: 131-159

Merland JJ, Natali J, Drouet L et al. (1982) Les angiomes et malformations vasculaires. Médicorama 255

Mibelli V (1889) Di una nuova forma di cheratosis, angiocheratoma. G Ital Mal Venereol 30: 285

Millender LH, Nalebuff EA, Kasdon E (1972) Aneurysms and thrombosis of the ulnar artery in the hand. Arch Surg 105: 686-690

Miller SH, Smith RL, Shochat SJ (1976) Compression treatment of hemangiomas. Plast Reconstr Surg 58: 573-579

Milling AP, Kinmonth MH (1977) False aneurysm of the ulnar artery. Hand 9/1: 57-59

Morestin H (1905) Anéurysme de l'arcade palmaire superficielle. Bull Mém Soc Anat Paris 80: 556

Moss Lee A, Stueber K, Hafiz MA (1982) Congenital hemangioendothelioma of the hand. J Hand Surg 7/1: 53-55

Neviaser RJ, Adams PA (1974) Vascular lesions in the hand. Clin Orthop 100: 111-119

Pack GT (1939) Symposium: Tumors of the hands and feet. Surgery 5: 1

Pack GT, Miller TR (1950) Hemangiomas. Angiology 1: 405

Palmari JT (1983) Subcutaneous hemangiomas of the hand. J Hand Surg 8/2: 201-204

Parkes-Weber F (1908) Haemangiectatic hypertrophies of the foot and extremity congenital or acquired. Med Presse 136: 261

Parkes-Weber F (1918) Haemangiectatic hypertrophy of limbs. Br J Child Dis 15: 13

Patel MR, Scririvasan KC, Pearlman HS (1978) Malig-

nant hemangioendothelioma in the hand: A case report. J Hand Surg 3: 585-588
Peled I, Iosipovich Z, Rousso M, Wexler MR (1980) Hemangioma of the median nerve. J Hand Surg 5/4: 363-365
Poirier RA, Stansel HC (1972) Arterial aneurysms of the hand. Am J Surg 124: 72-74
Posch JL (1956) Tumors of the hand. J Bone Joint Surg [Am] 38: 517
Posch JL (1976) Soft tissue tumors of the hand. In: Flynn JE (ed) Hand surgery, 2nd edn. William & Wilkins, Baltimore
Riche MC, Melki JP, Merland JJ (1981) L'angiographie thérapeutique, vol 1. Specia, Paris
Rotteleur G (1980) Le laser ionisé en thérapeutique dermatologique. Thèse pour le Doctorat en Médecine, Lille
Saijo M, Munro IR, Mancer K (1975) Lymphangioma: A longterm follow-up study. Plast Reconstr Surg 56: 642
Salyer WR, Salyer DC (1975) Intravascular angiomatosis: Development and distinction from angiosarcoma. Cancer 36: 995-1001
Sanchez A, Archer S, Levine NS, Buchanan RT (1982) Traumatic aneurysm of a common digital artery. J Hand Surg 7/6: 619-621
Sato S (1913) Über das cavernose Angioma an peripheren Nervensystem. Arch Klin Chir 100: 553
Sen R (1976) Hemangiopericytoma of the hand. Plast Reconstr Surg 57/6: 746-748
Sherwood B, Mori K, Hoffman S (1975) Aneurysm of the hand to Kaposi's syndrome. J Bone Joint Surg [Am] 57: 995-997
Smith J (1962) True aneurysm of traumatic origin in the palm. Am J Surg 104: 7
Soler JM, Muller HA, Trautlein JJ (1979) Posttraumatic aneurysms of the hand. IACEP 8: 232-234
Spittel JA (1958) Aneurysms of the hand and wrist. Med Clin North Am 42: 1007-1010
Srinivasan CK, Patel MR, Pearlman HS (in press) Malignant hemangioendothelioma in the bones
Stack G (1960) Tumors of the hand. Br Med J I: 517, 919-922
Sterling AP, Habermann ET (1975) Traumatic aneurysm of the radial artery. Hand 7/3: 294-296
Stewart FW, Treves N (1948) Lymphangiosarcoma in postmastectomy lymphedema. Cancer 1: 64
Stout AP (1943) Hemangioendothelioma: A tumor of blood vessels, featuring vascular endothelial cells. Ann Surg 118: 445-464
Stout AP, Lattes R (1967) Tumors of the soft tissues, series 2. The Armed Forces Institut of Pathology, Washington, p 145
Stout AP, Murray MR (1942) Hemangiopericytoma - a vascular tumor featuring Zimmerman's pericytes. Ann Surg 116: 26-33
Suzuki K, Takahashi S, Nakagawa T (1980) False aneurysm in a digital artery. J Hand Surg 5: 402-403
Taxy JB, Gray SR (1979) Cellular angiomas of infancy. Cancer 43: 2322-2331
Tropet Y, Merle M, Foucher G, Vichard P, Michon J (1982) Subcutaneous hemangiomas of the hand (In French). Ann Chir Plast 27/1: 68-71
Tseng-Tong K, Sayers CP, Rosai J (1976) Masson's "vegetant intravascular hemangioendothelioma" - a lesion often mistaken for angiosarcoma. Cancer 38: 1227-1236
Verdan C (1964) Traitement chirurgical des hémangiomes des membres et aspects particuliers de leur localisation à la main. Ann Chir Plast 9/4: 304-316
Waddell GF (1967) Haemangioma involving tendons. J Bone Joint Surg [Br] 49/1: 138-141
Wallerstein RO (1961) Spontaneous involution of giant hemangioma. Am J Dis Child 102: 233
Walsh EN, Becker SW (1941) Erythema palmare and naevus araneus. Arch Derm 44: 616-630
Walsh MJ, Conolly WB (1982) False aneurysms due to trauma to the hand. Hand 14/2: 77-181
Warner ED, Hoak JC, Fry GL (1971) Hemangioma-thrombocytopenia and anemia. The Kassabach-Meritt Syndrome in an animal model. Arch Pathol 91: 523-528
Wenger DR, Boyer DW, Sigurd SS (1980) Traumatic aneurysm of the radial artery in the anatomical snuff box. Hand 12/3: 266-270
Weinberg M Jr, Steiger Z, Fell EH (1960) Unusual congenital anomalies of the arteriovenous system. Surg Clin North Am 40: 67-74
Weisman PA (1959) Blood vessel tumors of the hand. Plast Reconstr Surg 23: 175-186
Whimster IW (1976) The pathology of lymphangioma circumscriptum. Br J Dermatol 94: 473
Wood VE (1982) Hemangioma with bone lesions. J Hand Surg 7/3: 287-290
Zweig J, Lie KK, Posch JL, Larsen RD (1969) Thrombosis of the ulnar artery following blunt trauma to the hand. J Bone Joint Surg [Am] 51: 1191

III. Tumours of Bone

Introduction

Osseous tumours of the hand are quite rare. In various series they represent from 1% (Leung 1981) to 5.4% (Bogumill et al. 1975) of all tumours of the hand. Present classifications (Dahlin 1973; Meary 1974; Mirra 1980) are based on the arrangement of Lichtenstein and Goldman (1964), which refers to both the cellular components of the tumour and the tissues to which they give rise. In terms of their cells, tumours may be cartilaginous, bony, fibrous or, much more rarely, vascular, neural, fatty, epidermal, histiocytic and haematopoietic. Some remain poorly labelled, and there are also the bone dystrophies which are pseudotumoral processes. This classification, which is somewhat inexact inasmuch as quite a number of tumours have several different cellular components, has the merit of clarity.
In the skeleton of the hand we may therefore envisage the following:

A. Cartilaginous tumours
  - chondroma, by far the most common of the bony tumours of the hand
  - chondroblastoma
  - osteochondroma
  - chondrosarcoma (see *G*)
  - chondromyxoid fibroma, containing both cartilaginous and fibrous cells

B. Bony tumours
  - osteoid osteoma
  - osteoblastoma
  - osteogenic sarcoma (see *G*), triply composed of osteoblasts, chondroblasts and fibroblasts

C. Fibrous tumours
  - fibrous dysplasias, which are not always classed with tumours
  - desmoid tumour (see p. 101)
  - fibrosarcoma (see G)
  - no cases seem to have been reported of non-ossifying fibroma in the hand

D. Other tumours. Apart from histiocytic tumours (eosinophilic granuloma), of which only case has been reported in the hand, each type of tumour is treated in the corresponding chapter:
  - vascular tumours (haemangioma, haemangiopericytoma, angiosarcoma
  - fatty tumours (lipomata)
  - epidermal tumours (epidermal cysts)
  - neural tumours (no case described in the hand)
  - chordomata (no case described in the hand)
  - completely separate tumours of haematopoietic origin (myeloma, reticulosarcoma) which will not be dealt with here

E. Unidentified tumours
  - giant-cell tumour (osteoclastic origin, according to Mirra 1980)
  - Ewing's sarcoma (see *G*)
  - no case of adamantinoma (of epidermal origin) has been reported in the hand

F. Bone dystrophies
  - solitary bone cysts
  - aneurysmal bone cysts
  - myositis ossificans

G. For clarity of exposition, we have artificially grouped the malignant tumours: osteosarcoma, fibrosarcoma, chondrosarcoma, Ewing's sarcoma, metastatic tumours.

Bibliography

Aegerter E, Kirkpatrick JA (1975) Orthopaedic diseases, 4th edn. Saunders, Philadelphia

Bogumill GP, Sullivan DJ, Baker GI (1975) Tumors of the hand Clin Orthop 108: 214-222

Bunnel S (ed) (1948) Tumors of the hand. In: Surgery of the hand, 2nd edn. Lippincott, Philadelphia, pp 861-913

Butler ED, Hamill JE, Seipel RS et al. (1960) Tumors of the hand. A ten year survery and report of 437 cases. Am J Surg 100: 293-302

Campanacci M (1975) Cartilaginous tumors of the skeleton of the hand. Chir Organi Mov 62: 483-490

Carroll RE (1966) Tumors of the hand skeleton. In: Flynn JE (ed) Hand surgery. Williams & Wilkins, Baltimore, p 1039

Coley BL, Higinbotham NL (1939) Tumors primary in the bones of the hands and feet. Surgery 5: 112-128

Cooley BL (1964) Tumors of the hand and forearm. In: Converse JM (ed) Reconstructive plastic surgery, vol 4. Saunders, Philadelphia London, pp 1740-1786
Dahlin DC (1973) Bone tumors, 2nd edn. Thomas, Springfield, p 285
Geschickter CF, Copeland MM (1949) Tumors of bone, 3rd edn. Lippincott, Philadelphia, p 810
Jaffe HL (1958) Tumors and tumorous conditions of the bones and joints. Lea & Febiger, Philadelphia
Leung PC (1981) Tumors of hand. Hand 13: 169-176
Lichenstein L, Goldman RL (1964) Cartilage tumors in soft tissues, particularly in the hand and foot. Cancer 17: 1203-1208
Mangini U (1967) Tumors of the skeleton of the hand. Bull Hosp Jt Dis Othop Inst 28: 61-103
Meary R (1974) Tumeurs bénignes osseuses et dystrophies pseudo-tumorales. Monographies des annales de chirurgie. Expansion Scientifique, Paris, p 118
Mirra JM (1980) Bone tumors. Diagnosis and treatment. Lippincott, Philadelphia, p 629
Rinaldi E (1976) Metacarpal loss treated by metatarsal substitution. Ital J Orthop Traumatol 2: 335-339
Shellito JC, Dockerty MB (1948) Cartilaginous tumors of the hand. Surg Gynecol Obstet 86: 465-472
Stein AH (1959) Benign neoplastic and non neoplastic destructive lesions in the long bones of the hand. Surg Gynecol Obstet 109: 189-197

Chondroma

Chondromata are the most common of the benign cartilaginous tumours. They preferentially affect the hand (45%-65% of cases, according to series: Geschickter and Copeland 1949, 52/110; Dahlin 1973, 56%; Jaffe 1958, 18/28; Mirra 1980, 75/153) and aetiologically constitute the most frequently found of the skeletal tumours of the hand (Carroll 1975, 90% of cases). Several major series of bony chondromata of the hand have been published, as follows: Takigawa (1971b), 110 cases; Jewusiak et al. (1971), 41; Tomeno and Forest (1974), 29; Noble and Lamb (1974a, b), 40 of 50 osteocartilaginous tumours of the hand; Alawneh et al. (1977), 33; They must be dinstinguished from solitary enchondroma, multiple enchondromata, and multiple enchondromatoses or Ollier's disease.

Solitary Enchondroma

This is by far the most common of the three types mentioned above in the hand: 68%-87%, depending on the series.

Incidence

Enchondroma may occur at any age, but it shows a marked predominance between 30 and 35 years. Takigawa (1971b) gives the average as 35 years, Tomeno and Forest (1974) 20-30 years, Mason (1937) before 30 years. Alawneh et al. (1977) give between 40 and 50 years but believe that the tumour often arises earlier and remains asymptomatic for a long time because of its poor growth potential. Men are slightly more affected than women. Noble and Lamb (1974a) find enchondroma three times as common as osteochondroma in the hand.

Clinical Features

The diagnosis of enchondroma may be made under three sets of circumstances. It may be revealed by a pathological fracture, which is the most frequent situation presented (Noble and Lamb 1974b, 75% of cases; Jewusiak et al. 1971, 59%; Alawneh et al. 1977, 54%; Eiken and Carstam 1966, 50%; Tomeno and Forest 1974, 30%). It may be a swelling that is indicative (Fig. 97a): 30%-57% in different series. Noble and Lamb (1974b) state that this swelling is always present at the time of the first examination, but it is not always in itself the motive for consultation, and its course is so slow that a period of several months to 20 years may elapse before the diagnosis is made. This swelling is of a hard, bony consistency and presents as a regular increase in the size of the bone. Different authors give different incidences of associated pain: Alawneh et al. (1977) state that it is always present, Noble and Lamb (1974b) that it occurs in only half the cases. Very occasionally, pain is the sole and revelatory feature. Finally, the enchondroma may be entirely asymptomatic and revealed only by radiography performed for some other reason (4%-20% of cases).

Radiology

The classical image is central, round and purely osteolytic. Initially, it respects the bony contours but in advanced forms the bone appears "blown-out" or distended, and the cortex is thinned at the level of the condroma, though it does not rupture unless there is a fracture (Fig. 97b). The cavity is sometimes dotted with calcification. However, the lesion may also be eccentric. Takigawa (1971b) classifies the appearances into five types (Fig. 98): central (the most common, 58%), eccentric (19%), combined (several lesions in the same bone, 21%), polycentric (11%; Fig. 99) and giant (3%). For Noble and Lamb (1974b) there are four types: central (27 cases), eccentric (6 cases), juxtacortical (3 cases) and giant (5 cases). The films sometimes show one or more other chondromata in the same hand (multiple chondromata) or even Ollier's disease (multiple chondromatosis).

Site

The chondroma is generally diaphyseal, only secondarily invading the metaphyses. In the hand it preferentially affects the ulnar fingers. Analysis of the five large series (Takigawa 1971b; Jewusiak et al. 1971; Tomeno and Forest 1974; Noble and Lamb

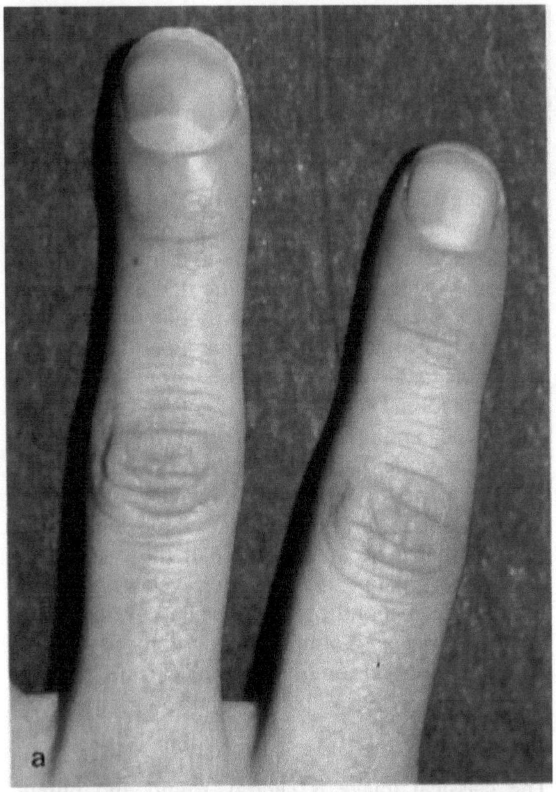

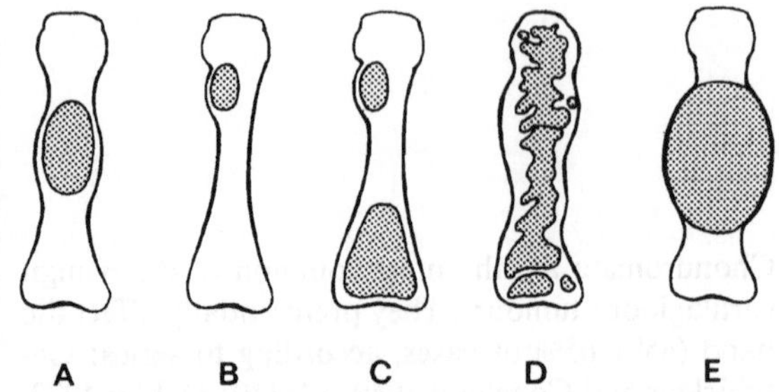

Fig. 98. Radiological classification of Takigawa (1971b). *A* central; *B* eccentric; *C* combined; *D* polycentric; *E* giant

1974b; Alawneh et al. 1977) shows the following distribution (Table 5): 5th finger, 36%; 4th finger, 21%; 3rd finger, 19%; 2nd finger, 16%; thumb, 5%. The segmental localisation in each ray is as follows: metacarpal, 23%; proximal phalanx, 45%; middle phalanx, 8%; carpus, 1%. Thus, the preferential localisation is the proximal phalanx, followed by the metacarpal and the second phalanx. Carpal lesions are exceptional (Takigawa 1971a, 1 case; Noble and Lamb 1974b, 1 case).

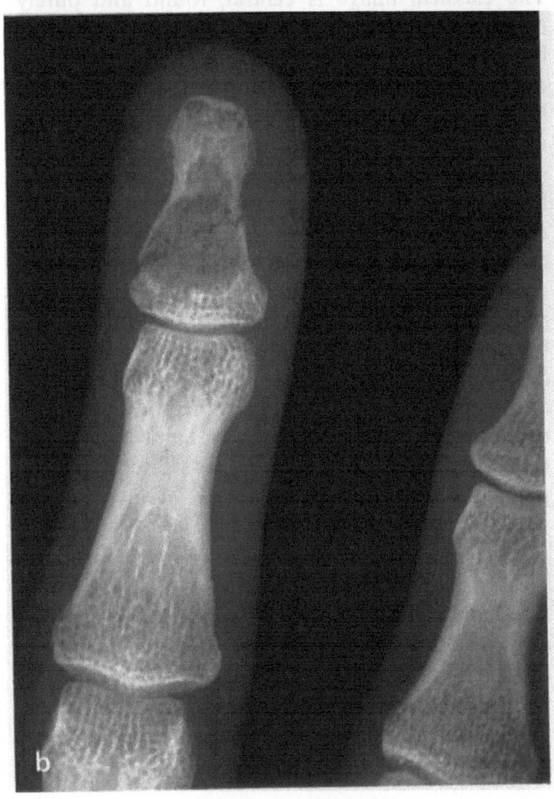

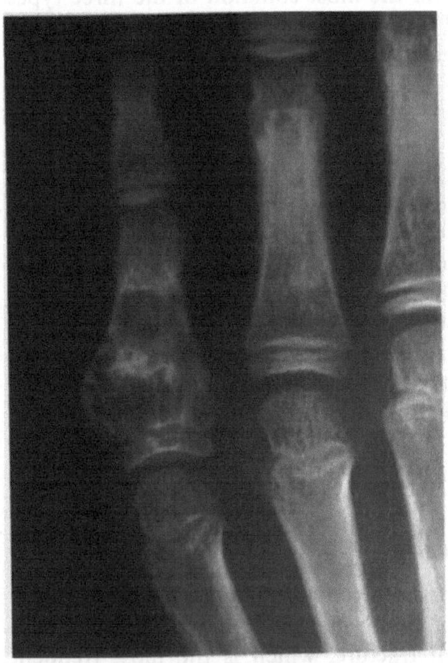

Fig. 99. Polycentric enchondroma

Fig. 97. ***a*** Chondroma of distal phalanx: note global hypertrophy, including the nail. ***b*** Same patient: note the pathological fracture *(left)*

Table 5. Distribution of solitary enchondroma in the hand

	Takigawa (1971b)	Alawneh et al. (1977)	Noble and Lamb (1974b)	Jewusiak et al. (1971)	Tomeno and Forest (1974)	Total	Percentage
Finger							
V	26	7	15	12	17	77	36
IV	17	6	10	3	9	45	21
III	18	7	6	4	7	42	19
II	10	8	10	4	3	35	16
I	3	5	3	0	1	12	5
Bone							
Distal phalanx	3	4	6	2	3	18	8
Middle phalanx	15	8	12	2	10	47	22
Proximal phalanx	36	15	19	11	15	96	45
Metacarpal	20	6	7	8	9	50	23
Carpus	1	0	1	0	0	2	1
						213	

Finally, Tomeno and Forest (1974) note that in the proximal and middle phalanges involvement tends to be proximal, while it is more often distal in the metacarpal.

Clinical Types

It is necessary to emphasise one particular type of solitary chondroma, the periosteal chondroma. Described for the first time by Lichtenstein and Hall (1952), it has since been reported by others (Jaffe 1956; Feinberg and Wilber 1956; Merlino and Nixon 1964; Cary 1965; Nosanchuck and Kaufer 1969; Rockwell et al. 1972) with some 40 published cases.

This is a chondroma that develops between the cortex of the bone and the periosteum. It most often affects the hand, but can occur also in the long bones (humerus, tibia) and the foot. It may appear at any age and is evidenced by a sometimes painful swelling. Exceptionally, it is revealed by injury. The radiographs show a regular cortical erosion at the chondroma (a dome-shaped impression) with sclerosis of the cortex, though the chondroma itself is hardly apparent, with small scattered calcifications. Occasionally, its outline is shown by a fine peripheral opacity (Fig. 100). Several views are required in order not to miss the cortical erosion. Treatment consists of excision en bloc, removing the entire tumour with the adjacent cortical lamella, so as not to leave any cartilaginous remnant that may be included in the cortical bone. Lichtenstein and Hall (1952) are content with curettage of the adjacent cortex after excision of the tumour.

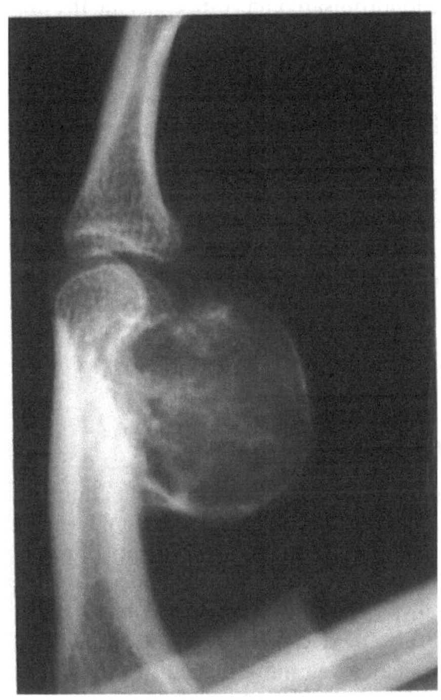

Fig. 100. Periosteal chondroma: note peripheral opacity, intratumoral calcification and erosion of diaphysial cortex

Pathology

Macroscopically, exploration shows a thinned cortex, sometimes very attenuated in "blown-out" forms. The material curetted is pearly or bluish-white, with variably remodelled lobules, in the midst of which are foci of calcification and mucoid zones.
The histological appearance is that of regular cartilaginous lobules, separated by poorly vascularised fibrous septa. The chondrocytes are normal in number and appearance, showing all the features of benign cells - small nuclei of poor chromatic density with few mitoses. They are scattered in a hyaline or myxoid stroma. The calcifications produce a granular appearance or even opaque streaks.
Periosteal chondroma presents as a rounded, lobulated, bluish-white swelling of firm consistency, sometimes adherent to the periosteum but easily detachable from the bone. The local cortex is very hard and sclerosed. The histology is much like that of intraosseous chondroma, but the chondrocytes may assume a pseudo-malignant appearance suggestive of chondromsarcoma, especially at the periphery of the tumour.

Aetiopathogenesis

Virchow (1875, cited by Alawneh et al. 1977) thought that a chondroma originated from cartilaginous remnants of the marrow. Bunnell 1948) and Geschickter and Copeland (1949) considered these cartilaginous residues to represent accessory articular cartilages. According to Bell (1941), this is a metaplasia derived from other connective tissues, a theory taken up by Noble and Lamb (1974a).

Differential Diagnosis

The radiographic appearance of osseous chondroma may be confused with that of an epidermoid cyst, a glomus tumour, a chondromyxoid fibroma, an osteoblastoma (in some very advanced forms) and, especially, a chondrosarcoma; the histological features may also be deceptive. Arriving at the diagnosis can be made even more difficult as the chondroma may become sarcomatous. The diagnosis must rest on a grouping of the clinical, radiological and histological evidence.
Periosteal chondroma may suggest an osteogenic exostosis, but even with numerous views the chondroma shows neither a base of implantation with bony trabeculae in continuity nor any cortical continuity with the healthy bone. It may also be confused with juxtacortical chondrosarcoma or even with myositis ossificans in the palm.

Development

The spontaneous course of enchondroma is slow and may evolve towards stabilisation of the lesions or, in some cases, spontaneous cure. However, chondroma may also sometimes degenerate into chondrosarcoma, though the diagnosis of such a change is often difficult. Classically, the chondroma becomes painful and suddenly increases in size. Radiographs show an increase in size of the lytic image, which becomes less confined. Cortical destruction may appear (which should not be taken for a pathological fracture in this painful context) and invasion of the soft parts. None of these signs is constant. The histology is not always decisive since not only is the distinction between these two tumours histologically difficult, but all intermediate stages exist between frankly benign chondroma and frankly malignant chondrosarcoma. Finally, the malignant change at the outset affects only a very limited area of the tumour, often missed at biopsy. Diagnosis must be based on collating the clinical, radiological and histological data. However, it should be noted that this degeneration is much rarer in the hand than at other sites of chondroma: Lansche and Spjut (1958), 1 case; Culver et al. (1975), 1 case. Takigawa (1971b) found no instance of malignant change in his series of 110 chondromata of the hand.

Treatment

Curettage is the treatment of choice for solitary enchondroma. Most authors combine this with packing the residual cavity, either with chips of cancellous bone, the most common method (Eiken and Carstam 1966; Takigawa 1971b) or with bank bone (Jewusiak et al. 1971). Some advise refraining from treatment and doing no more than applying a plaster splint in the case of fracture (Zifko 1966; Noble and Lamb 1974a, b). In very extensive forms, curettage is not as satisfactory as diaphyseal resection with reconstruction by a corticocancellous graft (Giannikas 1966).
Noble and Lamb (1974b) compared the results in 40 cases of abstention, curettage alone and curettage combined with cancellous packing. In the untreated cases (11) the result was uniformly excellent. With curettage alone (11 cases) minor symptoms persisted in 4 instances (pain, limitation of movement). When curettage was combined with cancel-

lous packing (18 cases) there were 4 instances with residual limitation of movement and 1 failure due to pseudarthrosis after fracture. However, for many others (Jewusiak et al. 1971), the theoretical risk of malignant change justifies surgical excision.
Recurrence after surgical treatment is not exceptional (Coley and Santoro 1947, 2 cases; Shellito and Dockerty 1948, 4 cases; Takigawa 1971b, 5 cases). It is generally associated with incomplete resection of the lesion and is evidenced by the reappearance of a lytic radiological image after some months. It calls for reoperation with careful histological examination of the specimen, as it must arouse the suspicion of an initial failure to recognise a chondrosarcoma.

Multiple Enchondromata

These are relatively common (Takigawa 1971b, 45%), often amounting to a double or triple localisation on the same bone ("combined" chondromata) or on the same hand ("multiple" chondromata) (Fig. 101). Their treatment is the same as that for each separate tumour (Mosher 1976). Some authors (Aegerter and Kirkpatrick 1975; Block and Burton 1977) have recorded multiple enchondromatoses strictly localised to the hand as true localised forms of Ollier's disease.

Multiple Enchondromatosis

Chondromata may form part of the picture of the rare multiple enchondromatosis of Ollier, characterised by the presence of a varying number of cartilaginous foci within bones of endochrondral origin. It is on the whole unilateral, and may be limited to two or three sites or affect an entire half of the body. Involvement of the bones of the hand is very common and troublesome (deformities, limitation of joint movements, pathological fractures). It may be associated with an ulnar club hand due to enchondroma of the lower end of the ulna. The radiographs show expansive lytic images, striped with often linear calcifications, developing from the epiphysis towards the diaphysis (Fig. 102).

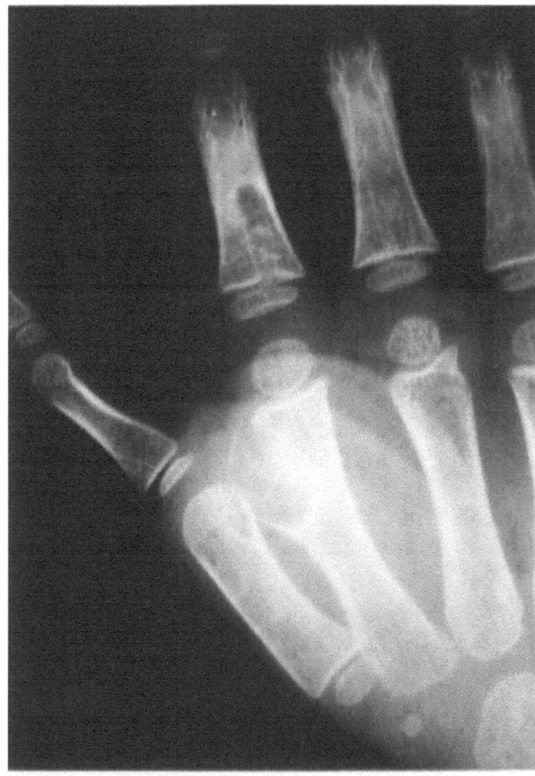

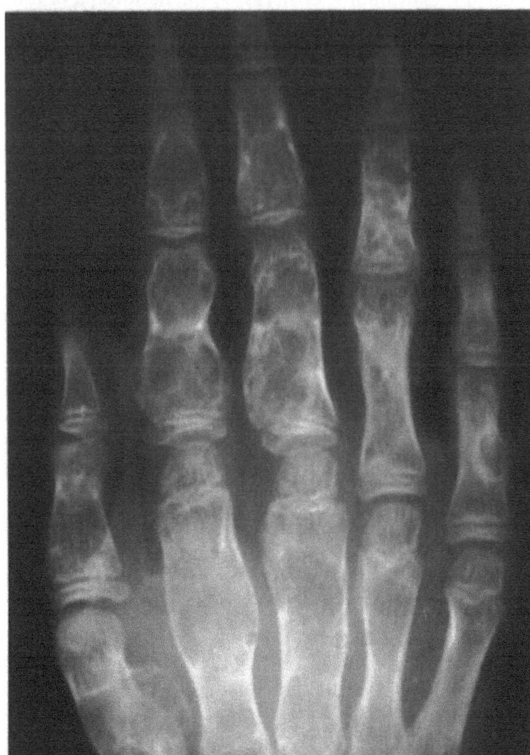

Fig. 101. Multiple chondromata: 2nd metacarpal (eccentric chondroma) and first phalanx of index finger *(above)*

Fig. 102. Multiple enchondromatoses in Ollier's disease *(right)*

The histological aspect is very much like that of solitary enchondroma, but is sometimes deceptive and resembles a chondrosarcoma. The risk of malignant degeneration is much higher than in solitary enchondroma (up to 50%). Such a change is indicated by the sudden appearance of pain and increase in size of a swelling. In the radiographs, the bone is blown-out, the cortex broken, and there is a periosteal reaction with spicules indicative of malignant change. Biopsy at the edge of the relevant tumour is called for. Whatever the circumstances, any clinical or radiological change in a lesion of Ollier's disease after the age of 20 must be considered malignant until proved otherwise. Treatment is difficult. Apart from lesions suspected of malignancy, it should always be limited to frankly troublesome chondromata, if possible waiting for skeletal maturation before intervening. In cases of severe deformity, it may be necessary to amputate (Jaffe 1958; Dahlin 1973).

Bibliography

Aegerter E, Kirkpatrick JA (1975) Orthopaedic diseases, 4th edn. Saunders, Philadelphia, p 116

Alawneh I, Giovanini A et al. (1977) Enchondroma of the hand. Int Surg 62/4: 218

Bell ET (1941) A textbook of pathology, 4th edn. Lea & Febiger, Philadelphia, p 276

Block RS, Burton RI (1977) Multiple chondrosarcomas in a hand: A case report. J Hand Surg 2: 310

Bunnell S (ed) (1948) Tumors of the hand. In: Surgery of the hand. 2nd edn. Lippincott, Philadelphia, pp 861-913

Carroll RE (1975) Tumors of the hand skeleton. In: Flynn JE (ed) Hand surgery, 2nd edn. Williams & Wilkins, Baltimore, pp 668-678

Cary GR (1965) Juxta cortical chondroma. A case report. J Bone Joint Surg [Am] 47: 1405

Coley BL, Santoro AJ (1947) Benign central cartilaginous tumors of bone. Surgery 22: 411

Culver JE, Sween DE, McCue FC (1975) Chondrosarcoma of the hand arising from a pre-existent benign solitary enchondroma. Clin Orthop 113: 128-131

Dahlin DC (1973) Bone tumors, 2nd edn. Thomas, Springfield, p 285

Eiken O, Carstam N (1966) Enchondroma in the hand. Acta Chir Scand [Suppl] 357: 148

Feinberg SB, Wilber MC (1956) Periosteal Chondroma: Report of two cases. Radiology 66: 383-386

Geschickter CF, Copeland MM (1949) Tumors of bone. 3rd edn. Lippincott, Philadelphia, p 810

Giannikas AC (1966) Treatment of metacarpal enchondromata. J Bone Joint Surg [Br] 48: 333-335

Habal MB, Snyder HH et al. (1973) Chondrosarcoma of the hand. Am J Surg 125: 775

Jaffe HL (1956) Juxta cortical chondroma. Bull Hosp Jt Dis Orthop Inst 17: 20

Jaffe HL (1958) Tumors and tumorous conditions of the bones and joints. Lea & Febiger, Philadelphia

Jewusiak EM, Spence KF, Sell KW (1971) Solitary benign enchondroma of the long bones of the hand. J Bone Joint Surg [Am] 53: 1587-1590

Lansche WE, Spjut HJ (1958) Chondrosarcoma of the small bones of the hand. J Bone Joint Surg [Am] 40: 1139

Lichtenstein L, Hall JE (1952) Periosteal chondroma. J Bone Joint Surg [Am] 34: 691-697

Mason ML (1937) Tumors of the hand. Surg Gynecol Obstet 64: 129-148

Matteson B (1975) Solitary benign enchodroma of the hand. Am J Orthop 68/4: 426-429

Merlino AF, Nixon JE (1964) Periosteal chondroma: Report of an atypical case and review of the literature. Am J Surg 107: 773

Mirra JM (ed) (1980) Bone tumors. Intra-osseous cartilage- and chondroid-producing tumors. Lippincott, Philadelphia, p 162

Mosher JF (1976) Multiple enchondromatosis of the hand. J Bone Joint Surg [Am] 58: 717-719

Mosher JF, Kettelkamp DB, Campbell CJ (1966) Intracapsular or para-articular chondroma. J Bone Joint Surg [Am] 48: 1561-1569

Noble J, Lamb DW (1974a) Enchondromata of the hand bones. Ann Chir 28: 855

Noble J, Lamb DW (1974b) Enchondromata of bones of the hand. A review of 40 cases. Hand 6: 275-284

Nosanchuk JS, Kaufer H (1969) Recurrent periosteal chondroma. J Bone Joint Surg [Am] 51: 375-380

Rockwell M, Saiter ET, Enneking WF (1972) Periosteal chondroma. J Bone Joint Surg [Am] 54: 102-108

Shellito JC, Dockerty MB (1948) Cartilaginous tumors of the hand. Surg Gynecol Obstet 86: 465-472

Takigawa K (1971a) Carpal chondroma. Report of a case. J Bone Joint Surg [Am] 53: 1601-1604

Takigawa K (1971b) Chondroma of the bones of the hand. A review of 110 cases. J Bone Joint Surg [Am] 53: 1591-1600

Tomeno B, Forest M (1974) Chondromes des membres. In: Meary R (ed) Tumeurs bénignes osseuses et dystrophies pseudo-tumorales. Monographies des annales de chirurgie. Expansion Scientifique, Paris, pp 21-26

Vaillant JM, Forest M (1974) Enchondromatose multiple. In: Meary R (ed) Tumeurs bénignes osseuses et dystrophies pseudo-tumorales. Monographies des annales de chirurgie. Expansion Scientifique, Paris, pp 27-30

Zifko B (1966) Behandlung von Frakturen in Chondromen der Hand. Monatsschr Unfallheilkd 69: 538

Chondroblastoma

Still sometimes known as "benign chondroblastoma", this is a relatively rare bony tumour (less than 1% of bony tumours according to Dahlin 1973), probably of cartilaginous origin. Though classically benign, several cases of pulmonary metastases have been reported. Location in the hand is exceptional.

Ewing, in 1923 (cited by Codman 1931), seems to have been the first to mention this lesion, which he termed "calcifying giant-cell tumor". Kolodny (1927) called it a "giant-cell tumor containing cartilage". We owe the first detailed description to Codman (1931), whose nine cases he named "epihyseal chondromatous, giant cell tumor". It was Jaffe and Lichtenstein, in 1942, who finally proposed the term of benign chondroblastoma.

Incidence

Chondroblastoma preferentially affects the long bones (humerus, femur) and more rarely the flat bones (ilium, ribs). Location in the hand is very rare. Coley and Santora (1947) reported one metacarpal lesion. Lichtenstein and Bernstein (1959) saw an "atypical" case in the first phalanx of the index finger, while Mangini (1964) published a case in the capitate. Sundaram (1966) found one lesion in the 3rd metacarpal in his six cases, while Schajowicz and Gallardo (1970) described two carpal and one metacarpal lesions. Neviaser and Wilson (1972) studied two phalangeal lesions. Mirra (1980), in his major series of 465 chondroblastomata, found 8 in the hand (2%).

Clinical Features

Chondroblastoma affects young subjects, mostly between the ages of 10 and 20 years, and males twice as often as females. It is generally evidenced in the hand by a painless swelling located anywhere in the metaphyseal diaphysis, bony in consistence and sometimes tender to pressure. Neviaser and Wilson (1972) report one case with spontaneous pain. This mode of presentation is entirely specific to the hand, for elsewhere the lesion remains small and confined to an apiphysis, is rarely palpable and tends to be painful and to limit joint motion. This last feature may also be found in the hand. There is not usually any relation to injury.

Radiology

Here again, the appearances of chondroblastoma are specific to the hand, taking the form of a clear, multilobular image situated in the metaphyseal diaphysis and often affecting the entire bone, including the epiphysis (or at least its metaphyseal aspect), with an overall increase in size of the bone and expansion of the thinned cortex (Fig. 103a). In one of the cases of Neviaser and Wilson (1972) the cortex was actually fractured (though in their other cases the image was small and round and limited to half an epiphysis). The lesion is bounded by a sharp line of sclerosed bone and is dotted with small calcified deposits. The epiphyseal plate remains active.

Pathology

Macroscopically, a chondroblastoma shows on section a central calcified nucleus, whitish and more or less necrotic, surrounded by a pink gelatinous zone, which separates it sharply from the surrounding healthy bone. But in practice the lesion is more often curetted, providing soft, reddish, semi-fluid tissue samples with greyish and haemorrhagic zones. Microscopically, the diagnosis rests on three features.

1. Two cell populations exist, composed of polyhedral cells, which are chondroblasts, and giant cells, rather different from those found in giant-cell tumours and met with more in the necrotic areas.
2. There are islands of hyaline or chondroid cartilage present away from the cellular areas. These occupy less than half the tumour, and if the opposite were the case it would suggest an enchondroma or a chondrosarcoma.
3. Calcified foci, the cardinal diagnostic feature, are found between the cells in the cellular areas. In practice, however, the diagnosis is often difficult, with resemblance to a giant-cell tumour or a chondrosarcoma.

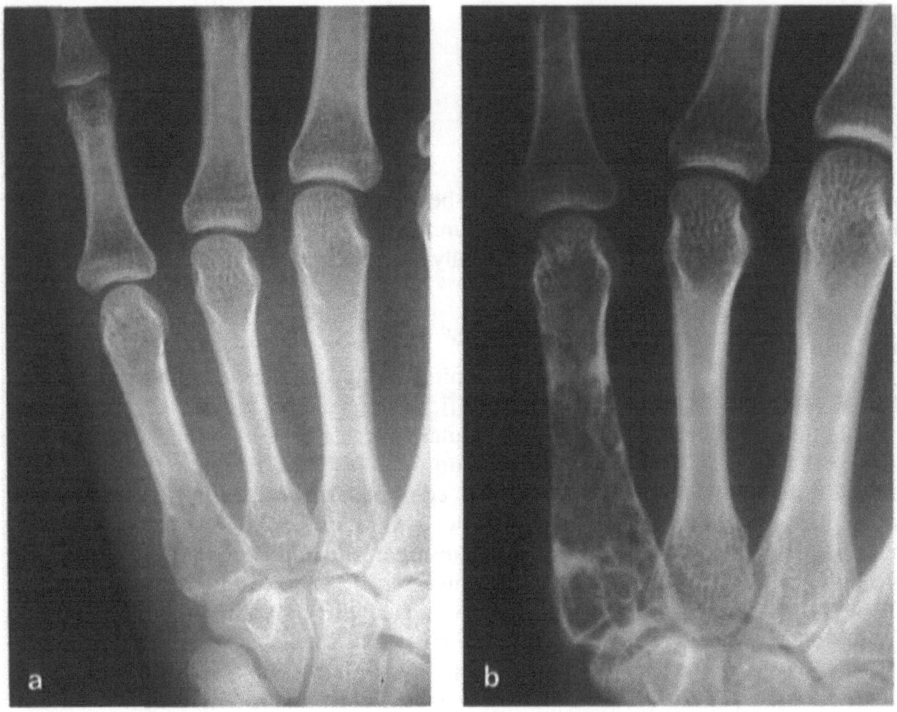

Fig. 103. ***a*** Chondroblastoma of 5th metacarpal. ***b*** Same patient: recurrence after curettage and grafting

Aetiopathogenesis

There are two opposing theories as to the origin of chondroblastoma. Codman (1931), followed by Willis (1948), regards it as a special form of giant-cell tumour. For Jaffe and Lichtenstein (1942), followed by Coley and Santora (1947) and Geschickter and Copeland (1949), the giant cells are secondary to haemorrhage and necrosis, and the tumour is of cartilaginous origin from the growth plate. Jaffe and Lichtenstein (1942) suggest that the tumour arises from the metaphyseal side of the epiphysis, while Geschickter and Copeland (1949), on the contrary, see it as starting at the epiphyseal side of the metaphysis. While this cartilaginous origin now seems generally accepted, it does not explain the development of a chondroblastoma in the flat bones. Dahlin (1956) mentions the possible relationships of chondroblastoma to chondromyxoid fibroma. Lichtenstein and Bernstein (1953) describe a number of borderline tumours between the two entities.

Development

The course of the tumour is usually slow and benign, but several cases have been reported of malignant degeneration with invasion of the soft parts and pulmonary metastases (1%-2% of cases). It is difficult to explain such a transformation when the lesion is histologically benign. Possibly, initial errors of diagnosis have mistakenly labelled a true chondrosarcoma as a chondroblastoma. It should also be noted that several of these cases of malignant change have occurred after therapeutic irradiation (Mirra 1980), casting the responsibility on the the irradiation rather than the original tumour. To our knowledge, no case of malignant change has been reported in the hand.

Differential Diagnosis

In the hand the radiographic appearances may suggest an enchondroma or aneurysmal cyst. The histology may be confused with that of a giant-cell tumour, a chondrosarcoma or a chondromyxoid fibroma.

Treatment

The treatment usually advised is curettage followed by packing with cancellous grafts. In the hand, the extent of the lesion may sometimes require disphyseal resection (including the metaphyseal aspect of the affected epiphysis), with reconstruction by a corticocancellous graft. Some authors (Valls et al. 1951; Kunkel et al. 1956) have advised radiotherapy after biopsy with apparently good results, Hatcher and Campbell (1951) reported a case of chondrosarcomatous degeneration after irradiation. Recurrences are rare, but not exceptional (some 5% of cases) and can occur in the hand: Neviaser and Wilson (1972), 1 case; and Sundaram (1966), 1 case (Fig. 103b, p. 170). They have been treated by further curettage and packing with spongy bone.

Bibliography

Barnes SN, Calandruccio RA et al. (1976) Chondroblastoma of bone: Complications and problems. J Bone Joint Surg [Am] 58: 734

Codman EA (1931) Epiphyseal chondromatous giant cell tumors of the upper end of the humerus. Surg Gynecol Obstet 52: 543

Coleman S (1966) Benign chondroblastoma with recurrent soft tissue and intra-articular lesions. J Bone Joint Surg [Am] 48: 1554-1560

Coley BL, Santora AJ (1947) Benign central cartilaginous tumors of bone. Surgery 22: 411

Dahlin DC (1956) Chondromyxoid fibroma of bone, with emphasis on its morphological relationship to benign chondroblastoma. Cancer 9: 195

Dahlin DC (1973) Bone tumors, 2nd edn. Thomas, Springfield, p 285

Geschickter CF, Copeland MM (1949) Tumors of bone, 3rd edn. Lippincott, Philadelphia, p 810

Hatcher CH, Campbell JC (1951) Benign chondroblastoma of bone - its histologic variations and a report of late sarcoma in the site of one. Bull Hosp Jt Dis 12: 411

Jaffe HL (1958) Tumors and tumorous conditions of the bones and joints. Lea & Febinger, Philadelphia

Jaffe HL, Lichtenstein L (1942) Benign chondroblastoma of bone. Am J Pathol 18: 969

Kolodny A (1927) Bone sarcoma. Surg Gynecol Obstet [Suppl] 44: 1

Kunkel MG, Dahlin DC, Young HH (1956) Benign chondroblastoma. J Bone Joint Surg [Am] 38: 817

Lichtenstein L, Bernstein D (1959) Unusual benign and malignant chondroid tumors of bone. Cancer 12: 1142-1157

Mangini U (1964) Chondroblastoma localized in the capitate bone. Bull Hosp Jt Dis Orthop Inst 25: 50

Mirra JM (ed) (1980) Bone tumors. Diagnosis and treatment. Chondroblastoma. Lippincott, Philadelphia, pp 219-233

Neviaser R, Wilson J (1972) Benign chondroblastoma in the finger. J Bone Joint Surg [Am] 54: 389-392

Schajowicz F, Gallardo H (1970) Epiphysial chondroblastoma of bone. A clinico-pathological study of 69 cases. J Bone Joint Surg [Br] 52: 205-226

Sundaram TKS (1966) Benign chondroblastoma. J Bone Joint Surg [Br] 48: 92-104

Valls J, Ottolenghi CE, Schajowicz F (1951) Epiphyseal chondroblastoma of bone. J Bone Jt Surg [Am] 33: 997

Willis RA (1948) Pathology of tumors. Butterworth, London, p 684

Wright JL, Sherman MS (1964) An unusual chondroblastoma. J Bone Joint Surg [Am] 46: 597-600

Osteochondroma (Solitary Osteogenic Exostosis)

Osteochondroma, sometimes referred to in the English language literature as "ecchondroma", is more familiar under the name of solitary osteogenic exostosis. It is a benign tumour arising from the growth plate and is so called because of its double nature, both bony and cartilaginous. Dahlin (1973) finds it responsible for about 4.5% of benign osseous tumours. It preferentially affects the metaphyses of the long bones (lower end of femur, upper end of humerus); though lesions in the hand are quite rare, it nevertheless represents the second most common cause of osseous tumours of the hand after chondroma (Carroll 1966). We do not include here either subungual exostosis or multiple exostoses, which will be described separately.

Clinical Features

Osteochondroma, which develops in bones during growth, is a lesion of young persons (10–30 years) and affects both sexes equally. It is evidenced in the great majority of cases as a painless swelling situated in the metaphyseal region of the long bones of the hand, usually distal in the metacarpals and proximal in the phalanges (base and proximal third), except for the very special case of osteochondroma of the tuft of the distal phalanx, which is discussed later. Carpal lesions are exceptional (Geschickter and Copeland 1949, 1 case in the scaphoid; Heiple 1961, 1 case in the trapezium; Medlar and Sprague 1979, 1 case in the scaphoid). The swelling may produce limitation of joint movement and even, when large, signs of vascular or nervous compression. Sometimes (10% of cases) the osteochondroma is entirely asymptomatic and discovered by chance during radiography for something else.

Radiology

Only the bony component of osteochondroma is visible radiologically, so that the image is much smaller than the clinical swelling. It presents as a bony excrescence, sometimes pedunculated (Fig. 104) and easy to see, sometimes sessile and requiring a number of views to distinguish its base of implantation and to differentiate it from an intraosseous tumour (Fig. 105). The trabeculae of spongy bone are continuous with those of the normal adjacent bone, sealing the diagnosis. The tumoral bone is often of normal density but may exhibit denser streaks of calcification. The cartilaginous cap may be visible in mature osteochondromata as a dotted line of microcalcifications suggesting its contour.

Pathology

The gross appearance of osteochondroma at surgical exploration is that of a whitish tumour with irregular, rounded contours, much larger than its radiographic image and covered with a sometimes adherent periosteum. This is the cartilaginous cap, which may conceal the tumour's base of implantation. When the specimen is opened, the peripheral cartilaginous zone is clearly seen, together with the bony base composed of an irregular spongiosa, surrounded by a fine cortex continuous with the cortex of the healthy bone.

Histologically, the diagnosis is made on examination of the cartilaginous cap. This is composed of cartilage cells of variable density; they may be binucleate but the nuclei always remain regular. In its deep part the cartilage is the site of intense vascular activity, with calcification and endochondral ossification leading to mature spongy bone in which calcified sclerosed zones persist. At the end of growth the osteochondroma undergoes changes; the cartilaginous cap thins and becomes less cellular and may even disappear. In every case the theoretical risk of chondrosarcomatous degeneration requires histological examination of the entire specimen, especially if this is large.

Aetiopathogenesis

While all authors agree that this is an osteogenic cartilaginous tumour, they differ as regards its origin. According to the theory of Virchow (1875), revived by Ambrosia and Ferguson (1968), it arises from the growth plate; Keith (1919) states that it originates from the periosteum, while it is interpret-

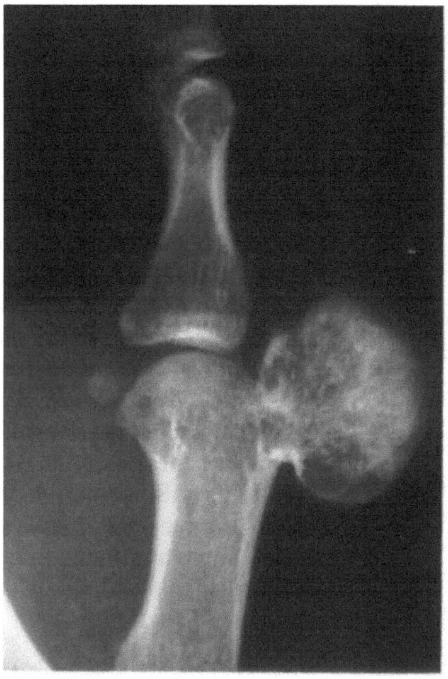

Fig. 104. Osteochondroma of 1st metacarpal: pedunculated form

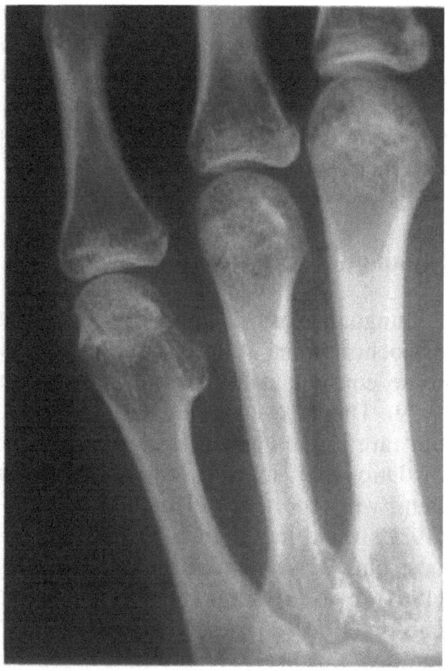

Fig. 105. Osteochondroma of 5th metacarpal: sessile form

ed by Jaffe (1958) as emanating from epiphyseal remnants.

Differential Diagnosis

The radiographic appearance may be suggestive of some other benign tumours of bone but visualisation of the base of implantation, the regular cortical dehiscence and the continuity of the trabeculae of the spongy bone virtually confirm the diagnosis. This is reaffirmed by the histological examination, while realising that this may sometimes suggest a periosteal chondroma or especially a chondrosarcoma. We shall refer to this later.

Treatment

It seems generally accepted, in the hand as elsewhere, that exostoses should only be operated on if they are a nuisance. This said, intervention is often more common in the hand because osteochondromata here soon become functionally troublesome. Insofar as it is possible, it is preferable to wait until the tumour has reached maturity before operating, in order to avoid recurrence. Excision must include the entire tumour with its base of implantation and with all the adjacent periosteum, without seeking to dissect the latter (for if this is accepted as the origin of the tumour, its preservation is a factor in recurrence).

Development

The spontaneous course of osteochondromata, as we have seen, is towards thinning and even disappearance of the cartilaginous cap at the end of growth. Chrisman and Goldenberg (1968), who were able to follow up two cases over periods of 26 and 40 years, report no progress in size of the radiographic image, with the appearance of ossification within the tumour. In 1% of cases (Jaffe 1958), however, there is evidence of malignant degeneration in the form of chondrosarcoma (or, exceptionally, osteosarcoma). One should be suspicious of an osteochondroma that occurs in adult life, is large from the outset, grows very rapidly or is painful. The radiographs are suggestive if they show peritumoral calcification or lacunae in the base of implantation (for these are cartilaginous foci penetrating in depth). Histological examination confirms the diagnosis in showing irregular nuclei with numerous binuclear forms and evidence of cytolysis, but there is often great difficulty in distinguishing between a benign exostosis and a very well-differentiated chondrosarcoma.

After operations osteochondromata may recur, and

although there are no figures valid for the hand, the proportion of recurrences seems to be of the order of 1%. We have remarked on the importance of operating for exostoses at the stage of maturity and of removing them with their periosteum to prevent recurrence.

Subungual Exostosis

Subungual exostosis represents localisation of an osteochondroma at the distal phalanx, even though some consider it a separate entity (Landon et al. 1979). These exostoses are very common in the toes but rare in the hand. They were initially described by Dupuytren in 1847, and the first lesion in the hand was reported by Hutchinson in 1857. Mason (1937) found less than a dozen cases in the literature, but 17 cases since been published: Loewenthal (1964), 1 case; Evison and Price (1966), 2 cases; Jacobson (1971), 2 cases; Matthewson (1977), 6 cases; Landon et al. (1979) found 5 cases in the hand of 44; finally, Ganzhorn et al. (1981) published 1 case which they classified separately.

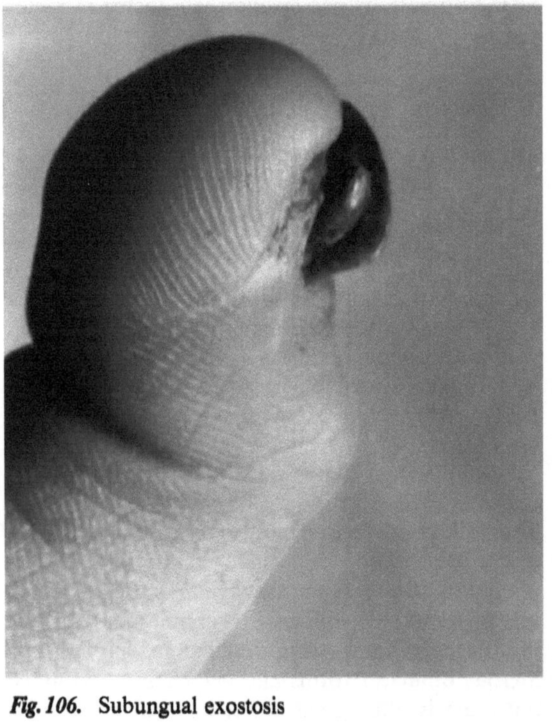

Fig. 106. Subungual exostosis

Clinical Features

Subungual exostosis may begin at any age (usually before 40 years) but advice may not be sought for a long time, perhaps several months to 20 years after the onset of symptoms. Matthewson (1977) states that it predominates between the ages of 10 and 30 years, but Dahlin (1973) found a mean age of 43 years in their 5 cases (contrasted with a mean age of 21 years for 39 lesions in the foot). The exostosis presents as a solitary swelling that is painful and elevates the tip of the nail or one of its margins and may bring about dystrophy or even ulceration of the nail (Ganzhorn et al. 1981) (Fig. 106). It may be pigmented, suggestive of melanoma. If there is ulceration, there is often associated chronic infection. It may also be periungual, respecting the integrity of the nail. Subungual exostosis predominates in the thumb and index finger, and there is a history of antecedent trauma in a quarter of the cases (Dahlin 1973).

Radiology

Diagnosis is made by the radiographic demonstration of a sessile or pedunculated excrescence of the tuft of the distal phalanx, made of trabecular bone. These trabeculae are continuous with those of the phalangeal tuft, which is itself intact. The exostosis is bounded by a cortex continuous with that of the healthy bone, save at its distal extremity, where there is no definite boundary between the spongy bone and the soft parts (Fig. 107a, b). Subungual exostosis always originates from the tuft, except in the case of Ganzhorn et al. (1981), in which it originated from the metaphysis; this led these authors to believe that, although it ulcerated the nail, it was not a true subungual exostosis but an osteochondroma of the distal phalanx.

Aetiopathogenesis

The gross and histological appearances of subungual exostosis are identical with those of osteochondroma; yet Landon et al. (1979), noting that it is never found in multiple exostoses, regard it as a separate entity. Its precise aetiology remains unexplained, although definite preceding injury is found in a certain number of cases.

Differential Diagnosis

Before the radiographic appearances are available, certain other diagnoses may come to mind: botryomycoma, onychomycosis, epithelioma of the nail bed, subungual wart, glomus tumour, or even melanoma. The films will verify the true diagnosis.

Treatment

This is the same as for metaphyseal osteochondroma, the only problem being the approach to the tumour. Matthewson (1977) advises removal of the nail, then careful elevation of the nail bed in continuity with the matrix in order to excise the exostosis. When ulceration of the nail is extensive, it may be preferable to excise the remains of the nail bed with the matrix and replace them with a full-thickness skin graft. Cure is normally to be expected if care has been taken to excise the entire cartilaginous cap; however, reccurrence was not uncommon in Dahlin's series of subungual exostoses (11%). No case of sarcomatous degeneration at this particular site has been reported to our knowledge.

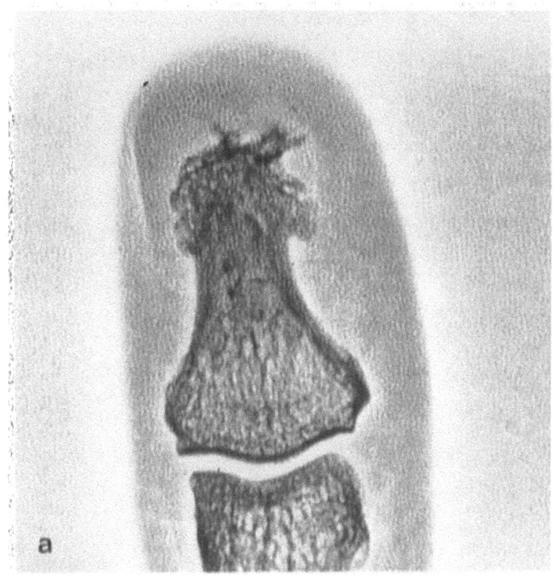

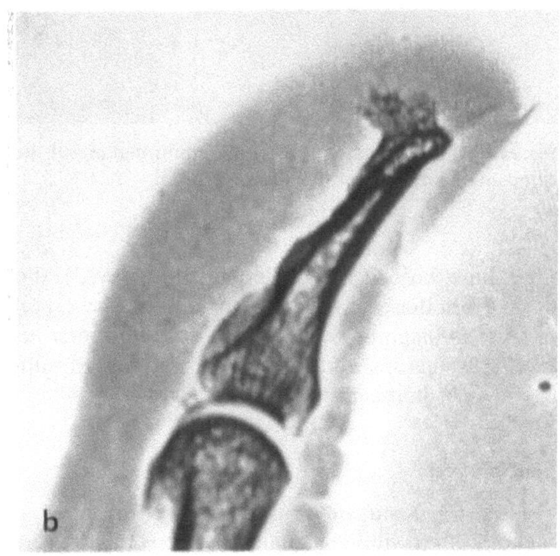

Fig. 107a, b. Subungual exostosis. ***a*** Note in frontal view that the bony trabeculae are continuous with those of the tuft. ***b*** Lateral view, showing absence of definite demarcation from soft tissues

Multiple Exostoses

Here, the multiplicity of the exostoses is the characteristic feature. This is a rare disorder of a hereditary nature, with autosomal dominant transmission. The exostoses appear in early childhood. They are situated at the extremities of bones of endochondral origin, are bilateral, more or less symmetrical and may be associated with lack of stature or various deformities.

Clinical Features

Involvement of the upper limb is very common, with special localisation to the lower ends of the ulna and radius (85% and 80% of cases, respectively, according to Shapiro et al. 1979), thus causing the typical deformity of radial club hand with restricted pronation, supination and luxation of the radial head (Bessel-Hagen disease). Involvement of the hands is somewhat less common. Apart from metaphyseal protuberances identical with those of the solitary exostosis, there may exist shortening and overall broadening of the hand, but ill effects on joint mobility are exceptional. Lesions in the carpus are extremely uncommon.

Radiology

The radiographic appearances are a little different from those of solitary exostoses; they occupy a larger part of the metaphysis and are more irregular, and they are often associated with global shortening and broadening of several metacarpals and phalanges (Fig. 108).

Pathology

Macroscopically, the exostoses are often almost entirely composed of cartilage, with very little extraosseous bone (Mirra 1980). The histological fea-

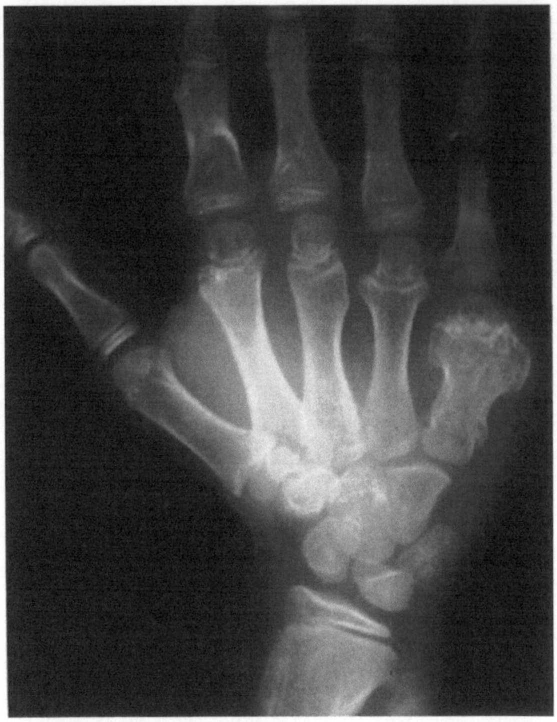

Fig. 108. Multiple exostoses: note multiphalangeal lesions and shortening of the 5th metacarpal

tures have been well detailed by Jaffe (1943) and differ from those of solitary exostoses in the extent of their cartilaginous cap, the presence of numerous calcified outcrops and the sometimes marked paucity of cells in the cartilage.

Development

The exostoses and deformities cease to progress at the end of growth. Malignant change is much more common than with solitary exostoses: from 11% (Jaffe 1943) to 33% (Vaillant and Forest 1974) of cases. It likewise takes the form of chondrosarcoma, which must routinely be suspected if there is pain, increase in size or a changed radiographic appearance - especially in the adult. Such degeneration seems more frequent in the trunk than the limbs (Vaillant and Forest 1974).

Treatment

The treatment of the various expressions of the exostoses conforms to the same rules as for solitary exostoses. The deformities may require specific treatment (osteotomy of the radius, elongation of the ulna, resection of the radial head). Given the high risk of malignant change, surveillance of the exostoses must be very regular. Any suggestive sign of such change should lead to wide biopsy or, better, to initial radical excision, given the difficulties of histological diagnosis, which calls for minute examination of the whole of the cartilaginous cap.

Bibliography

Ambrosia R de, Ferguson AB (1968) The formation of osteochondroma by epiphyseal cartilage transplantation. Clin Orthop 61: 103-115

Baran R (1978) Exostoses sous-unguéales. Ann Dermatol Venereol 105: 1075-1076

Bennett RG, Gammer S (1973) Painful callus of the thumb due to phalangeal exostosis. Arch Dermatol 108: 826

Bessel Hagen F (1891) Ueber Knochen- und Gelenkanomalien, insbesondere bei partiellem Riesenwuchs und bei multiplen cartilaginären Exostosen. Arch Klin Chir 41: 420, 505, 749, 969

Carroll RE (1966) Tumors of the hand skeleton. In: Flynn JE (ed) Hand surgery. Williams & Wilkins, Baltimore, p 1039

Chrisman OD, Goldenberg RR (1968) Untreated solitary osteochondroma. J Bone Joint Surg [Am] 50: 508

Dahlin DC (1973) Bone tumors, 2nd edn. Thomas, Springfield, p 19

Dupuytren G (1847) On the injuries and diseases of bone. Sydenham Society, London (Translated and edited by F Le Gros Clark)

Evison G, Price CHG (1966) Subungueal exostosis. Br J Radiol 39: 451-455

Forest M, Chitour S (1974) Exostoses ostéogéniques solitaires. In: Meary R (ed) Tumeurs bénignes osseuses et dystrophies pseudo-tumorales. Monographies des annales de chirurgie. Expansion Scientifique, Paris, pp 14-17

Ganzhorn RW, Bahri G, Horowitz M (1981) Osteochondroma of the distal phalanx. J Hand Surg 6: 625

Geschickter CF, Copeland MM (1949) Tumors of the bone, 3rd edn. Lippincott, Philadelphia, p 810

Heiple KG (1961) Carpal osteochondroma. J Bone Joint Surg [Am] 43: 861-864

Jacobson S (1971) The comparative pathology of the tumors of bone. Thomas, Springfield, p 94

Jaffe HL (1943) Hereditary multiple exostosis. Arch Pathol 36: 335

Jaffe HL (1958) Tumors and tumorous conditions of the bones and joints. Lea & Febinger, Philadelphia

Keith A (1919) Studies on the anatomical changes which accompany certain growth disorders of the human body. J Anat 54: 101

Landon GC, Johnson KA, Dahlin DC (1979) Subungual exostoses. J Bone Joint Surg [Am] 61: 256-259

Loewenthal K (1964) Subungual exostosis on a forefinger. NY State J Med 64: 2691

Mason ML (1937) Tumors of the hand. Surg Gynecol Obstet 64: 129

Matthewson MH (1978) Subungueal exostoses of the fingers. Br J Dermatol 98: 187-189

Medlar R, Sprague H (1979) Osteochondroma of the carpal scaphoid. J Hand Surg 4: 150

Mirra JM (1980) Bone tumors, diagnosis and treatment. Lippincott, Philadelphia, p 532

Morton KS (1964) On the question of recurrence of osteochondroma. J Bone Joint Surg [Br] 46: 723-725

Posch JL (1956) Tumors of the hand. J Bone Joint Surg [Am] 38: 517-539

Shapiro F, Simon S, Glimcher M (1979) Hereditary multiple exostoses. J Bone Joint Surg [Am] 61: 815

Solomon L (1974) Chondrosarcoma in hereditary multiple exostosis. South Afr Med J 48: 671

Vaillant JM, Forest M (1974) Maladie exostosante. In: Meary R (ed) Tumeurs bénignes osseuses et dystrophies pseudo-tumorales. Monographies des annales de chirurgie. Expansion Scientifique, Paris, p 18

Virchow R (1875) Über Entstehung des Enchondroma und seine Beziehung zur Enchondrosis und Exostosis Cartilaginea. Monatsberichte der Kgl. preuss. Akademie der Wissenschaften, S 760

Chondromyxoid Fibroma

Chondromyxoid fibroma is a benign tumour of cartilaginous origin which produces simultaneously fibrous, chondroid and myxoid tissue. It is a very rare tumour, forming only 0.4% of all bone tumours according to Mirra (1980). It was characterised by Jaffe and Lichtenstein in 1948 and may raise difficult problems of diagnosis from certain other tumours of cartilaginous origin.

Incidence

The preferential location of chondromyxoid fibroma is at the knee (upper end of tibia). It is extremely rare in the hand, there having been only 2 cases in the series of Meary (1974, 9 cases), and 3 cases in the series of Mirra (1980, 100 cases).

Clinical Features

The lesion occurs mainly in the second and third decades of life, but is sometimes seen before the age of 10 years. Both sexes are equally affected. In the long bones it remains metaphyseal and often unicortical, but the entire bone is generally affected in the case of the small tubular bones of the hand. It often manifests as a painless bony swelling, more rarely with pain of moderate intensity.

Radiology

In the hand, chondromyxoid fibroma presents as a rounded lacunar image occupying the whole diameter of the bone, whose outline may be deformed to produce a fusiform expansion. It erodes the cortex from within outwards, forming notches which may be outlined by marginal sclerosis and, at worst, the cortex may be destroyed (Fig. 109). The lacunar image may be uniform, but more often it produces a trabeculated appearance related to the peripheral lobulations.

Pathology

Biopsy yields a fairly homogeneous, yellowish-white tissue of cartilaginous consistency. The adjacent spongy bone is often sclerosed. There may be

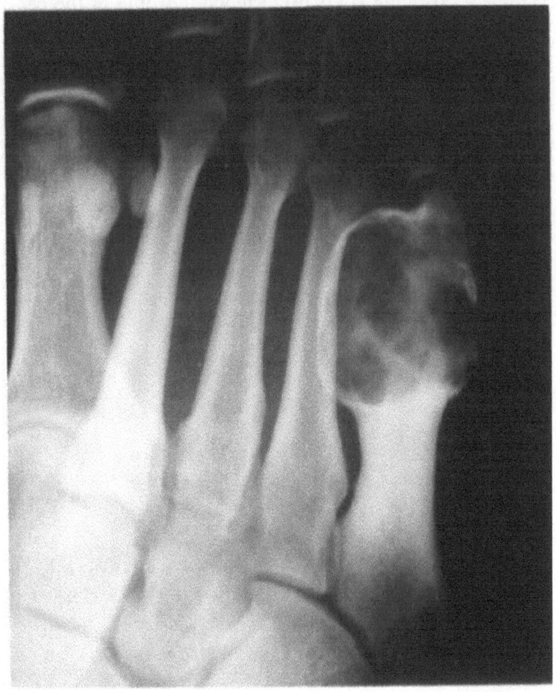

Fig. 109. Chondromyxoid fibroma of the 5th metatarsal: note cortical erosion

infraction of the cortex, with invasion of the soft parts. Histological examination confirms the presence of three cell populations - chondroid, myxoid and fibrous - in varying proportions. The chondroid tissue is arranged in lobules, with stellate or rounded cells at the periphery, separated by myxoid or fibrohyaline areas. Between the lobules there is a very cellular tissue composed of fusiform and oval cells, macrophages and plasmocytes. These cells may sometimes exhibit major nuclear anomalies, but there is little mitotic activity. The proportion of each component may vary greatly from one region of the tumour to another, calling, as always, for examination of the entire specimen.

Differential Diagnosis

The radiological appearance of chondromyxoid fibroma in the hand may be primarily suggestive of an enchondroma, less often of an aneurysmal cyst or solitary cyst. Histological analysis excludes these diagnoses but sometimes raises difficult problems of distinction from (a) chondroblastoma, the more so since the two may be combined as we shall see, and especially (b) chondrosarcoma when atypical nuclei are numerous, a diagnosis that could lead regrettably to amputation.

Development

The spontaneous course of chondromyxoid fibroma is towards slow increase in size. Malignant change (which has never been observed in the hand) has been reported only in two cases in the form of chondrosarcoma (Jaffe and Lichtenstein 1948; Iwata and Coley 1958) and in one case as fibrosarcoma (Meary 1974) in an irradiated patient. Recurrence after surgical treatment usually occurs after incomplete curettage. It may take another form, that of chondroma (Meary 1974) or, particularly a chondroblastoma, a diagnosis all the more difficult to make since the two tumours are very similar histologically and may, classically, occur together. Scaglietti and Stringa (1961) and Ralph (1962) were struck by the greater frequency of recurrence in very young subjects (under the age of 15 years).

Treatment

Curettage, followed or not by packing with cancellous grafts, remains the simplest treatment for this benign lesion but, as we have seen, it carries a risk of recurrence, especially in young subjects. This is why, for the latter, resection en bloc followed by reconstruction with a cortico cancellous graft is preferable.

Bibliography

Dahlin DC, Wells AH (1953) Chondro-myxoid fibroma of bone. J Bone Joint Surg [Am] 35: 831-834

Iwata S, Coley B (1958) Report of 6 cases of chondromyxoid fibroma of bone. Surg Gynecol Obstet 107: 571

Jaffe HL, Lichtenstein L (1948) Chondromyxoid fibroma of bone: A distinctive benign tumor likely to be mistaken especially for chondrosarcoma. Arch Pathol 45: 541-555

Languepin A, Forest M (1974) Fibromes chondromyxoïdes. In: Meary R (ed) Tumeurs bénignes osseuses et dystrophies pseudo-tumorales. Expansion Scientifique, Paris, pp 38-42

Lichtenstein L, Bernstein D (1959) Unusual benign and malignant chondroid tumors of bone. Cancer 12: 1142-1156

Meary R (1974) Tumeurs bénignes osseuses et dystrophies pseudo-tumorales. Monographies des annales de chirurgie. Expansion Scientifique, Paris, p 118

Mirra JM (ed) (1980) Bone tumors. Diagnosis and treatment. Chondromyxoid fibroma. Lippincott, Philadelphia, pp 234-243

Rahimi A, Dahlin DC et al. (1972) Chondromyxoid fibroma: A clinico-pathologic study of 76 cases. Cancer 30: 726-736

Ralph LL (1962) Chondromyxoid fibroma of bone. J Bone Joint Surg [Br] 44: 7-24

Scaglietti O, Stringa G (1961) Myxoma of bone in childhood. J Bone Joint Surg [Am] 43: 67-80

Schajowicz F (1971) Chondromyxoid fibroma of bone. J Bone Joint Surg [Br] 53: 198-216

Osteoid Osteoma

This is a benign tumour of bony origin, characterised particularly by its painful symptomatology and by its limited potential for growth. It is a relatively common lesion (Dahlin 1973, 10% of benign bone tumours), but its localisation in the hand is quite rare, assessed at 6%-10% by various authors (Jackson et al. 1977; Freiberger et al. 1959; Chaouat et al. 1976; Braun et al. 1979). Haber et al. (1965) found only 1 instance in over 2000 tumours of the hand.

The earliest known reference seems to have been that of Heine in 1927. It was Jaffe who gave a detailed description of this lesion in 1935, following this 10 years later with the report of a series of 62 cases (Jaffe 1945). Carroll (1953) was the first to take an interest in lesions in the hand and published six personal cases of this type. Since then, apart from a certain number of isolated cases, several series have been published: Dunitz and Lipscomb (1957), four cases; Tubiana et al. (1978), two carpal lesions; Braun et al. (1979), ten cases; Leclerq and Goutallier (1981), nine cases; Descamps et al. (1984), three cases, two of them carpal.

Clinical Features

Osteoid osteoma is a lesion of the young (10-30 years) but may occur in the elderly (Tubiana et al. 1978, one patient of 75 years) and affects males twice as often as females. Though there is a predilection for the long bones, it may be found virtually anywhere in the skeleton. In the hand it preferentially involves the phalanges (50%), especially the proximal and distal, the second being curiously exempt, then the metacarpals (20%-25%) and the carpus (25%-30%). The scaphoid is the most common of the carpal bones to be affected.

Pain is the classic symptom, typically continuous, very intense, localised, predominantly nocturnal, relieved by aspirin and nonsteroidal anti-inflammatory agents (Tubiana et al. 1978), but it may be absent (Lawrie et al. 1970) or rather diffuse. In the hand swelling is virtually constant in the phalangeal sites (Braun et al. 1979), but may be absent in the carpus and metacarpals. It consists of a hypertrophy of the soft parts, producing the appearance of a "teat" in the proximal phalanx and of a "barrel" in the middle phalanx (Fig. 110), while the terminal phalanx may have a clubbed appearance (Fig. 111). For a long period this swelling may remain the only feature, preceding the development of pain by several months or even years and posing numerous diagnostic problems. Carroll (1953) was the first to describe the pain provoked at a precise point, the importance of which will be referred to later. Herndon et al. (1974) reported a case of carpal tunnel syndrome due to an osteoid osteoma of the hamate bone.

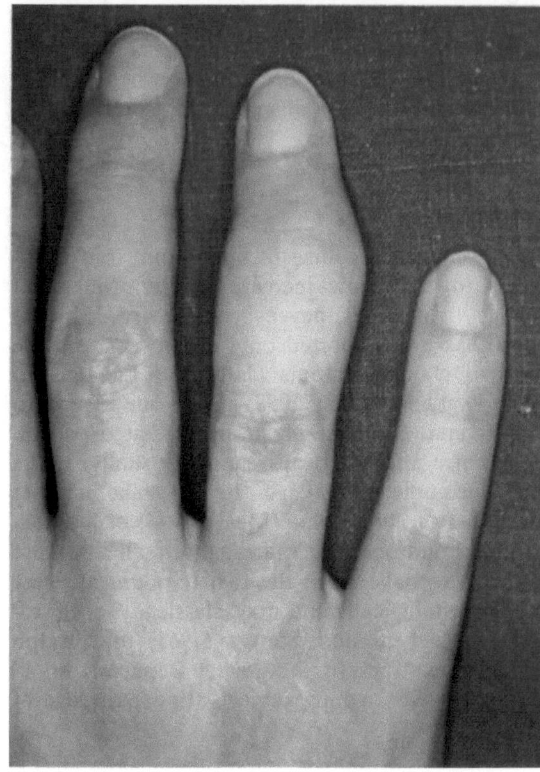

Fig. 110. Osteoid osteoma of middle phalanx: characteristic "barrel" appearance

Fig. 111. Osteoid osteoma of distal phalanx (compare with Fig. 110): global hypertrophy of phalanx and pseudo-clubbing

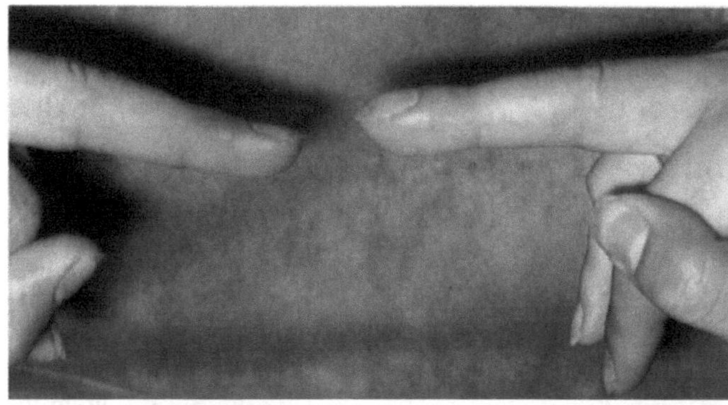

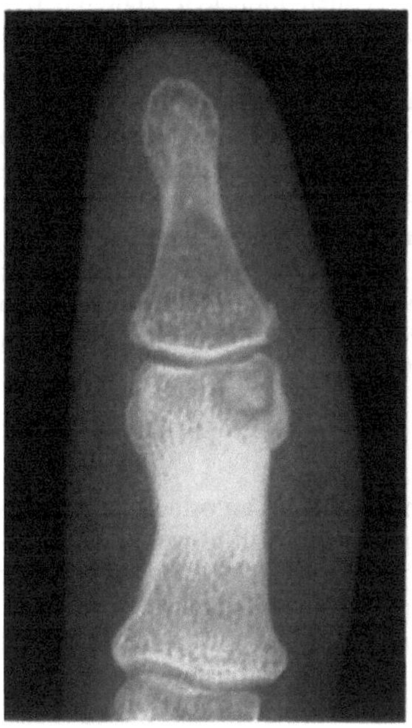

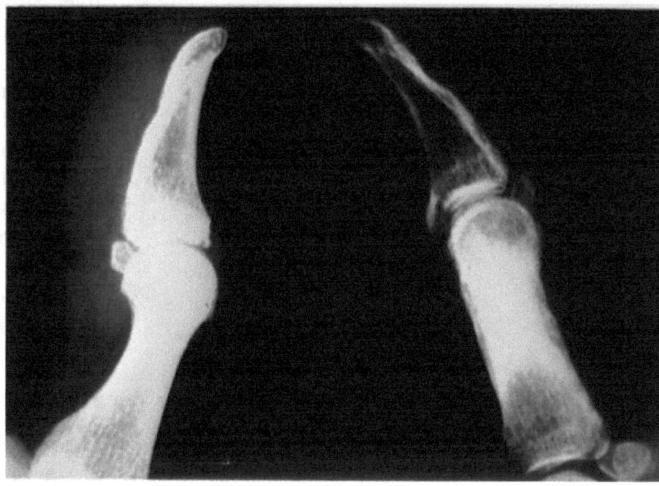

Fig. 113. Osteoid osteoma of first phalanx of thumb: comparative films do not show the nidus, but global hypertrophy of the phalanx

Fig. 112. Osteoid osteoma of middle phalanx: classic radiographic appearance of "nidus" *(left)*

Radiology

It is the discovery of the radiographic nidus that confirms the diagnosis of osteoid osteoma, but it is sometimes very difficult to locate this nidus in the hand. Typically, *cortical* osteoid osteoma is evidenced by a central, rounded, radiolucent zone of a diameter less than 1 cm, the nidus (Fig. 112). This is surrounded by a dense zone of sclerosis with localised cortical thickening, the osteosclerotic reaction sometimes being such as to obscure the nidus. Several views around the bone are then required. Within the nidus there may be a central calcified image, the nucleus, producing the classic roundel or rosette appearance. Sometimes there is a periosteal reaction opposite the lesion. However, if the osteoma is *cancellous* (carpus), the diagnosis may be much more difficult, for the nidus may be absent and the sclerosis is less dense and especially less thick, even reduced to a narrow, opaque band. It is even more difficult with a *subperiosteal* osteoma, for there is little or no peripheral sclerosis, and the nidus raises the periosteum. Finally, certain appearances are very deceptive: global hypertrophy of the bony segment (Braun et al. 1979) (Fig. 113) or condensation that is not peripheral but at a distance from the nidus. In our series of nine cases (Leclercq and Goutallier 1981), the radiological, nidus was absent six times. In difficult cases, help may be afforded by tomography which allows localisation of the nidus. In our experience, radiographic enlargements have repeatedly proved useful.

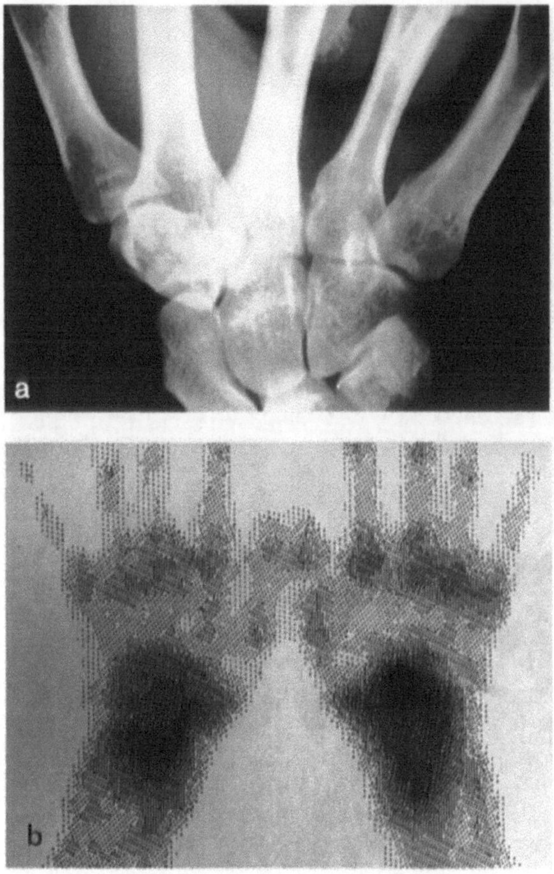

Fig. 114. ***a*** Osteoid osteoma of the base of the 3rd metacarpal: simple metaphyseal thickening. ***b*** Same patient: scintigraphy confirms site of lesion *(right)*

Scintigraphy

Bone scintigraphy with technetium gives a zone of intensification at an osteoid osteoma (Fig. 114) and is doubly important. Although nonspecific, it confirms the bony origin of the lesion. It is also valuable in localisation, particularly in the carpus (Bohne et al. 1975) where it sometimes permits precise identification of the affected bone (Tubiana et al. 1978). It is sometimes negative (Leroy et al. 1980, two cases).

Arteriography

This has been advised by Lindblom et al. (1960) and O'Hara et al. (1975) in the diagnosis of osteoid osteoma. It gives an appearance of a constant hypervascularisation of the soft parts, with early opacification of the nidus, but this hypervascularisation is not always found at the nidus itself. Arteriography sometimes proved negative in our experience (one case) as in that of others (Tubiana et al. 1978). At all events, it is justified only when the other complementary investigations are negative.

After all this assessment, one may have no more than a suspicion of the diagnosis, without precise arguments in favour of osteoid osteoma (two cases in our experience). The clinical symptomatology is profuse, but ordinary X-ray examinations and tomograms are negative or nonspecific, while scintigraphy, even if positive, is not decisive. The arteriogram may be difficult to interpret or may be negative. Surgery is then indicated, centered over the painful area. If the diagnosis of osteoid osteoma has been put forward, the endeavour should be to perform resection of the suspicious zone in its entirety, rather than simple biopsy. It is with lesions in the proximal part of the metacarpus and in the carpus that the search for a precisely localised tender point, a search made patiently with a blunt-tipped instrument, millimeter by millimeter, proves so valuable, for it helps to decide the exact site of the lesion and how best to approach it (Leclercq and Goutallier 1981).

Pathology

The gross appearance of an osteoid osteoma is that of a rounded, reddish-brown (mulberry-like; Fig. 115) or else greyish focus, rarely exceeding 1 cm in diameter, surrounded by a zone of sclerosed bone that may be very thick in cortical regions and so hinder identification of the nidus.

Microscopically, in the midst of young connective tissue with abundant fibroblasts and capillaries, there are differentiated bony cells, osteoblasts and osteoclasts; there is also osteoid tissue arranged in trabeculae that calcify progressively during maturation of the osteoma. The perifocal sclerosis consists of compact haversian bone.

Aetiopathogenesis

It was formerly considered that osteoid osteoma was a minor form of osteomyelitis (Compere 1932; Brailsford 1953), but since the work of Jaffe (1945) all authors agree that it is a true neoplasm, benign and of bony origin. It is difficult to account for the severe pain characteristic of this lesion. It seems to be related to an increase of tension and of oedema within the tumour. Schulman and Dorfman (1970) have demonstrated the existence of perivascular nerve endings within the nidus itself.

Development

Several authors report cases of tumours corresponding to osteoid osteoma which spontaneously regress after a course of 15 years (Sherman 1947; Moberg 1951; Vickers et al. 1959), but no opinion can be reached in the absence of histological confirmation.

Differential Diagnosis

We have remarked on the frequent difficulty in diagnosing osteoid osteoma. In our experience from 6 months to 5 years have elapsed between the onset of symptoms and diagnosis (average, 20 months). It is clear that numerous other diagnoses may come to mind: glomus tumour, in the absence of radiographic evidence; osteitis, because of the sclerosis; or Brodie's abscess, because of the cavity in the bone. The histological appearance may suggest a benign osteoblastoma (Jackson et al. 1977; p. 197).

Treatment

The diagnosis of osteoid osteoma may be difficult, but its treatment is simple and well established. Curettage often risks recurrence (39 cases to be found in the literature), and resection of the lesion in its entirety is preferable with removal of the whole of the nidus. We have seen that clinical examination permits very precise localisation of the tumour, and this is assisted by preoperative radiographic guidelines. Excision of a small slab of bone including the nidus with the aid of a hammer and chisel seems the technique of choice. Radiography of the excised specimen will confirm that it includes the entire nidus. The immediate postoperative course is marked by the spectacular cessation of the pain, and any persistence of painful phenomena must suggest incomplete excision. Further, the return of pain signals the recurrence (or the persistence) of the tumoral process. It would seem that, in every case in which ablation has been formally complete, cure has been obtained.

In certain juxtarticular sites, it may be necessary to remove part of the cartilaginous surface with the osteoid osteoma, particularly with condylar lesions (as in our one case of resection of the medial condyle of the proximal phalanx of the thumb, with residual limitation of mobility of the interphalangeal joint).

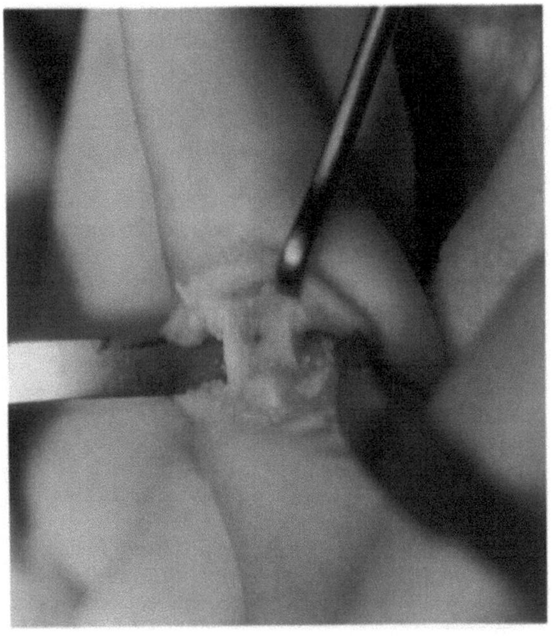

Fig. 115. Osteoid osteoma: gross appearance showing "mulberry-like" focus

Bibliography

Azar CA (1978) Painless osteoid osteoma in phalanx of finger. Bull Hosp Jt Dis Orthop Inst 39: 86-91

Bohne V, Levine DB et al. (1975) Scintimetric diagnosis of osteoid osteoma of the carpal scaphoid bone. Clin Orthop 107: 156-158

Bordelon RL, Cracco A et al. (1975) Osteoid osteoma producing premature fusion of the epiphysis of the distal phalanx of the big toe. J Bone Joint Surg [Am] 57: 120

Brailsford JF (1953) The radiology of bones and joints, 5th edn. Williams & Wilkins, Baltimore

Braun S, Chevrot A, Tomeno B et al. (1979) A propos des ostéomes ostéoïdes phalangiens. Rev Rhum Mal Osteoartic 46: 225-233

Carroll RE (1953) Osteoid osteoma in the hand. J Bone Joint Surg [Am] 35: 888-893

Carroll RE (1966) Tumors of the hand skeleton. In: Flynn JE (ed) Hand surgery. Williams & Wilkins, Baltimore, pp 1033-1035

Chaouat Y, Kanovitch B et al. (1976) L'ostéome ostéoïde. A propos de 5 observations. Rev Rhum Mal Osteoartic 43: 297-303

Compere EL (1932) Streptococcus viridans osteomyelitis. J Bone Joint Surg 14: 264

Dahlin DC (1973) Bone tumors, 2nd edn. Thomas, Springfield, p 285

Debeyre F, Touzard R (1974) Traitement chirurgical de 48 cas d'ostéome ostéoïde. Acta Orthop Belg 40: 34-44

Descamps L, Jarsaillon B, Schuster P, Vergnat C (1984) Les ostéomes osteoïdes de la main et du pied - 8 observations. Compte rendu de société. Rev Chir Orthop 70: 347

Dunitz NL, Lipscomb PR (1957) Osteoid osteoma of the hand and wrist. Am J Surg 94: 65-69

Eideken J, de Palma AF et al. (1966) Osteoid osteoma. Roentgenographic emphasis. Clin Orthop 49: 201-206

Fotoohi Giam G, Bora W (1978) Osteoid osteoma of the carpal bones. J Hand Surg 3: 280-283

Freiberger RH, Loitman BS et al. (1959) Osteoid osteoma. A report on 80 cases. AJR 82: 194

Gagnon JH (1978) Osteoid osteoma in cancellous bone. J Can Assoc Radiol 61: 399-403

Giannikas A, Papachristou G et al. (1977) Osteoid osteoma of the terminal phalanges. Hand 9: 295-300

Golding JSR (1954) The natural history of osteoid osteoma. J Bone Joint Surg [Br] 36: 218-227

Grundberg AB (1977) Osteoid osteoma of the thumb: Report of a case. J Hand Surg 2: 266-267

Haber MH, Alter AH, Wheelock MC (1965) Tumors of the hand. Surg Gynecol Obstet 121: 1073-1080

Heine J (1927) Einheilender Knochensequester an der Grundphalanx des Ringfingers. Arch Klin Chir 146: 737

Herndon JH, Eaton RG et al. (1974) Carpal tunnel syndrome: An unusual presentation of osteoid osteoma of the capitate. J Bone Joint Surg [Am] 56: 1715-1718

Hundley JD (1976) Osteoid osteoma of the trapezium. First case report of Rx-demonstrated progression in the trapezium. Clin Orthop 116: 170-172

Jackson RP, Reckling FW et al. (1977) Osteoid osteoma and osteoblastoma. Clin Orthop 128: 303-311

Jaffe HL (1935) Osteoid osteoma: A benign osteoblastic tumor composed of osteoid and atypical bone. Arch Surg 31: 709-728

Jaffe HL (1945) Osteoid osteoma of bone. Radiology 45: 319

Jaffe HL, Mayer L (1932) An osteoblastic osteoid tissue forming tumor of a metacarpal bone. Arch Surg 24: 550-564

Lamb DW, Del Castillo F (1981) Phalangeal osteoid osteoma in the hand. Hand 13: 291-295

Lawrie TR, Aterman K et al. (1970) Painless osteoid osteoma. J Bone Joint Surg [Am] 52: 1357-1363

Leclercq C, Goutallier D (1981) Communication personnelle. Congrès de la Société française de Chirurgie de la Main, Luxembourg

Leroy V, Couturaud M et al. (1980) La scintigraphie osseuse est-elle si fiable dans la recherche de l'ostéome ostéoïde? Rev Rhum Mal Osteoartiç 47: 53-56

Lindbom A, Lindvall N et al. (1960) Angiography in osteoid osteoma. Acta Radiol [Oncol] 54: 327-333

Martorell F (1964) Sudek's atrophy due to osteoid osteoma of the carpal navicular. Angiologia 16: 170-172

Moberg E (1951) The natural course of osteoid osteoma. J Bone Joint Surg [Am] 33: 166-170

Norman A (1978) Persistance or recurrence of pain: A sign of surgical failure in osteoid osteoma. Clin Orthop 130: 263-266

O'Hara JP, Tegtmeyer C et al. (1975) Angiography in the diagnosis of osteoid osteoma of the hand. J Bone Joint Surg [Am] 57: 163-166

Phelip X, Mouries D et al. (1978) Les manifestations articulaires de l'ostéome ostéoïde. Essai de classification. Rhumatol 30: 219-226

Raspall JT, Llort MC (1935) Une lésion rare du scaphoïde carpien. Rev Chir Orthop 22: 53-57

Rigault P, Mouterde P et al. (1974) Osteome ostéoïde chez l'enfant. A propos de 29 cas. Rev Chir Orthop 61: 627-646

Rosborough D (1966) Osteoid osteoma: Report of a lesion in a terminal phalanx of a finger. J Bone Joint Surg [Br] 48: 485-487

Rosenfeld K, Bora W et al. (1973) Osteoid osteoma of the hamate. J Bone Joint Surg [Am] 55: 1085-1087

Schajowicz F, Aiello CL, Slullitel I (1972) Cystic and pseudocystic lesions of the terminal phalanx with special reference to epidermoid cyst. Clin Orthop 68: 84-92

Schulman L, Dorfman HD (1970) Nerve fibers in osteoid osteoma. J Bone Joint Surg [Am] 52: 1351-1356

Sezurier E, Simonin JL et al. (1976) Osteome ostéoïde avec synovite. A propos de 2 observations. Rev Rhum Mal Osteoartic 43: 521-526

Sherman MS (1947) Osteoid Osteoma. J Bone Joint Surg 29: 918

Sherman M, McFarland G Jr (1965) Mechanism of pain in osteoid osteoma. South Med J 58: 163-166

Sim FH, Dahlin DC et al. (1975) Osteoid osteoma: Diagnostic problems. J Bone Joint Surg [Am] 57: 154-158

Spinner M et al. (1972) Osteoid osteoma of the hamate. Bull Hosp Jt Dis Orthop Inst 33: 8-14

Sullivan M (1971) Osteoid osteoma of the fingers. Hand 3: 75-80

Swee R, McLeod RA et al. (1979) Osteoid osteoma: Detection, diagnosis and localisation. Radiology 130: 117-123

Tubiana R, Menkes CJ et al. (1978) Localisation carpienne de l'ostéome ostéoïde. Rev Rhum Mal Osteoartic 45: 133-135

Verheugen P (1969) Ostéome ostéoïde du trapézoïde. J Belg Rhum Med Phys 24: 316-321

Vickers CW, Pugh DC et al. (1959) Osteoid osteoma: A 15 years follow-up of un untreated patient. J Bone Joint Surg [Am] 41: 357-358

Benign Osteoblastoma

Benign osteoblastoma, specified as such by Lichtenstein in 1956, bears much clinical and histological resemblance to osteoid osteoma. Diagnosis can often be made from the radiological appearances. Despite the appellation "benign", several cases of malignant transformation have been described.

Previously called osteogenic fibroma or ossifying fibroma (Lichtenstein 1952) or giant osteoid osteoma (Dahlin and Johnson 1954), benign osteoblastoma was characterised simultaneously by Lichtenstein (1956) and Jaffe (1956). Several small series have been published since, but it is not a very common lesion (1% of primary tumours of bone). Its preferential site is in the spine (40%). Jackson et al. (1977), in a review of the English language literature, found 184 cases of which 15 (8%) were in the hand. Lichtenstein and Sawyer (1964) published 4 cases in the hand, Mosher and Peckham (1978) 1 case and Chaise and Witvoet (1983) 1 case.

Clinical Features

In the hand this lesion occurs predominantly in the metacarpals (70%). Carpal (18%) and phalangeal (12%) lesions are rarer. Males are affected two or three times as often as females, and usually between the ages of 10 and 30 years. The chief clinical features are pain, often less defined and intense than that of osteoid osteoma and without an exquisitely tender point and without nocturnal exacerbation, and a palpable, expansive swelling that may increase to segmental hypertrophy of the bone.

Radiology

Radiography shows a circumscribed lytic image, sometimes granular, that may become quite dense. It expands, but normally remains bounded by a layer of cortical bone (absent in the case of Chaise and Witvoet 1983). It is usually diaphyseal but rapidly extends to both ends of the bone until the entire bone is invaded. There is no cortical thickening or periosteal reaction.

Pathology

Surgical exposure reveals as well-circumscribed, osseous tumour, surrounded by a layer of young cortical bone and sometimes by a thickened periosteum. When opened, the tumour is red and friable and dotted with calcifications and haemorrhages. There are sometimes several confluent multicentric foci. Microscopic examination reveals richly vascularised, connective tissue that contains a great number of osteoblasts without atypical nuclei or excess of mitoses, some osteoclasts, osteoid seams forming lamellar bone of irregular architecture and some giant cells.

Differential Diagnosis

Benign osteoblastoma and osteoid osteoma are closely allied in several aspects and have been long confused. However, there are some features that usually make it possible to distinguish them. Pain is less intense in osteoblastoma and is not worse at night, and there is no very localised tender point. The swelling is purely osseous in osteoblastoma, while it lies more in the soft parts in osteoid osteoma. Radiologically, the diameter of osteoid osteoma rarely exceeds 1 cm, while osteoblastoma is bigger; the reactive sclerosis is much denser and more extensive in osteoid osteoma. In osteoblastoma, the histological examination typically shows many more osteoblasts, more profuse vascularisation, more regular osteoid seams and less systematised calcification. Histologically the diagnosis may be extremely difficult, and the distinction must often be made on joint anatomical, radiological and clinical evidence.

The other diagnoses to be eliminated include aneurysmal cyst, also expansive with some radiological resemblances, and cartilaginous tumours (enchondromata). In all cases the operative appearance and the histological findings correct the diagnosis.

Clinical Types

Yoshikawa et al. (1977) reported two cases of vitamin D deficiency induced by a benign osteoblastoma, one situated on the 4th metacarpal. These "rachitic osteomalacias" associated with a neoplasm have been known since a report by Prader et al. in 1959, but these seem to have been the first cases in which osteoblastoma was involved.

Treatment

The two types of treatment normally advised are curettage (with or without cancellous packing) and resection in entirety (with or without bone grafting), which seems the sole guarantee of definitive cure. Of 184 cases collated from the literature (including all sites), Jackson et al. (1977) noted 9% recurrences, all appearing after curettage. In the hand, bony reconstruction may be accomplished by use of an iliac or a metatarsal graft. The only indication for radiotherapy occurs when surgical access to the lesion is dangerous (e.g. spine).

Development

Over the past 15 years (since Mayer 1967), several reports have appeared in the literature of the malignant transformation of a benign osteoblastoma into an osteosarcoma. This may consist of local malignancy without metastasis (Mayer 1967; Schajowicz and Lemos 1976; Jackson 1977) or a full osteosarcomatous picture with metastases (Lichtenstein 1972; Seki et al. 1975).

Bibliography

Chaise F, Witvoet J (1983) Les ostéoblastomes bénins à la main. Ann Chir 2: 270

Dahlin DC, Johnson EW Jr (1954) Giant osteoid osteoma. J Bone Joint Surg [Am] 36: 559

Jackson RP, Reckling FN et al. (1977) Osteoid osteoma and osteoblastoma. Clin Orthop 128: 303-311

Jaffe HL (1956) Benign osteoblastoma. Bull Hosp Jt Dis Orthop Inst 17: 141

Lichtenstein L (ed) (1952) Osteogenic fibroma of bone. In: Bone tumors. Mosby, St. Louis, p 82

Lichtenstein L (1956) Benign osteoblastoma. A category of osteoid and bone forming tumor other than classical osteoid osteoma which may be mistaken for giant cell tumor or osteogenic sarcoma. Cancer 9: 1044

Lichtenstein L (ed) (1972) Bone tumors, 4th edn. Mosby, St. Louis, p 118

Lichtenstein L, Sawyer W (1964) Benign osteoblastoma. Further observations and report of 20 additional cases. J Bone Joint Surg [Am] 46: 755-765

Marsh B et al. (1975) Benign osteoblastoma: Range of manifestations. J Bone Joint Surg [Am] 57: 1-9

Mayer L (1967) Malignant degeneration of so-called benign osteoblastoma. Bull Hosp Jt Dis Orthop Inst 28: 4

Merryweather R, Middlemiss J (1980) Malignant transformation of osteoblastoma. J Bone Joint Surg [Br] 62: 381-384

Mosher J, Peckham A (1978) Osteoblastoma of the metacarpal: A case report. J Hand Surg 3: 358

Prader A, Illig R, Uhlinger E, Stalder G (1959) Rachitis infolge Knochentumors. Helv Paediatr Acta 14: 554

Schajowicz F, Lemos C (1976) Malignant osteoblastoma. J Bone Joint Surg [Br] 58: 202

Seki T, Fukoda H, Ishii Y et al. (1975) Malignant transformation of benign osteoblastoma. J Bone Joint Surg [Am] 57: 424

Tomeno B (1974) Osteomes ostéoïdes et ostéoblastomes. In: Meary R (ed) Tumeurs bénignes osseuses et dystrophies pseudo-tumorales. Monographies des annales de chirurgie. Expansion Scientifique, Paris, pp 44-55

Yoshikawa S, Nakamura T et al. (1977) Benign osteoblastoma as a cause of osteomalacia. J Bone Joint Surg [Br] 59: 279-286

Fibrous Dysplasia

As its name indicates, fibrous dysplasia is not a tumour properly speaking. It rather represents an invasion of the bone by connective tissue of bony origin. It exists in two forms: monostotic (a solitary lesion), extremely rare in the hand, and polyostotic, in which the skeleton of the hand is very often affected.

Monostotic Forms

To our knowledge, only two cases have been reported in the hand: a metacarpal lesion described by Ross and Vitale (1955) and a phalangeal lesion reported by Mirra (1980). No localisation in the hand was reported in the series of Schlumberger (1946), 67 cases; Harris et al. (1962), 13 cases; or the Meary group (Mechin and Forest 1974), 32 cases.

Clinical Features

The lesion is revealed by a regular swelling of the bone (Ross and Vitale 1955), pain or pathological fracture. It tends to affect young adults (20-30 years), but the patient of Ross and Vitale (1955) was aged 6 years at the first consultation.

Radiology

Radiography displays an intraosseous lacunar image that is usually very clear but may sometimes have a "frosted glass" appearance, associated with the presence of fine intracavitary bony seems. This sharply outlined image is bounded by a thinned cortex. It respects the growth plate but may invade the epiphyses in the adult. Except in the case of a fracture, there is no periosteal reaction.

Differential Diagnosis

The lacunar image is most likely to simulate a solitary cyst, but an aneurysmal cyst may also come to mind if the trabeculation is marked. Biopsy confirms the diagnosis.

The relations between fibrous dysplasia and ossifying fibroma remain very ill-defined. Although the ossifying fibroma affects the mandible almost exclusively, histological examination shows the presence of osteoblasts (absent in fibrous dysplasia), and its course is more aggressive (recurrence and spread), many authors consider the two as one and the same lesion. Mirra (1980), for instance, regards ossifying fibroma as an early form of fibrous dysplasia.

Pathology

Beneath a thinned but intact cortex biopsy reveals a greyish-white heterogeneous tissue. Haemorrhagic or oedematous remodelling leading to cystic cavities is rare in the monostotic form. Histological findings show the predominance of fusiform or oval cells in sheets that build an irregular connective tissue framework. In the midst of this there are found osteoid outcrops "embedded like marbles in the collagen tissue" (Mechin and Forest 1974). Some cartilaginous foci are occasionally seen at the periphery.

Development

The course is usually benign, the rare osteosarcomatous transformations (1%) affecting mainly the cranial and maxillary lesions.

Treatment

In the absence of any dysfunction, it may be enough to perform a diagnostic biopsy. The severity of the pain or swelling may justify surgical treatment, either curettage or resection and grafting (Ross and Vitale 1955).

Polyostotic Forms

The polyostotic form of fibrous dysplasia is usually discovered in the first or second decade of life and is manifested most often by a pathological fracture and also by pain and skeletal deformities (genu valgum, unequal leg length). The bony lesions may be the only features or may be associated with distur-

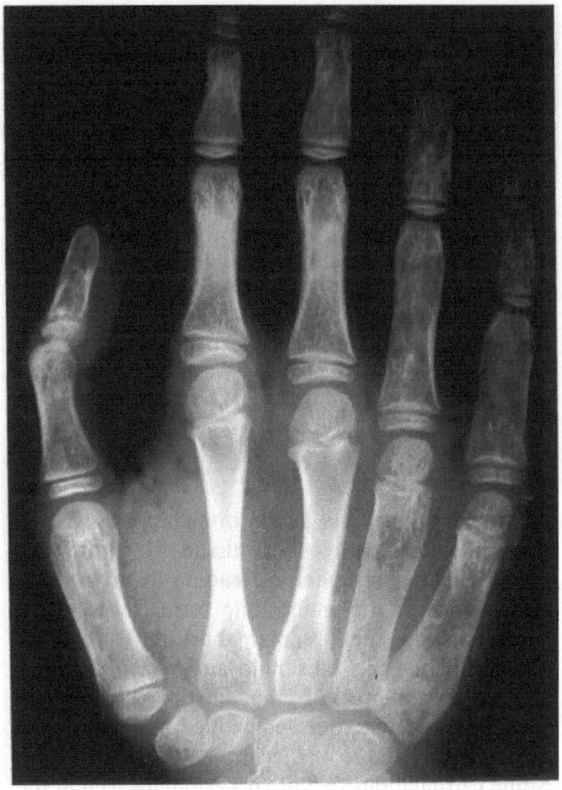

Fig. 116. Fibrous dysplasia: polyostotic form. Note length of the third and fourth rays

bances of skin pigmentation (café-au-lait spots), early skeletal maturation and precocious puberty in girls, the whole constituting the syndrome of Albright et al. (1937). Involvement of the hand is only exceptionally the initial presenting feature in polyostotic fibrous dysplasia and has little effect on the prognosis, which is chiefly linked to the progressive deformities of the limbs and trunk. It deserves mention, though, because of its frequency and characteristic appearance.

The detailed studies of Harris et al. (1962) and of Mirra (1980) assess the distribution of hand lesions as follows: 50%-55% in the metacarpals, 35%-40% in the phalanges, 8%-10% in the carpus. These three sites may or may not be concomitant. The radiographic appearances are quite typical, showing global broadening of the metacarpal and phalangeal diaphyses, with the very characteristic "frosted glass" appearance of the bony texture (Fig. 116).

Bibliography

Albright F, Butler AM et al. (1937) Syndrome characterized by osteitis Fibrosa Disseminata, areas of pigmentation and endocrine dysfunction, with precocius puberty in females. N Engl J Med 216: 727-746

Harris WH, Dudley R, Barry RJ (1962) The natural history of fibrous dysplasia. J Bone Joint Surg [Am] 44: 207-233

Lichtenstein L, Jaffe HL (1942) Fibrous dysplasia of bone. Arch Pathol 33: 777-816

Mechin JF, Forest M (1974) Fibromes ossifiants et dysplasie fibreuse. In: Meary R (ed) Tumeurs bénignes osseuses et dystrophies pseudo-tumorales. Monographies des annales de chirurgie. L'Expansion Scientifique, Paris, pp 65-75

Mirra JM (1980) Bone tumors. Diagnosis and treatment. Lippincott, Philadelphia, p 122

Ross D, Vitale C (1955) Monostotic fibrous dysplasia of the metacarpal: A case report. J Bone Joint Surg [Am] 37: 196-200

Schlumberger HG (1946) Fibrous dysplasia of single bones (Monostotic fibrous dysplasia). Mil Surg 99: 504-527

Giant-Cell Tumour

This osseous tumour remains a mystery in many respects. In particular, it poses the problem of its barely predictable course, with the possibility of malignant transformation. Giant-cell tumours have long been known and were clearly defined by Jaffe et al. in 1940, who established a classification in three grades. Very many cases have been published since, in particular the series of Coley et al. in 1958 (108 cases), Goldenberg in 1970 (222 cases), Dahlin et al. in 1970 (195 cases) and Meary et al. in 1975 (85 cases). Localisation of this tumour in the hand is quite rare. Averill et al. (1980) collected 20 large series, making a total of 1288 tumours, of which 31 were situated in the hand (2%). They presented a personal series of 28 cases, the largest recently published. Gupta et al. (1980) mentioned 6 metacarpal lesions in the series of 108 giant-cell tumours of Reddy et al. (1974).

Clinical Features

Giant-cell tumours most often affect young adults (20–40 years), with a slight female predominance. They are very rare before the age of 15 years (Minguella 1982). The average age seems to be a little earlier in the hand than at other sites (Averill et al. 1980). The tumour is usually revealed by a localised, dull, persistent pain, which may be unleashed by exertion. Swelling is less often the primary feature, but is often present at the time of first examination. Averill et al. (1980) note that patients with this tumour in the hand sought advice on average markedly earlier than those with lesions elsewhere. The swelling is hard, of bony consistence and tender to pressure. It may be accompanied by a minor restriction of joint movement. Sometimes the presenting feature is a pathological fracture. Distribution of the tumour in the hand occurs essentially in the long bones: phalanges 50%, metacarpals 45%. Carpal lesions are very rare (Fitzpatrick and Bullough 1977). Averill et al. (1980) are impressed by the frequency of multifocal lesions in afflictions of the hand (18%) and advise a routine search for these by bone scintigraphy using technetium.

Radiology

Although no absolutely pathognomonic image exists (Goldenberg et al. 1970), several features are suggestive of giant cell tumour (Fig. 117). There is a lytic radiolucent image with sharp boundaries but without peripheral sclerosis. This is epiphyseal and expansive, extending in a bipolar fashion towards the metaphysis and the articular cartilage, expanding the diameter of the bone. The cortex is often thinned, even perforated or fractured. This is a tumour with little osteogenic capacity, and intratumoral calcification is rare. The classic image of "soap bubble" trabeculation should suggest recurrence. Owing to its rapid course, it is common to find at first consultation that the tumour has already extended over half the epiphyseal surface; this is particularly true for the hand where it usually occupies the entire epiphysis. Periosteal reaction is

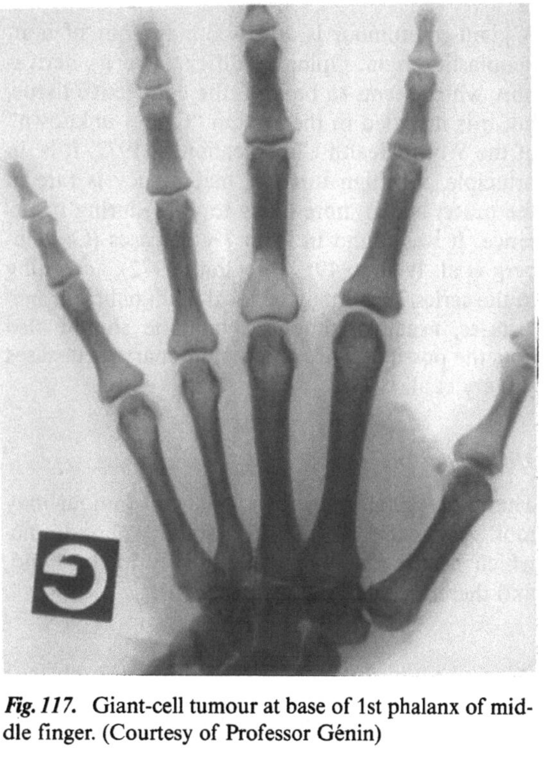

Fig. 117. Giant-cell tumour at base of 1st phalanx of middle finger. (Courtesy of Professor Génin)

trivial or absent. In the carpus diagnosis is difficult, as there is merely a radiolucent image without increase in size of the bone.

Pathology

Macroscopically, the thinned periosteum covers a brown or dark-red friable tissue of vascular appearance, with yellowish-grey zones of necrosis. Histologically, there is a meshwork of round, oval or fusiform cells and giant-cells in a more or less loose intercellular stroma. Jaffe et al. (1940) drew up a classification in three grades for prognostic purposes. In grade I the stroma is very vascular and loose, there are few or no mitoses and numerous giant cells. Conversely, in grade III the stroma is cellular, with pleomorphic cells; there are many mitoses and the giant cells are smaller and less numerous. Grade II is intermediate between grades I and III. This classification is difficult to apply since numerous intermediate forms exist (I-II). In particular, there is no true correlation with the clinical course, and malignant change is possible in all three grades.

Aetiopathogenesis

A giant-cell tumour is an osseous tumour of non-neoplastic origin. Opinions differ as to its derivation, which seems to be from the connective tissue, but it is included in the section "Origin unknown" of the World Health Classification of 1972. It is, in principle, a benign tumour; malignancy is rare at the outset and is more likely to arise during recurrence. It was found in from 7% of cases (Goldenberg et al. 1970) to 19% (McGrath 1972), according to the series. There may be local malignancy or metastases, usually pulmonary, but one should also note the possibility of benign pulmonary metastases (Meary et al. 1975).

Differential Diagnosis

The radiological image of a giant-cell tumour may look very much like that of a solitary chondroma, which occurs rather more frequently in the hand, and therefore will often be evoked first.

Treatment

Several methods of treatment have been suggested: curettage, curettage and packing with cancellous bone, simple resection, resection with bone grafting, amputation and radiotherapy. Though curettage has the advantage of being simple, it carries a high risk of recurrence, probably related to inadequate excision. The recurrence rate was 85% in the series of Goldenberg et al. (1970) after curettage, whose indications are therefore limited to small tumours with an intact cortex and a history of less than 3 months. Marcove et al. suggest supplementing curettage with cryotherapy, instilling liquid nitrogen into the residual cavity (Marcove et al. 1968), a technique that seems to give satisfactory results only after several years' experience (Marcove et al. 1978) and carries the risks of transient necrosis and pathological fracture. It was used once in the hand for a recurrence in a phalanx; the tumour did not recur again, but there was a pathological fracture 2 months after cryotherapy. Curettage followed by bone grafting calls for rigorous technique. The curettage must be complete, freshening the walls of the cavity, which is filled by packing with cancellous and corticocancellous grafts. According to the series, this technique yields 30%-45% of recurrences, with one exception - the 13 lesions of the hand in the series of Averill et al. (1980), of which 11 recurred.

Resection consists of total ablation of the tumour with its walls, and in the hand this means resection of the entire skeletal segment. One must endeavour not to open the tumour during the excision.

Whether or not followed by bone grafting, this treatment carries a much lower risk of recurrence than curettage (Meary et al. 1975, 1/20; Averill et al. 1980, 3/7). Though it may be very mutilating at other sites, reconstruction poses less of a problem in the hand. A bone graft may be used provided there is no cortical infraction, and several workers have grafted into the hand the phalanx of a toe or the corresponding metatarsal (Smith and Millender 1979; Minguella 1982). We should also note the giant-cell tumour of the lunate reported by Fitzpatrick and Bullough (1977), treated by resection of the proximal row of the carpus, the preoperative diagnosis having been Kienböck's disease. The amputation of a digital ray or segment is reserved for forms manifestly malignant from the outset (Gupta et al. 1980) or for recurrences (see below).

Radiotherapy has been abondoned by most authors and is never indicated in the hand, this tumour being hardly radiosensitive.

Development

Besides the possible local complications, such as infection and fracture, the most common complication is recurrence, which appears as often in the hand as elsewhere (30%-35%). We have seen that it is much more prevalent after curettage than after resection. This diagnosis will be suspected if there is a return of pain and the reappearance of a radiolucent image with lysis of the bone grafts. It justifies more aggressive treatment, resection or amputation. In the series of Averill et al. (1980) the three recurrences after curettage treated by further curettage all suffered recurrence for a second time. Recurrence particularly raises the pressing problem of malignancy (Fig. 118), since the appearance of metastases seems to be promoted by recurrence. The lung metastases are often solitary initially and may be treated by lobectomy. To conclude, we must note the caution expressed by Averill et al. (1980) in their analysis of their 28 cases of giant-cell tumour of the hand, of which 16 recurred. They state that, whatever the histological grade, the tumours must be regarded as locally aggressive.

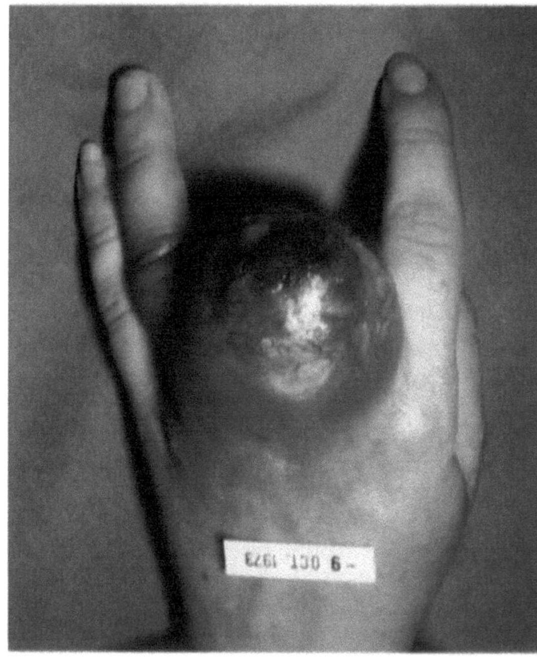

Fig. 118. Giant-cell tumour: malignant change after recurrence and amputation. (Courtesy of Professor Génin)

Bibliography

Averill RM, Smith RJ, Campbell CJ (1980) Giant cell tumors of the bones of the hand. J Hand Surg 5: 39

Campbell C, Akbarnia B (1975) Giant cell tumor of the radius treated by massive resection and tibial bone graft. J Bone Joint Surg [Am] 57: 982-986

Coley BL, Higinbotham ML, Kogure T (1958) Giant cell tumor of bone. Am J Surg 96: 479

Dahlin DC, Cupps RE, Johnson EW (1970) Giant cell tumor: A study of 195 cases. Cancer 25: 1051

D'Alonzo RT, Pitcock JA, Millford LW (1972) Giant cell reaction of bone. J Bone Joint Surg [Am] 54: 1267-1271

Fitzpatrick DJ, Bullough PG (1977) Giant cell tumor of the lunate bone: A case report. J Hand Surg 2: 269

Goldenberg R, Campbell CJ, Bonfiglio M (1970) Giant cell tumor of bone. An analysis of 218 cases. J Bone Joint Surg [Am] 52: 619-664

Gupta S, Kumar A, Gupta I (1980) Giant cell tumor of the first metacarpal bone. Hand 12: 288-292

Jaffe HL, Lichtenstein L, Portis R (1940) Giant cell tumor of bone. Its pathological appearance, grading supposed variance. Arch Pathol 30: 993

Marcove RC, Miller TR, Cahen WC (1968) The treatment of primary and metastatic bone tumors by repetitive freezing. Bull NY, Acad Med 44: 532

Marcove RC, Weis LD et al. (1978) Cryosurgery in the treatment of giant cell tumors of bone. Clin Orthop 134: 275-289

McGrath PJ (1972) Giant cell tumor of bone. An analysis of fifty-two cases. J Bone Joint Surg [Br] 54: 216-229

Meary R, Merle d'Aubigne R et al. (1975) Tumeurs à cellules géantes. 80 observations suivies. Rev Chir Orthop 61: 391-413

Minguella J (1982) Giant cell tumor of metacarpal in a child of unusual site and age. Hand 14: 93-96

Mirra JM (1980) Bone tumors. Diagnosis and treatment. Lippincott, Philadelphia, p 629

Reddy CRM, Rao PS, Rajakumari K (1974) Giant cell tumors of bone in South India. J Bone Joint Surg [Am] 56: 617

Roberts A, Long J, Wickstrom J (1976) A metacarpal giant cell tumor, a sternal osteoblastoma and a pubic osteogenic sarcoma in the same patient. South Med J 69: 660

Schobinger R, Stoll H (1957) The arteriographic picture of benign bone lesions containing giant cells. J Bone Joint Surg [Am] 39: 953

Sim F, Dahlin D, Beabout J (1977) Multicentric giant cell tumor of bone. J Bone Joint Surg [Am] 59: 1052-1060

Smith JA, Millender LH (1979) Treatment of recurrent giant-cell tumor of the digit by phalangeal excision and toe phalanx transplant. J Hand Surg 4: 164

Smith RJ, Mankin HJ (1977) Allograft replacement of the distal radius for giant cell tumor. J Hand Surg 2: 299

Srivastase T, Tuli J et al. (1975) Giant cell tumor of metacarpal. Indian J Cancer 12: 164-169

Thompson A, Turner-Warwick K (1955) Skeletal sarcomata and giant-cell tumor. J Bone Joint Surg [Br] 31: 266-303

Vistnes L, Vermuelen W (1975) The natural history of a giant cell tumor of bone. J Bone Joint Surg [Am] 57: 865

Solitary Bone Cyst

Solitary bone cyst, sometimes called benign or idiopathic bone cyst, is a lesion of childhood and adolescence, localised in the great majority of cases (75%) to the upper ends of the humerus and femur. It is extremely rare in the hand. Among the large series of solitary cysts (Spence et al. 1969; Meary 1974; McKay and Nason 1977), we found only a single case in the hand (Mirra 1980: 1/195 patients). Likewise, only one case has been reported in the large series of tumours of the hand (Bogumill et al. 1975). Carroll (1966) seems to have treated several but does not specify the number and, to our knowledge, only two other cases have been published (Miyakawa 1961; Ewald 1972). Therefore, a brief discussion will suffice.

Solitary cysts occur most often before the age of 20, with a maximum incidence between 9 and 14 years. They affect boys more than girls. They are often revealed by pain, a progressive increase in size or even pathological fracture.

Radiology

The films generally provide the diagnosis. There is a well-circumscribed, metaphyseal, lacunar image, not involving the growth plate, expansive, eroding the cortex and unilocular, with the classical appearance of the "base of an eggcup" at the junction between cyst and diaphysis. Bony septa traversing the cavity are clearly seen. In the hand it often invades the entire bone, leaving only a thinned but well-defined cortex and the juxta-epiphyseal subchondral bone. Carroll (1966) states that the lesion preferentially affects the distal third of the metacarpals and the shaft of the two proximal phalanges. It is often eccentric. These images may suggest an aneurysmal cyst (but this is more eccentric, more of a "blowout", and may infringe on the growth-plate) or an enchondroma (no intracavitary septa).

Pathology

Macroscopically, the periosteum is intact. Beneath an often very thin cortex there is a yellowish-brown fluid. The cyst walls, sometimes supported by bony septa, are lined with a soft grey tissue not forming a true membrane. The histological examination confirms the diagnosis. In the midst of a connective tissue with abundant fibroblasts there are fibrinous deposits, zones of absorption of the cancellous bone and zones of osteogenesis in contact with the cortex. Cohen (1960, 1970) postulates a disturbance of the venous circulation as the origin of this lesion.

Treatment

The classical treatment of these cysts combines evacuation of the cavity and packing it with iliac bone, bank-bone or even plaster of Paris. Some prefer conservatism, especially in latent forms and when the site of the lesion is very close to the growth plate. Simple aspiration of the cyst has been suggested. McKay and Nason (1977) advise subtotal resection without grafting and a simple plaster splint. The gratifying results obtained by injection of corticoids (methyl prednisolone) into the cyst should be noted (Scaglietti et al. 1979). The two cases reported in the literature were treated, the one by two curettages and packing at a 7-month interval, the other by subtotal resection of the phalanx and a massive iliac graft.

The course of these cysts may be towards recurrence, especially in the so-called active cysts. In all cases, restitution *ad integrum* is rare; a definitive bony "scar" often persists.

Bibliography

Baker DM (1970) Benign unicameral bone cyst. A study of 45 cases, with long term follow-up. Clin Orthop 71: 140

Bogumill GP, Sullivan DJ, Baker GI (1975) Tumors of the hand. Clin Orthop 108: 214-222

Boseker EH, Bickel WH, Dahlin DC (1968) A clinico-pathologic study of simple unicameral bone cysts. Surg Gynecol Obstet 127: 550

Carroll RE (1966) Tumors of the hand skeleton. In: Flynn JE (ed) Hand surgery. Williams & Wilkins, Baltimore, p 1039

Cohen J (1960) Simple bone cysts. Studies of cystfluid in six cases with a theory of pathogenesis. J Bone Joint Surg [Am] 42: 609

Cohen J (1970) Etiology of simple bone cysts. J Bone Joint Surg [Am] 52: 1493
Cohen J (1977) Unicameral bone cysts: Synthesis. Orthop Clin North Am 8: 715-736
Elandaloussi M, Padovani JP et al. (1982) Que peut-on attendre des injections intracavitaires de corticoïdes dans le traitement des kystes osseux essentiels de l'enfant? Rev Chir Orthop [Suppl 2] 68: 107-113
Ewald FC (1972) Bone cyst in a phalanx of a two-and-a-half year old child. J Bone Joint Surg [Am] 54: 399-401
Garceau GJ, Gregory CF (1954) Solitary unicameral bone cysts. J Bone Joint Surg [Am] 36: 267
McKay DW, Nason SS (1977) Treatment of unicameral bone cysts by subtotal resection without graft. J Bone Joint Surg [Am] 59: 515-519
Meary R (1974) Tumeurs bénignes osseuses et dystrophies pseudo-tumorales. Monographies des annales de chirurgie. Expansion Scientifique, Paris, p 118
Mechin JF, Forest M (1974) Kystes essentiels des os et images kystiques de nature indéterminée. In: Meary R (ed) Tumeurs bénignes osseuses et dystrophies pseudo-tumorales. Monographies des annales de chirurgie. Expansion Scientifique, Paris, pp 78-82
Mirra JM (1980) Bone tumors. Diagnosis and treatment. Lippincott, Philadelphia, p 629
Miyakawa G (1961) Replacement of the shaft of the phalanx with iliac bone. J Bone Joint Surg [Am] 43: 905-907
Neer CS, Francis KC et al. (1973) Current concepts on the treatment of solitary unicameral bone cyst. Clin Orthop 97: 40
Nelson JP, Foster RJ (1976) Solitary bone cyst with epiphyseal involvement. Clin Orthop 118: 147
Peltier LF, Jones RH (1978) Treatment of unicameral bone cysts by curettage and backing with plaster of Paris. J Bone Joint Surg [Am] 60: 820-822
Platt H (1930) Cysts of the long bones of the hand and foot. Br J Surg 18: 20-37
Porat S, Lowe J, Rousso M (1978) Solitary bone cyst in the infant radius. Clin Orthop 135: 132-136
Scaglietti D, Marchetti PG, Bartolozzi P (1979) The effects of methyl-prednisolone acetate in the treatment of bone cysts. J Bone Joint Surg [Br] 61: 200
Spence KF, Sell KW, Brown RH (1969) Solitary bone cyst: Treatment with freeze-dried cancellous bone allograft. J Bone Joint Surg [Am] 51: 87

Aneurysmal Bone Cyst

Aneurysmal bone cyst is a rare bony dystrophy that may occur in any part of the skeleton, with a predilection for the long bones and spine. Lesions of the hand are rare: 3/50 in the series of Lichtenstein (1957), 1/33 in the Meary group series (Mechin and Forest 1974), 4/134 in the series of Dahlin (1978) and none in the series of Koskinen et al. (1976). Fuhs and Herndon (1979) found 17 of 516 cases (3%) reported in the literature, to which they added 2 personal cases. Alnot et al. (1983), reporting a case, collected 12 metacarpal lesions from the literature.

The tumour was distinguished by Jaffe and Lichtenstein in 1942, clarifying the frequent confusion between aneurysmal cyst and giant-cell tumour, haemangioma, ossifying haematoma, etc. Dominogk (1971) found 344 cases in the world literature.

Clinical Features

This is a lesion of young patients, two-thirds of whom are less than 20 years old, but it may also be found at a ripe age (one patient aged 72 in the series of Slowick et al. 1968). Both sexes are equally affected. Pain is often the presenting symptom, either spontaneous or on movement; otherwise, attention is attracted by a pathological fracture. Examination may disclose increased local warmth over a swelling that is palpable and tender, with minor restriction of joint movement. Very often the cyst is asymptomatic, being discovered during an X-ray examination for some other reason. The three cases of aneurysmal cyst of the hand reported by Chalmers (1981) exhibited a painful swelling progressing over 6 months.

Radiology

The cyst has a predilection for the zones of slow growth in the long bones. It is metaphyseal and respects the epiphysis as long as the growth plate is patent; it rarely affects the diaphysis. In the hand it is preferentially located in the distal part of the metacarpals and the proximal ends of the phalanges (Carroll 1966). It very rarely affects the carpus (one case in the series of Fuhs and Herndon, 1979; one case in that of Barbieri 1984).

Radiological diagnosis is sometimes difficult in the hand and may be aided by film enlargements. An aneurysmal cyst is characterised by the following features (Fig. 119): a cystic appearance, radiolucent, classically eccentric, though often placed centrally in the small bones and occupying the entire width of the bone; a rapidly expansive image that may produce ballooning; a thinned cortex, blown-out but never destroyed (save in cases of pathological fracture); a trabecular appearance of the interior of the cyst with numerous intracavitary septa. This appearance is regarded by Sherman and Soong (1957) as that of a mature cyst, the first stage being entirely cystic; sometimes there is intracystic and subperiosteal calcification. Alnot et al. (1983) report a juxtacortical lesion in a metacarpal (see Fig. 119).

Arteriography is advised by several authors (Koskinen et al. 1976, 11 cases) and shows abnormal vascularisation at the periphery of the tumour with sinusoidal vessels of variable calibre, retention of contrast medium and, according to some (Lindbom et al. 1961), the presence of arteriovenous shunts. It has the theoretical importance of distinguishing an aneurysmal cyst from a solitary cyst and especially from a giant-cell tumour.

Pathology

The gross appearance of an aneurysmal cyst is that of "a sponge filled with blood". It consists of cavities filled either with blood, soft brown or yellowish tissue, or occasionally solid fibrous tissue.

The cyst is lined with a distinct membrane, unlike the giant-cell tumour. The cortex is often reduced to a fine bony shell, while the periosteum remains intact.

Microscopically, the cavities are separated by seams of fibrious tissue containing macrophages full of haemosiderin, multinucleated giant cells and osteoid cells (with signs of osteogenesis). The septa are covered with endothelial cells but, unlike haemangioma, there are no muscle or elastic cells. Several authors (Koskinen et al. 1976) have noted the existence of another benign tumour, either adjacent (chondromyxoid fibroma, nonossifying fibroma) or subjacent (fibrous dysplasia).

Aetiopathogenesis

The initial anomaly would seem to be an alteration in intraosseous haemodynamics (Lichtenstein 1950) with arteriovenous disequilibrium (Lindbom et al. 1961), possibly related to another pre-existing tumour, such as Biesecker et al. (1970) found in 32% of cases of aneurysmal bone cyst.

Differential Diagnosis

There was for a long time confusion between the giant-cell tumour and aneurysmal cyst, the latter being regarded as a variant of the former. Distinction is possible on several grounds: an aneurysmal cyst is not initially epiphyseal, it affects a younger population than the giant cell tumour, the cortex is never perforated, the gross appearance on section of the tumour is different, and the course is always benign. The alternative, and often suggested, diagnosis is enchondroma, but the peroperative and histological appearances allow differentiation.

Treatment

Several methods have been proposed: biopsy only, curettage, curettage plus cancellous packing, resection with or without bone grafting and radiotherapy. Conservatism is sometimes justified in certain stable lesions when surgical access is difficult and after biopsy. Curettage is variously assessed. Chalmers (1981) makes it the treatment of choice for this benign lesion in the hand adding, when necessary, packing with spongy bone. This technique has the inconvenience of profuse operative bleeding, which is difficult to control until the entire cyst has been resected. Others (Koskinen et al. 1976) consider curettage sometimes inadequate and a factor in the occasionally large recurrence rates. Barbieri (1984) stresses that, in the hand, bony destruction is sometimes such as to require resection en bloc followed by reconstruction with a massive corticocancellous graft (Verdan et al. 1976; Burkhalter et al. 1978) or even amputation of a ray (followed, in Barbieri's case, by transposition of the adjacent ray). Radiotherapy has provoked several cases of postirradiation sarcoma (Tillman et al. 1968) and should not be employed in the hand.

The recurrence rate is quite high, assessed by Chalmers (1981) from the literature at around 30%. It is higher after curettage than after resection en bloc,

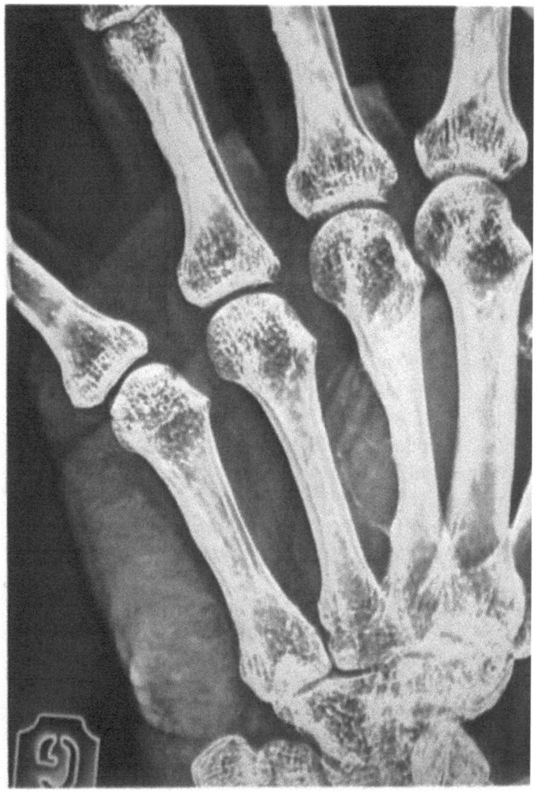

Fig. 119. Aneurysmal cyst of 3rd metacarpal, eccentric. (Courtesy of Professor Alnot)

though this does not prevent Chalmers (1981) from treating his cases of recurrence after curettage by a second curettage, on the grounds that resection en bloc is too mutilating a treatment for this benign lesion in a region as accessible as the hand. One may also refer to a case of a lesion in the 5th metacarpal, reported by Slowick et al. (1968), which regressed spontaneously after two recurrences.

Bibliography

Alnot JY, Badelon O, Grossin M, Bocquet J (1983) Kyste anévrysmal «juxta-cortical» du troisième métacarpien. Ann Chir 2: 358

Barbieri CH (1984) Aneurysmal bone cyst of the hand. An unusual situation. J Hand Surg 9/B: 89

Besse BE, Dahlin DC et al. (1956) Aneurysmal bone cysts. Additional considerations. Clin Orthop 7: 93-102

Biesecker SC, Marcove RA, Hovos AG, Miké V (1970) Aneurysmal bone cyst. A clinicopathological study of 66 cases. Cancer 26: 615

Burkhalter W et al. (1978) Aneurysmal bone cysts occur-

ring in the metacarpals: A report of three cases. J Hand Surg 3: 579
Carroll RE (1966) Tumors of the hand skeleton. In: Flynn JE (ed) Hand surgery. Williams & Wilkins, Baltimore, p 1039
Chalmers J (1981) Aneurysmal bone cysts of the phalanges. A report of three cases. Hand 13: 296-300
Clough J, Price C (1968) Aneurysmal bone cysts. Review of twelve cases. J Bone Joint Surg [Br] 50: 116-127
Dahlin DC (ed) (1978) Aneurysmal bone cysts. In: Bone tumors, 3rd edn. Thomas, Springfield, p 370
Dahlin DC, Besse BE et al. (1955) Aneurysmal bone cysts. Radiology 64: 56-65
Dominogk GW (1971) Aneurysmal bone cyst. 7 personal cases and 344 cases in literature. Langenbecks Arch Klin 328: 153
Fuhs S, Herndon J (1979) Aneurysmal bone cyst involving the hand: Two cases. J Hand Surg 4: 152
Jaffe HL, Lichtenstein L (1942) Solitary unicameral bone cyst, with emphasis on the roentgen picture, the pathologic appearance and the pathogenesis. Arch Surg 44: 1004
Koskinen EV, Visuri TI et al. (1976) Aneurysmal bone cyst. Evaluation of resection and of curettage in 20 cases. Clin Orthop 118: 136-146
Lichtenstein L (1950) Aneurysmal bone cyst. A pathological entity commonly mistaken for giant cell tumor and occasionally for hemangioma and osteogenic sarcoma. Cancer 3: 279
Lichtenstein L (1957) Aneurysmal bone cysts. Observations on fifty cases. J Bone Joint Surg [Am] 39: 873-882
Lindbom A, Soderberg G et al. (1961) Angiography of aneurysmal bone cyst. Acta Radiol 55: 12
Mechin JF, Forest M (1974) Kystes anévrysmaux. In: Meary R (ed) Tumeurs bénignes osseuses et dystrophies pseudo-tumorales. Monographies des annales de chirurgie. Expansion Scientifique, Paris, pp 83-89
Schobinger R, Stoll H (1957) The arteriographic picture of benign bone lesions containing giant cells. J Bone Joint Surg [Am] 39: 953
Sherman RS, Soong KY (1957) Aneurysmal bone cyst: Its roentgen diagnosis. Radiology 68: 54
Slowick F, Campbell CJ et al. (1968) Aneurysmal bone cyst. An analysis of 13 cases. J Bone Joint Surg [Am] 50: 1142-1151
Stein AH (1959) Benign neoplastic and non neoplastic destructive lesions in the long bones of the hand. Surg Gynecol Obstet 109: 189-197
Tillman BP, Dahlin DC et al. (1968) Aneurysmal bone cyst. An analysis of 95 cases. Proc Mayo Clin 43: 478
Verdan C, Loosli RM, Bozic C (1976) Diaphysectomie totale du 5e métacarpien pour kyste anévrysmal. Reconstitution par greffe osseuse du péroné. Ann Chir 30: 935

Malignant Tumours of Bone*

Malignant tumours of bone indicate malignant neoplasms originating from bony tissue, to which must be added other tumours located in bone but of differing histological origin (such as angiosarcoma). This definition excludes tumours of the haematopoietic tissue, such as myeloma and leukaemia. Primary malignant tumours must be distinguished from secondary or metastatic malignant lesions.

Primary Malignant Tumours

In the hand primary malignant tumours are, in any event, very rare (around 3% in Dahlin's 1973 series, less than 1% in that of Thomson and Turner-Warwick 1955). In 1973, Bryan et al. reported 126 malignant tumours of the hand, wrist and forearm, of which only 16 were osseous. Kendall et al. (1969) found only 3 bony lesions in 78 primary malignant tumours of the hand. These may be categorised as follows: osteogenic sarcoma, malignant cartilaginous tumours, Ewing's sarcoma - a rather separate group of ill-defined histological origin - and fibrosarcoma, extremely rare in the hand. The cartilaginous tumours, by far the most common (70%), are also the least threatening. Osteogenic sarcoma (25%) has an extremely grave prognosis when intraosseous, much more so than when the lesion is juxtacortical in site. The catastrophic prognosis of Ewing's sarcoma (4%) has improved with the advent of new programmes of combined chemotherapy and radiotherapy. Rather special to the hand is the *parosteal* localisation of osteogenic sarcoma and chondrosarcoma, which will be dealt with elsewhere.

Metastatic Tumours

These are rare and are generally evidence of the terminal stage of a cancer, the most common original site of which is bronchopulmonary.

Clinical Features

Two classic features characterise malignant bone tumours: pain, often moderate, of no especial character; and swelling, solid, hard, generally regular, not very tender and progressively enlarging. The related skin and soft parts are moderately inflamed or normal. The particular features of individual tumours will be discussed in the corresponding chapters.

Radiology

It is often the X-ray appearances that arouse suspicion of a diagnosis of malignancy, such as the following characteristic features: an osteolytic lesion, destruction of the cortex, periosteal reaction and invasion of the soft parts. The boundaries of the tumour are often ill-defined. Again, the image is not always a purely lytic one and may include zones of osteosclerosis. In every case, the presence of radiological signs suggestive of malignancy must lead to histological examination as soon as possible.

Differential Diagnosis

However, the radiological appearance is not always so suggestive; doubt may arise under the following conditions. A *lytic* image may suggest (a) an infection, and thus bacteriological, sampling must be routine, just as histological sampling is essential in osteitis, (b) a chondroma (and, as will be seen, diagnosis is often difficult in chondrosarcoma), (c) an aneurysmal cyst and (d) much more rarely, other benign osteolytic tumours (giant-cell tumour, chondroblastoma) or even fibrous dysplasia.

Osteosclerosis may suggest (a) an osteochondroma, so that several views should be taken to seek its base of implantation, (b) even the dense type of osteoblastoma and (c) in the case of juxtacortical sclerosing lesions (parosteal sarcoma): osteochondroma, myositis ossificans, ossifying haematoma, even gout.

Histological examination is essential *in every case*, as will be discussed. Histological analysis does not always lead to a certainty in diagnosis, however; this can only be confirmed by collation of the clini-

* This chapter was written in collaboration with Professor Jean Génin (Head, Department of Surgery) and Dr. Gilles Missenard (Surgical Consultant in Orthopaedics) of the Institut Gustave Roussy (Villejuif, Paris).

cal, radiological and histological data by an experienced team of collaborators. Only on this basis can the treatment programme be decided.

Treatment

The treatment of malignant bone tumours of the hand is difficult. The gravity of the prognosis and the small number of known cases make it difficult to lay down definite rules of management. The current tendency is to combine local and general treatment. Local treatment consists of wide surgical excision, limited to a finger or a ray when the tumour is phalangeal, more difficult to quantify when the lesion is more proximal. Reconstructive surgery, when necessary, must always be secondary and cautious. General treatment consists of antimitotic chemotherapy, often initiated prior to surgery and in some cases (Ewing's sarcoma) combined with radiotherapy.

Bibliography

Basora J, Fery A (1975) Metastatic malignancy of the hand. Clin Orthop 108: 182-186
Bryan RS, Dobyns JH, Pritchard D, Soule EH, Linscheid RL (1973) Malignant tumors of the hand, wrist and forearm. J Bone Joint Surg [Am] 55: 878
Clifford R, Kelly A (1955) Primary malignant tumors of the hand. Plast Reconstr Surg 15: 227-232
Dahlin DC (1973) Bone tumors, general aspects and data of 3987 cases, 2nd edn. Thomas, Springfield
Geschickter CF, Copeland MM (eds) (1949) Tumors of bone, 3rd edn. Lippincott, Philadelphia, p 185
Kendall TE, Robinson DW, Masters FW (1969) Primary malignant tumors of the hand. Plast Reconstr Surg 44: 37-40
Thomson AD, Turner-Warwick RT (1955) Skeletal sarcomata and giant cell tumor. J Bone Joint Surg [Br] 37: 266-303
Wilson P, Lance E (1965) Surgical reconstruction of the skeleton following segmental resection for bone tumor. J Bone Joint Surg [Am] 47: 1629-1656
Wu KK, Guise ER (1978) Metastatic tumors of the hand: A report of six cases. J Hand Surg 3: 271-276

Osteogenic Sarcoma

Osteogenic sarcoma, still sometimes called osteosarcoma or osteoblastic sarcoma, is a rare malignant bone tumour, and localisation in the hand is exceptional. Its very grave prognosis justifies even mutilating surgical treatment, together with chemotherapy.
The term of osteogenic sarcoma has long been the subject of controversy (Thomson and Turner-Warwick 1955), some authors using it to designate a sarcoma forming bone and others a sarcoma situated in bone. Dahlin and Coventry (1967) clearly defined it as being a sarcoma producing osteoid material, whatever the other components of the tumour. In fact, there are often associated chondrosarcomatous or fibrosarcomatous elements. Its situation in the hand is very rare: 0%-2% in various series (Price 1961, 0/88; Weinfeld and Dudley 1962; 2/94; Dahlin and Coventry 1967, 3/600; Weiss et al. 1978, 0/50). Carroll (1957) reported 10 cases, 4 personal, Drompp (1961) a bilateral case, Fleegler et al. (1980) 2 cases; there are perhaps 20 in all. In terms of site, there is a distinction between skeletal sarcoma, juxtacortical sarcoma (see p. 205) and sarcoma of the soft parts.

Clinical Features

This is a tumour of young patients, occurring most often in the second decade. When it affects elderly persons, it is generally secondary to Paget's disease. It involves the metacarpals and phalanges with equal frequency and exhibits the usual features of malignant osseous tumors (see p. 197).

Radiology

The films show lesions that are often very advanced. They combine diffuse bony remodelling (osteolysis and osteosclerosis), cortical destruction, bony proliferation invading the adjacent soft parts with "grass-fire" appearances and ill-defined boundaries. If the diagnosis of osteogenic sarcoma arises at this stage, the review is completed by a skeletal survey and a search for pulmonary metastases. Tomodensitometry can be used to define the precise boundaries of the tumour.

Pathology

The gross appearance is variable. There is a solid, manifestly bony tumour, eroding or destroying the

cortex and invading the periosteum and soft parts.
The histological appearance is usually characteristic, showing a hypercellular tissue formed of poorly differentiated, pleomorphic, osteoblastic cells, irregularly arranged, sometimes in spicules, and osteoid tissue. The presence of cartilage cells or fibroblasts does not exclude the diagnosis, for these two cell types are often noted in osteogenic sarcoma, and the tumour may be classified in terms of the dominant histological differentiation as osteoblastic (55%), fibroblastic (23%), and chondroblastic (22%). To these may be added the anaplastic osteogenic sarcomata, with little or no osteoid tissue, and the polymorphous variety. The degree of differentiation permits a prognostically based classification; that of Broders (1940), in four stages, is the one most used.

Aetiopathogenesis

Though osteosarcoma is most often primary, it may be secondary to Paget's disease (3%) or osteochondroma (0.3% no case reported in the hand) or to irradiation (2%). A traumatic origin has also been suggested.
In 1972, Haverbush et al. formally confirmed the possible localisation of Paget's disease in the hand. The literature contains two cases of malignant change: that of Carroll (1957) with multiple osteogenic sarcomata, one in the 5th metacarpal; and that of Drompp (1961) with bilateral degeneration (proximal phalanx of right middle finger and of left index) finger in a bony lesion labelled Paget's disease but without histological confirmation. In this particular type the sarcoma affects elderly subjects, and the prognosis is especially grave. Concerning irradiation as a cause, two cases of osteogenic sarcoma, have been reported, one after irradiation of the hand for arthrosis (Aub et al. 1952) and one after occupational irradiation in a dentist (Carroll 1957). One should also mention the possibility of bony dissemination with multiple localisations (osteosarcomatosis), including the hand (Drompp 1961).

Differential Diagnosis

After reviewing the clinical and radiological features, one may be led to consider the possibility of other sarcomata situated in bone, i.e. chondrosarcoma and fibrosarcoma. Histological examination of the specimen will establish the diagnosis. Reference has also been made (Dahlin and Coventry 1967) to the histological resemblances between very differentiated osteogenic sarcoma (grade 1) and fibrous dysplasia.

Treatment

Biopsy is essential in every case. The biopsy site must be carefully chosen in relation to the radiographic and CT scan data and the macroscopic appearance. It is essential for the sample to be examined by a forewarned pathologist.
Some authors (e.g. Carroll 1957) advise local radical resection, but this runs the risk of rapid recurrence. In the hand it seems better to perform amputation, limited to the finger for a distal lesion or extended to the ray if it is proximal. In all cases the excision must be made wide of the proximal limit of medullary extension of the tumour, precisely determined by a scanner.
Radiotherapy, whether followed or not by radical surgery, seems to give 5-year survival results slightly superior to those of amputation alone in large series (Trifaud and Meary 1972). It is less profitable in the hand, where there is a risk of leaving a nonfunctioning finger, and where amputation of a single ray preserves excellent functional levels.
Pulmonary metastases, when they are solitary, may be surgically resected. If multiple, they become the province of chemotherapy. Finally, there has been a growing interest in recent years in the quasi-prophylactic treatment of dissemination before it becomes perceptible, using supplementary chemotherapy.
The treatment programme currently employed for osteogenic sarcoma (Institut Gustave Roussy) begins with chemotherapy for 1-2 months. If the tumour reacts favourably (no increase in size under chemotherapy), surgical resection is employed on the basis outlined above. If the tumour is resistant, or if surgical resection is unsatisfactory, local radiotherapy is given, combined or not with chemotherapy in relation to the histological grade. The overall prognosis of osteogenic sarcoma is grave. Although the series of these tumours in the hand are very small, it does seem that the course of the distal forms is a little less dire, with a 10-year survival rate on the order of 30%. Local recurrence is common, and dissemination occurs by the blood stream, with metastases in the lungs and bones.

Osteogenic Sarcoma of the Soft Parts

This is a very rare malignant tumour, representing less than 1% of sarcomata of the soft parts and around 4% of osteogenic sarcomata (Dahlin and Coventry 1967). It is a tumour of middle life (average 47 years), preferentially localised in the thigh. Its prognosis is very grave (only 2 long-term survivals in the 26 cases of Allan and Soule 1971). Spread is by blood and lymphatic pathways. To our knowledge, only two cases have been reported in the hand: one at the wrist (Khoo and Balachandran 1966) in a man (60 years old and one in the index finger Allan and Soule 1971) in a woman aged 33.

It presents as a very hard swelling of the soft parts, of recent appearance, which grows rapidly. Clinical examination reveals nodules or even spicules at the periphery of the tumour. The films show a sometimes calcified opacity, not in contact with the adjacent skeleton, which is intact. These features may evoke other sarcomata of the soft parts, a mesenchymoma or myositis ossificans, but the histology establishes the diagnosis, for the tumour possesses the same characteristics as bony osteogenic sarcoma, with neoplastic cells that are always very poorly differentiated (grades 3 or 4 of Broders 1940).

The treatment of choice seems to be wide amputation, combined in certain cases with supplementary radiotherapy. Chemotherapy seems to have no effect on these tumours. The course is marked by local recurrences and by blood and lymphatic spread. The 5-year survival rates do not exceed 8% (Allan and Soule 1971; Fine and Stout 1956).

Bibliography

Allan CJ, Soule EH (1971) Osteogenic sarcoma of the somatic soft tissues. Cancer 27: 1121-1133

Aub JC, Evans RD, Hempelmann LH, Martland SH (1952) The late effects of internally-deposited radioactive materials in man. Medicine (Baltimore) 31: 221

Broders AC (1940) The microscopic grading of cancer. In: Pock GT, Livingston EM (eds) Treatment of cancer and allied diseased, vol 1. Hoeber, New York, p 19

Carroll RE (1957) Osteogenic sarcoma in the hand. J Bone Joint Surg [Am] 39: 325-331

Carroll RE, Godwin JK, Watson WL (1952) Osteogenic sarcoma of phalanx after chronic roentgen-ray irradiation. Cancer 9: 753-755

Clark CE (1952) Osteogenic sarcoma of the finger. Am J Surg 83: 112-114

Dahlin D, Coventry M (1967) Osteogenic sarcoma. A study of 600 cases. J Bone Joint Surg [Am] 49: 101-110

Drompp BW (1961) Bilateral osteosarcoma in the phalanges of the hand. J Bone Joint Surg [Am] 43: 199-204

Farr GH, Huvos AG (1972) Juxta cortical osteogenic sarcoma. J Bone Joint Surg [Am] 54: 1205-1216

Fine G, Stout AP (1956) Osteogenic sarcoma of the extrasqueletal soft tissues. Cancer 9: 1027-1043

Fleegler EJ, Marks KE et al. (1980) Osteosarcoma of the hand. Hand 12: 316-322

Geschickter CF, Copeland MM (1951) Parosteal osteoma of bone. A new entity. Ann Surg 133: 790

Haverbush TJ, Wilde AH, Phalen GS (1972) The hand in Paget's disease of bone. J Bone Joint Surg [Am] 54: 173

Khoo TK, Balachandran N (1966) Osteogenic sarcoma of soft tissue. Singapore Med J 7: 30-35

Lorentzon R, Larsson SE et al. (1980) Parosteal (juxtacortical) osteosarcoma. J Bone Joint Surg [Br] 62: 86-92

Mangini U (1967) Tumors of the skeleton of the hand. Bull Hosp Jt Dis Orthop Inst 28: 61

McKenna RJ, Schwinn OP et al. (1966) Sarcomata of the osteogenic series. An analysis of 552 cases. J Bone Joint Surg [Am] 48: 1-26

Moseley JE, Bass MH (1956) Sclerosing osteogenic sarcomatosis. Radiology 66: 41-45

Price CH (1961) Osteogenic sarcoma. An analysis of survival. J Bone Joint Surg [Br] 43: 300-313

Stark HH, Jones FE, Jernstrom P (1971) Parosteal osteogenic sarcoma of a metacarpal bone. J Bone Joint Surg [Am] 53: 147-153

Thomson AD, Turner-Warwick RT (1955) Skeletal sarcomata and giant cell tumor. J Bone Joint Surg [Br] 37: 266-303

Trifaud A, Meary R (1972) Pronostic et traitement des sarcomes ostéogéniques. Rapport du 74ème congrès Français de Chirurgie. Masson, Paris

Van der Heul RO, JR von Ronnen (1967) Juxtacortical osteosarcoma. J Bone Joint Surg [Am] 49: 415-439

Weinfeld MS, Dudley HR (1962) Osteogenic sarcoma. A follow-up study of 94 cases. J Bone Joint Surg [Am] 44: 269-276

Weiss AB, Adams G et al. (1978) Osteosarcoma: A review of 50 patients. Clin Orthop 135: 137-147

Malignant Cartilaginous Tumours

These are malignant osseous tumours originating in cartilaginous tissue. Their slow course and essentially local malignancy give them a quite favourable prognosis. They include *chondrosarcoma* and, added in recent years, *aggressive chondroma* which, despite its histological resemblance to chondroma, exhibits the same clinical, radiological and developmental features as chondrosarcoma.

It was in 1930 that Phemister distinguished chondrosarcoma from the other bony sarcomata. Its localisation in the hand is rare (1%-2% in different

series), and it was mentioned for the first time in the hand by Lichtenstein and Jaffe in 1943 when they reported a lesion in the proximal phalanx of the index finger. The world literature now contains around 70 cases in the hand, the largest series being that of Roberts and Price (1977), composed of 19 cases.
Depending on the relation of the site of the lesion to the bone, malignant cartilaginous tumours are classed either as *skeletal,* which is most common (85%), or *extraskeletal,* including juxtacortical lesions (10%) (see p. 206) and those of the soft parts (5%) (see p. 114).

Clinical Features

As stated by Roberts and Price (1977), "any painful swelling of the hand having recently increased in size in a subject over the age of 50 is very suggestive of chondrosarcoma" (Fig. 120a). Indeed, chondrosarcoma, like aggressive chondroma, is rare before the age of 40 years. It affects both sexes equally. The slowly developing swelling is generally the presenting feature and may reach a considerable size. The accompanying pain is often moderate.

Radiology

This shows the usual radiographic features of malignant bone tumours (see p. 202). The lytic image is often of very large size (Fig. 120b.). There is sometimes calcification within the interior of the tumour or in the adjacent soft parts. A metacarpophalangeal site is also very suggestive, most of these lesions being situated on the proximal two-thirds of the first phalanx or (a little more rarely) on the distal half of the metacarpal. Lesions of the middle phalanx are exceptional. Two cases have been reported in the carpus (Dahlin and Salvador 1974; Granberry and Bryan 1978), none in the terminal phalanx. The few arteriographs performed have shown vascular anomalies with retardation of the venous return.

Pathology

Chondrosarcoma

Macroscopically, chondrosarcoma presents as a bluish-white, lobulated tumour of cartilaginous appearance, exhibiting softer, greyish, gelatinous zones. The tumour infiltrates the bone, ruptures the cortex and invades the soft parts.
The histological diagnosis is extremely difficult because of the very great resemblance between chon-

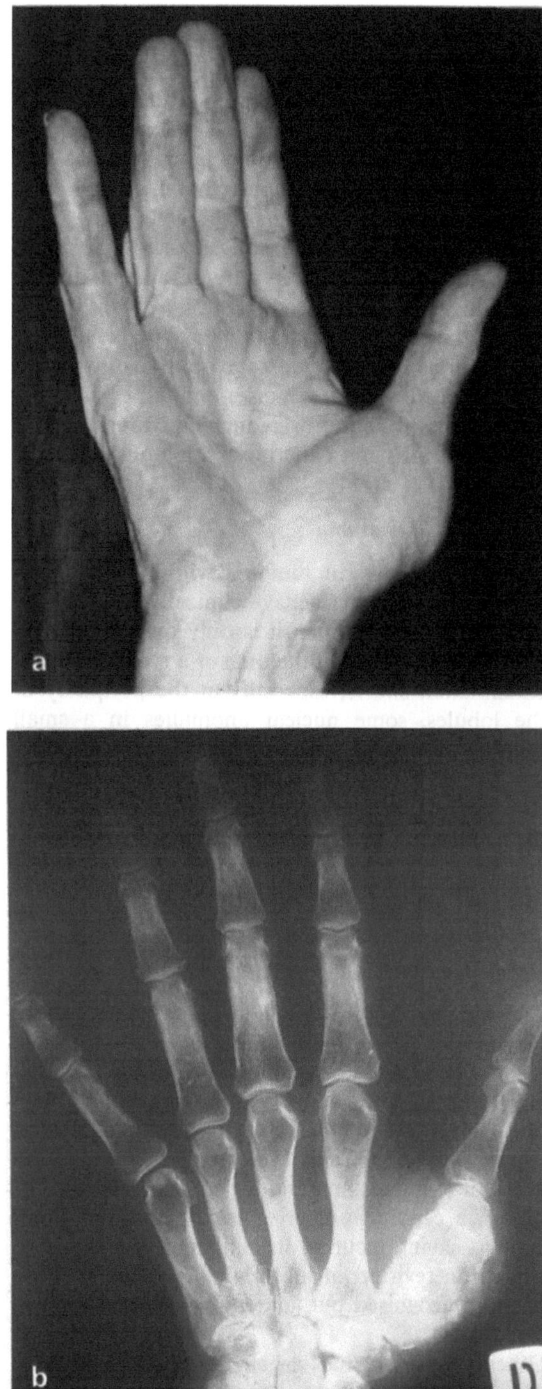

Fig. 120a, b. Chondrosarcoma of 1st metacarpal. ***a*** Clinical appearance. ***b*** Same patient: radiographic appearance, showing cortical destruction and invasion of soft parts. (Courtesy of Professor Génin)

drosarcoma and chondroma. Typically, chondrosarcoma is composed of numerous, binucleate or multinucleated, polymorphous, cartilage cells. These nuclei are "bizarre", plump and swollen, irregular and immature, with visible mitoses. The cells invade the bony medulla and adjacent soft parts. Chondroma, on the other hand, consists of a tissue with few cells, the cartilage cells being regular, well-differentiated, the nucleus solitary, small, dense and round, with very rare mitoses, and there is no invasion of adjacent tissues. However, it is known that chondromata may be hypercellular (in active forms), that chondrosarcoma is often well-differentiated with rare mitoses and, especially, that there exists in chondrosarcomata juxtaposition of chondromatous and chondrosarcomatous zones.

Aggressive Chondroma

A certain number of tumours whose histology is that of a chondroma behave clinically like a chondrosarcoma. Quite often it is only on carefully reviewing the sections at the time of a recurrence that examination shows, especially at the periphery of the lobules, some nuclear anomalies in a small number of cells (less than 5%) or even invasion by the chondrocytes of the vascular sheaths. It is therefore essential in every case to have the entire specimen examined by a forewarned pathologist and to collate the histological findings with the clinical and radiological data (see Differential Diagnosis).

Aetiopathogenesis

There are two contrasting theories. For some, chondrosarcoma is a degeneration of a chondroma. Several clinical cases seem to confirm this hypothesis, in particular those of Lansche and Spjut (1958) and Culver et al. (1975), showing the juxtaposition in the same bone, and even in the same tumoral zone, of a chondroma and a chondrosarcoma. For others, a certain number of lesions labelled chondroma, having recurred in malignant form, are revealed on revision of the original sections to have been unrecognised but authentic primary chondrosarcomata.

Others consider that degeneration of a solitary chondroma is altogether exceptional; the series of 110 enchondromata of the hand of Takigawa (1971) contains no case of malignant change. Again, Dahlin's (1973) series of 331 chondrosarcomata (taking all sites together) records 10% of secondary forms, related to either an osteochondroma (solitary or multiple) or to multiple enchondromatosis, but never to a solitary enchondroma. Malignant change does seem more frequent in multiple enchondromata (Block and Burton 1977) or the enchondromatoses (with a gloomier prognosis in this last instance).

Differential Diagnosis

The clinical and radiological features may suggest the possibility of other primary malignant tumours, a myositis ossificans or even certain types of fibrous dysplasia. The histological examination will provide the correct diagnosis. However, as we have seen, the essential problem remains that of chondroma, for the histology may be indefinite. In such cases one must rely on the clinical and radiological data, and any "chondroma" must be regarded as malignant until proved otherwise if it occurs after the age of 50 years, if the swelling is large and/or painful, if the radiograms show cortical infraction and/or invasion of the soft parts, and if the lesion recurs after curettage and grafting. In practice, it is very often on this last occasion that the diagnosis is rectified, casting doubt on the diagnosis of secondary degeneration of a "benign" chondroma. One should also mention the importance attached by Gottschalk and Smith (1963) to preoperative radioactive labelling, the count after excision of the tumour permitting differentiation of chondrosarcoma, which fixes more of the isotope than normal cartilage, from chondroma, which fixes less.

Treatment

All authors seem agreed that amputation is justified for chondrosarcoma; this is transmetacarpal in the rare cases of lesions of the middle phalanx, or excision of the entire digital ray in lesions of the metacarpal and proximal phalanx. Should the lesion seem to be invading the adjacent rays, the excision must be extended; it is very rare to be compelled to perform amputation at the wrist or of the forearm. In every case in which the diagnosis is made histologically after curettage and grafting of a lesion labelled benign, amputation should be immediate according to the rules given above. Whenever the preoperative diagnosis is uncertain, the excision-biopsy must be carefully planned with regard to a possible repetition for an amputation. The latter will have to be made en bloc, well away from the biopsy scar.

Insofar as one is dealing with elderly subjects for the most part, elaborate reconstructive surgery is rarely justified, but it is theoretically to be contem-

plated, given the overall favourable prognosis of chondrosarcoma. Although the published series are too small for any assertion, it seems that chondrosarcoma carries a better prognosis in the hand than at other sites. The series of Roberts and Price (1977, 19 cases) reports no deaths, while in the major series of osseous chondrosarcomata of Dahlin and Henderson (1956) only 41% of patients considered as correctly treated were alive after 10 years.
This is a tumour of essentially local malignancy, as demonstrated by recurrences after inadequate excision (Lansche and Spjut 1958; Patel et al. 1977) and by the extraskeletal recurrences after amputation (Trias et al. 1978; Patel et al. 1977). Metastases are classically rare, but the literature does show an incidence of around 10%, mainly pulmonary. They usually occur after one or more local recurrences.

Bibliography

Barnes R, Catto M (1966) Chondrosarcoma of bone. J Bone Joint Surg [Br] 48: 729

Block R, Burton R (1977) Multiple chondrosarcomas in a hand: A case report. J Hand Surg 2: 310

Coley BL, Higinbotham NL (1954) Secondary chondrosarcoma. Ann Surg 139: 547-559

Cruickshank AH (1945) Chondrosarcoma of a phalanx with cutaneous metastases. J Pathol 57: 144-145

Culver JE, Sweet DE, McCue FC (1975) Chondrosarcoma of the hand arising from a pre-existent benign solitary enchondroma. Clin Orthop 113: 128-131

Dahlin DC (1973) Bone tumors, 2nd edn. Thomas, Springfield

Dahlin DC, Henderson ED (1956) Chondrosarcoma. A surgical and pathological problem. J Bone Joint Surg [Am] 38: 1025-1038

Dahlin DC, Salvador AH (1974) Chondrosarcoma of the bones of the hand and feet. Cancer 34: 755-760

Goldenberg RR, Cohen P, Steinlauf P (1967) Chondrosarcoma of the extra-squeletal soft tissue. A report of 7 cases and review of the literature. J Bone Joint Surg [Am] 49: 1487

Gottschalk RG, Smith RT (1963) Chondrosarcoma of the hand. J Bone Joint Surg [Am] 45: 141-150

Granberry W, Bryan W (1978) Chondrosarcoma of the trapezium: A case report. J Hand Surg 3: 277-279

Habal MB, Snyder HH et al (1973) Chondrosarcoma of the hand. Am J Surg 125: 775

Jokl P, Albright JA, Goodman AH (1971) Juxta cortical chondrosarcoma of the hand. J Bone Joint Surg [Am] 53: 1370-1376

Lansche WE, Spjut H (1958) Chondrosarcoma of the small bones of the hand. J Bone Joint Surg [Am] 40: 1139-1145

Lester JG (1976) Two interesting tumors of the hand. J Bone Joint Surg [Br] 58: 386

Lichtenstein L, Jaffe HL (1943) Chondrosarcoma of bone. Am J Pathol 19: 553-590

Patel M, Pearlman HS et al (1977) Chondrosarcoma of the proximal phalanx of the finger. J Bone Joint Surg [Am] 59: 401-403

Phemister DB (1930) Chondrosarcoma of bone. Surg Gynecol Obstet 50: 216

Roberts P, Price C (1977) Chondrosarcoma of the bones of the hand. J Bone Joint Surg [Br] 59: 213-221

Sbarbaro JL, Straub LR (1960) Chondrosarcoma in a phalanx. Report of a case. Am J Surg 100: 751

Takigawa K (1971) Chondrosarcoma of the bones of the hand. A review of 110 cases. J Bone Joint Surg [Am] 53: 1591-1600

Trias A, Basora J et al (1978) Chondrosarcoma of the hand. Clin Orthop 134: 297

Wu K, Kelly AP (1977) Periosteal (juxtacortical) chondrosarcoma: Report of a case in the hand. J Hand Surg 2: 314

Fibrosarcoma

Fibrosarcoma is a very rare malignant tumour (2%-4% of primary malignant bone tumours) whose differentiation is purely fibroblastic. It may be primary or secondary to Paget's disease, fibrous dysplasia, irradiation or even chronic osteomyelitis. Localisation in the hand is quite exceptional (Dahlin and Ivins 1969; Larsson et al. 1976; Campanacci and Olmi 1977; André et al. 1983).

Clinical Features

Fibrosarcoma affects all ages, with a slight predominance around the age of 50 years. It does not exhibit any specific clinical features within the general context of malignant tumours (see p. 197). It is sometimes revealed by fracture.

Radiology

The initial site of the lesion is generally metaphyseal, but in the hand it rapidly invades the entire bone. The usual radiographic signs are found, but the lytic image is often multilocular, and the periosteal reaction may be absent (see p. 197).

Differential Diagnosis

The diagnosis of malignancy can usually be made on the clinical and radiographic data. Possible alternatives include other malignant bony tumours and osteomyelitis, so that bacteriological testing must be routine.

Pathology

Fibrosarcoma develops at the expense of the bone marrow. It consists of purely fibroblastic proliferation, without osteogenesis or chondrogenesis, made up of fusiform cells often grouped in bundles and surrounded by a collagen mesh of variable density. Atypical nuclei and abnormal mitoses are common. The histological diagnosis is sometimes difficult for the following reasons: the presence of osteoblastic cells suggests a diagnosis of osteogenic osteosarcoma of fibroblastic type; there is also the possibility of an exceptional fibrosarcoma of the soft parts secondarily involving the bone; finally, one may think of a more recent histological entity, malignant histiocytofibroma of bone. Study of the cell enzymes (Jefree and Price 1965) may help to differentiate fibrosarcoma from other sarcomata of bone.

Development

Spread is via the blood stream, with the appearance of pulmonary and bony metastases. The prognosis for survival is grave, with 5-year rates variously assessed at between 25% and 35% (André et al. 1983). It is even gloomier after the age of 40 years, and with the secondary forms of fibrosarcoma.

Treatment

In its distal sites amputation remains the treatment of choice, combined with chemotherapy. Radiotherapy is classically rather ineffective and should be reserved for cases where surgery is contraindicated.

Bibliography

André S, Tomeno B, Forest M, Carlioz A (1983) Fibrosarcome des os. Rev Chir Orthop 69: 107

Campanacci M, Olmi R (1977) Fibrosarcoma of bones. Study of 114 cases. Ital J Orthop Traumatol 3: 199

Dahlin DC, Ivins JC (1969) Fibrosarcoma of bones. Cancer 23: 35

Jefree GM, Price CHG (1965) Bone tumors and three enzymes. J Bone Joint Surg [Br] 47: 120

Larrson SE, Lorentzon R, Boquist L (1976) Fibrosarcoma of bone. Study of all cases recorded in the Swedish Cancer Register from 1958 to 1968. J Bone Joint Surg [Br] 58: 412

Ewing's Sarcoma

This was described by Ewing in 1921 and is the most common of the malignant bone tumours of childhood. Although situated within the bone, Ewing's sarcoma (or malignant endothelioma) is not an osseus tumour properly speaking as it does not arise from bone cells. Its origin is unknown. Although it formerly had a very sinister reputation, its prognosis has greatly improved since the advent of new compounds in chemotherapy.

Incidence

Localisation of Ewing's sarcoma in the hand is very rare: not a single case in the series of 165 cases of Dahlin et al. (1961) and only 2 cases in the 177 cases of Genin et al. (1976, cited in Genin et al. 1980). There are 6 other cases in the English language literature: Dick et al. (1971), 2 cases; Pandey (1972), 1 case; Bogumill et al. (1975), 1 case; Bhansali and Desay (1963), 1 case; and Dreyfuss et al. (1980), 1 case.

Clinical Features

Ewing's sarcoma occurs in the first two decades of life, affecting boys rather more often than girls. In addition to the usual clinical features of malignant tumours of the hand (pain and swelling), there are local inflammatory features that are sometimes preponderant and systemic - low-grade fever, wasting, leukocytosis and increased sedimentation-rate - which are classical sources of diagnostic error. These are uncommon with lesions of the hand (Dick et al. 1971).

Radiology

The classical "onion skin" appearance is a combination of osteolysis, cortical destruction and layers of new periosteal bone parallel to the cortex. There are, however, other and less typical appearances: osteosclerosis, a "grass fire" periosteal reaction at right angles to the shaft and a tumour of the soft parts that is barely visible if not calcified.

Pathology

The gross appearance is that of a soft grey or whitish tumour, with zones of liquefaction corresponding to areas of necrosis that may suggest a bone infection.
Histological examination shows a hypercellular tis-

sue formed of very anaplastic, small, round cells, impossible to relate to any specific cell line, with large, regular nuclei arranged in nests or strands. There is little or no intercellular connective tissue. These very undifferentiated cells are almost always of grade 4 in Broders' (1940) classification, but they may have a different appearance; in particular, they may be larger and less regular. The diagnosis is always difficult, and special forms of staining may be useful.

Aetiopathogenesis

The absence of cell differentiation prevents determination of the histological origin of Ewing's sarcoma. Dreyfuss et al. (1980) reported a lesion of the first metacarpal that was probably related to repeated trauma.

Differential Diagnosis

The clinical and radiographic features may suggest other malignant bone tumours, an eosinophilic granuloma, an aneurysmal cyst and, above all, an osteomyelitis, because of the inflammatory clinical and haematological signs. The danger consists of treating a lesion whose gross appearance is also deceptive as an infection without taking a biopsy. Conversely, bacteriological samples should be routinely taken at the time of biopsy. In some cases the histology may be evocative of metastasis (Dahlin et al. 1961).

Treatment

Surgery was formerly the treatment of choice in Ewing's sarcoma. It has now been almost entirely replaced by radiotherapy as local treatment, but applied to the entire bone, and systemic chemotherapy, which has completely transformed the prognosis of this tumour. However, amputation may come into the picture in certain distal lesions of the fingers (Dick et al. 1971). In this disease, as with other bone tumours, it is essential for treatment to be directed by a collaborative team of specialised physicians and surgeons.

Development

The prognosis, which used to be disastrous, with 50% of deaths in the first year and less than 15% survival rate at 5 years (Dahlin et al. 1961) is now much less gloomy: the Institut Gustave Roussy reports a 5-year survival rate of 40% (Génin et al. 1980). Spread occurs locally, by extension throughout the same bone (hence the need to irradiate the entire bony segment) and also via the blood stream, with remote bony and pulmonary metastases.

Bibliography

Bhansali SK, Desai PB (1963) Ewing's sarcoma. Observations on 107 cases. J Bone Joint Surg [Am] 45: 541-552

Bogumill GP, Sullivan DJ, Baker GI (1975) Tumors of the hand. Clin Orthop 108: 214-222

Broders AC (1940) The microscopic grading of cancer. In: Pack GT, Livingstone EM (eds) Treatment of cancer and allied diseases, vol 1. Hoeber, New York, p 19

Dahlin DC, Coventry MB et al (1961) Ewing's sarcoma. A critical analysis of 165 cases. J Bone Joint Surg [Am] 43: 185-192

Dick HM, Francis KC et al (1971) Ewing's sarcoma of the hand. J Bone Joint Surg [Am] 53: 345-348

Dreyfuss UY, Auslander L et al (1980) Ewing's sarcoma of the hand following recurrent trauma. Hand 12: 300-303

Dryer RF, Buckwalter JA, Flatt AE, Bonfiglio M (1979) Ewing's sarcoma. J Hand Surg 4: 372

Ewing J (1921) Diffuse endothelioma of bone. Proc NY Pathol Soc 21: 17

Génin J, Schweisguth O et al (1980) Sarcomes d'Ewing. In: Encyclopedie médico-chirurgicale. Appareil locomoteur, tome 2. Techniques. Paris, p 14030 D 40

Pandey S (1972) Ewing's sarcoma of the third metacarpal. Int Surg 57: 984

Pritchard DJ, Dahlin DC et al (1975) Ewing's sarcoma. A clinicopathological and statistical analysis of patients surviving 5 years or longer. J Bone Joint Surg [Am] 57: 10-16

Juxtacortical Sarcoma

Juxtacortical sarcoma was defined only relatively recently (Geschickter and Copeland 1951); despite its rarity, it is relatively common in the hand. It deserves a separate description because of its particular clinical characteristics and because its global prognosis is better than that of other bone sarcomata, thus justifying more conservative treatment. There are three forms: parosteal osteosarcoma, periosteal osteosarcoma and juxtacortical chondrosarcoma.

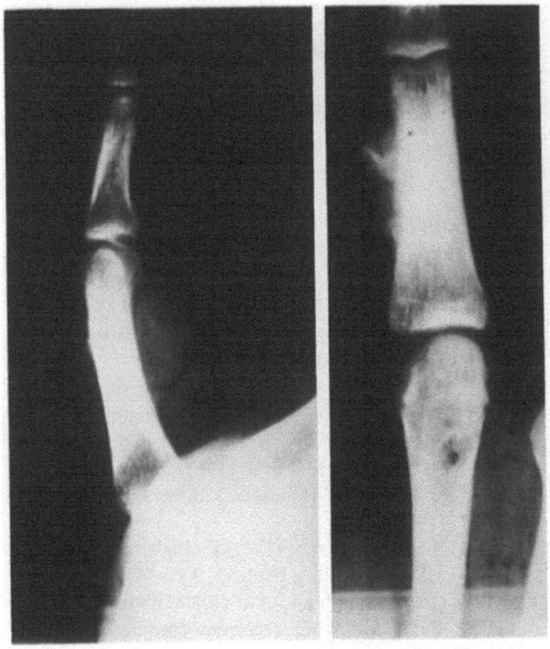

Fig. 121. Juxtacortical osteosarcoma (parosteal) of 1st phalanx. (Courtesy of Professor Génin)

Parosteal Osteosarcoma

Taking all sites together, juxtacortical osteosarcoma ("parosteal" in the English language literature) represents around 3% of all osteogenic sarcomata. To our knowledge four cases have been reported in the hand (Mangini 1967; Stark et al. 1971; Fleegler et al. 1980). The average age of predilection (30 years) is somewhat higher than for intraosseous osteogenic sarcoma.

The clinical picture is modified by the eccentric site of the tumour, which is well-defined to palpation. Radiographs show a juxta-osseous mass at the metaphysis, dense and lobulated, tending to surround the shaft and attached to the cortex by a broad base of implantation (Fig. 121). The gross appearance is that of a smooth, white, encapsulated tumour, very hard but with softer areas, enveloping the diaphysis while respecting the cortex. Moreover, the medulla is intact, even at the level of the base of implantation. Histological examination confirms the diagnosis of osteogenic sarcoma. Cartilaginous and fibroblastic cells are less common than in intraosseous osteogenic sarcoma. On the other hand, the bony tissue is very well developed, forming irregular seams. There is little infiltration of the adjacent soft tissues.

The alternative diagnoses suggested by the radiographic appearances include ossifying haematoma, myositis ossificans and osteogenic exostosis. Only rarely is malignancy diagnosed before operation (in which case the question of juxtacortical chondrosarcoma would arise.

Treatment is difficult to plan in view of the small number of known cases. However, it seems that en bloc resection of the tumour with the adjacent cortical bone and the invaded soft parts may lead to cure (Fleegler et al. 1980). Adjuvant chemotherapy is a possibility. Amputation is required after recurrence (Stark et al. 1971). The prognosis seems better than that for intraosseous osteogenic sarcoma. The four patients with this tumour in the hand were alive and without evidence of spread at 2, 3 and 4 years (two cases) later. However, late metastases may occur.

Periosteal Osteosarcoma

This tumour was presented by Farr and Huvos in 1972 and formally defined by Unni et al. (1976b). The distinguishing features from parosteal osteosarcoma are: its sessile rather than pedunculated form; the radiographic appearance of thickening of the bony cortex, associated with invasion of the soft parts by a tumour that is often less dense, more irregular in outline and not encircling the shaft; and its different histological features - cartilaginous islets without mature bony tissue and containing few malignant osteoblastic cells. Like parosteal osteosarcoma, the prognosis of periosteal osteosarcoma is better than that for intraosseous osteosarcoma. Given the small number of reported cases, no plan of treatment can be affirmed. To date, no case in the hand has been reported.

Juxtacortical Chondrosarcoma

This is an extremely rare lesion; only fewer than ten cases can be collected from the literature (Jokl et al. 1971; Lester 1976; Wu and Kelly 1977; Vichard et al. 1980).

It presents as a swelling of the soft parts, of solid consistency, not mobile over the bony plane and increasing progressively in size over several months. The radiographs display a juxtacortical radiolucent zone hollowing out a saucer-shaped notch in the cortex, which is sometimes sclerosed at this level but never invaded. The lesion may be dotted with scattered opacities. Surgical exploration shows a bluish-white subperiosteal lesion that is well-circumscribed and does not invade the bone or the soft parts. Amputation usually leads to cure. Simple excision is followed by local recurrences, but their development is relatively slow.

Bibliography

Farr GH, Huvos AG (cf biblio Sarcome ostéogénique)
Fleegler EJ, Marks KE et al (cf biblio Sarcome ostéogénique)
Geschickter CF, Copeland MM (1951) Parosteal osteoma of bone. A new entity. Ann Surg 133: 790
Jokl P, Albright JA, Goodman AH (cf biblio Chondrosarcome)
Lester JG (cf biblio Chondrosarcome)
Mangini U (1967) Tumors of the skeleton of the hand. Bull Hosp Jt Dis Orthop Inst 28: 61
Stark HH, Jones FE, Jernstrom P (cf biblio Sarcome osteogénique)
Unni KK, Dahlin DC, Beabout JW, Ivins JC (1976a) Parosteal osteogenic sarcoma. Cancer 37: 2466
Unni KK, Dahlin DC, Beabout JW (1976b) Periosteal osteogenic sarcoma. Cancer 37: 2476
Vichard P, Pinon P et al (cf biblio Chondrosarcome)
Wu K, Kelly AP (cf biblio Chondrosarcome)

Metastatic Tumours

Metastatic tumours of the hand are very rare (0.1% of all metastases). The first instance, a patient suffering from cancer of the breast with a secondary deposit in a metacarpal, was described by Handley in 1906. In 1907, De Massary et al. reported the case of a metastatic tumour of the soft parts of the hand secondary to cancer of the kidney. Numerous detailed studies have been made since, among which are those of Kerin (1953), Tubiana and Proux (1965), Bouvier et al. (1971) and Asencio and Hafdi (1982). In 1983 Kerin made an exhaustive review of the 123 cases published in the literature.

Tumoral metastases in the hand are essentially bony, more rarely cutaneous or in the soft tissues. Although very few, their incidence is probably underestimated in view of the long-established dogma that metastases are very rare below the knee and exceptional beyond the elbow, which has not encouraged the routine performance of distal radiography when assessing neoplasms. Most authors are agreed that the lesion is rare, that there is a marked predominance of bony lesions (Kerin 1983 found only 16 metastases in the soft parts in 123 cases) and that, among the primary tumours, bronchopulmonary carcinoma is responsible in 42% of cases. In 12% there is a carcinoma of the breast, almost always an adenocarcinoma, and in 12% a renal carcinoma. The alimentary tract (colon), parotid, prostate, oropharynx, uterus, testicle, liver, skin (melanoma) and lymphosarcoma form a heterogeneous group responsible for 30% of metastases in the hand (Kerin 1983). The lesions occur most often in males (66%) at the average age of 55 years, as is the case with bronchopulmonary carcinoma. They are equally common on either side; 66% of cases are found in the phalanges (with a very marked distal predominance: 52% in the terminal phalanx, 30% in the proximal and 18% in the middle phalanges), 17% in the metacarpus and 17% in the carpus. Multiple forms are relatively infrequently found (25%), and bilateral cases have been reported, while some authors find the thumb and middle finger more often involved. Metastases of the soft parts are much rarer (13%); they are most often situated on the dorsum of the hand and may secondarily invade by contiguity the local bony structures.

Clinical Features

The clinical polymorphism of these lesions accounts for the diagnostic difficulties. The metastasis may occur in the context of a known primary tumour or, not exceptionally, it may be solitary and appear as the presenting feature of the malignant process. In most cases, there is pain of varying intensity which may be spontaneous or provoked, acute or chronic, usually stubborn and tenacious, and frequently associated with a firm, palpable swelling.

Inflammatory features - redness, warmth and dysfunction - may dominate the clinical picture, suggesting a rheumatic or infective disorder (osteitis, osteoarthritis, whitlow), but there is no portal of entry and no pus (Fig. 122). A pathological fracture may be the starting point for an enquiry into causation. Finally, in some cases a hitherto asymptomatic lesion may be discovered at routine radiographic examination.

Lesions of the soft parts are evidenced by the appearance of first one and then several nodules, which are small, hard, well-defined and covered with normal skin, but which then break down, become purplish and ulcerate. At worst, the final appearance is that of an infiltrated, indurated and diffuse patch.

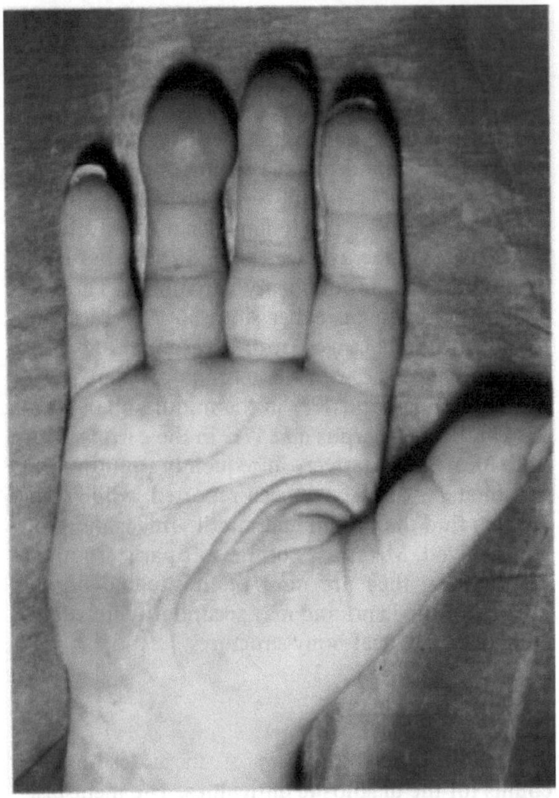

Fig. 122. Bony metastasis from breast cancer: note very inflammatory appearance. (Courtesy of Professor Génin)

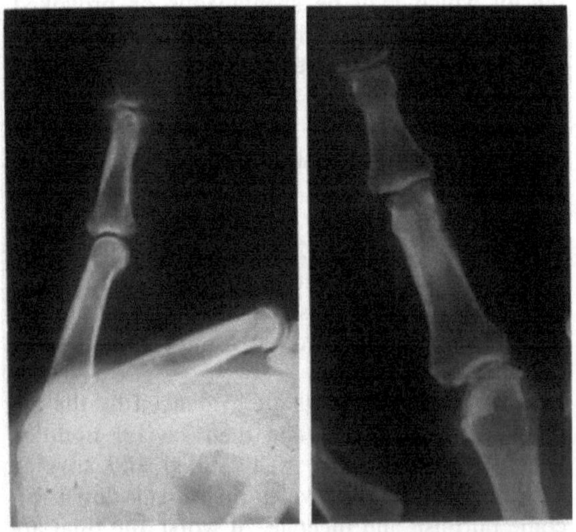

Fig. 123. Bony metastasis from oesophageal cancer: note virtually complete osteolysis of distal phalanx, but preserving subchondral bone. (Courtesy of Professor Génin)

Radiology

When the bone is invaded, radiographic examination reveals the various aspects of malignant osteolysis. Nearly always, the images are of an osteolytic nature and irregular aspect, extending to virtually the entire bone but with preservation of the joint spaces (Fig. 123). At its maximum, there is complete disappearance of a phalanx or metacarpal. Spread to the soft parts is associated with rupture of the cortex. Osteoporosis of the adjacent bony segments is also frequently noted.

Histology

Histological examination constitutes the essential element in the aetiological enquiry. It may make it possible to relate the metastatic lesion to an already recognised cancer or to arouse suspicion of an initial lesion and aid the search for the primary tumour (Kerin 1983 reports 22 cases in which metastases were thus revelatory in a series of 123 patients). The result of the histological examination may help to rectify a certain number of diagnosis initially evoked by the clinical polymorphism and the rarity of metastases in the hand. Thus, the clinical features may suggest a whitlow, an osteitis, an arthritis or a glomus tumour, while the radiographic features may suggest an osseous chondroma or a giant cell tumour.

Pathology

Different authors have suggested different mechanisms relating to the mode of origin of these tumours. Bronchopulmonary cancers are well-known for their tendency to spread to bone (10% of all osseous metastases, but Depass and Roswitt 1958 found only 2 cases of bony metastases in the hand in 800 bronchopulmonary cancers). The primary lesions are more often adenocarcinomata than anaplastic or epidermoid cancers. Spread is by the blood stream following neoplastic irruption into the pulmonary capillaries. Localisation in the fingers is explained by the rich vascularity and the local circulatory slowing (Tubiana and Proux 1965).
Bony metastases from breast cancer usually occur later, several months or even years after the initial lesion. Spread takes place via the internal mammary trunk and then the intrapulmonary and alveolar vessels. The most common origin is an adenocarcinoma. With colonic cancers, spread is via the portal system and then the inferior vena cava. As for cutaneous digital metastasis secondary to a le-

sion of the floor of the mouth, Kominoth and Florange (1977) state that this results from spread in two stages, first by the lymphatics to the mediastinal nodes, and then via the blood stream from the mediastinum to the digital dermis.

Treatment

Prognostically, the occurrence of a metastasis indicates the terminal phase of the neoplastic process. The average duration of life varies from 68 days (Greene 1957) to 3½ months (Asencio and Hafdi 1982). This rapidly fatal outcome influences the treatment programme, for only palliative measures are in order and no essay at reconstruction is permissible. Some authors propose radiotherapy to counter pain, possibly combined with chemotherapy; however, most will opt for amputation, with the aim of relieving the patient and improving his comfort, amputation that Basora and Fery (1975) suggest can be done on a finger under local anaesthesia.

Bibliography

Agansi M, Rzetelni V, Axer A (1981) Osteochondritis dissecans of the carpal scaphoid. J Hand Surg 6: 351

Altner RC, Singh SK (1981) An unusual case of ectopic ossification in a finger. J Hand Surg 6: 142

Asencio G, Hafdi CH (1982) Métastases osseuses au niveau de la main. Revue générale à propos de trois cas. Ann Chir 1/2: 97-102, 137-145

Basora J, Fery A (1975) Metastatic malignancy of the hand. Clin Orthop 108: 182-186

Benkeddache Y, Gottesman M (1982) Skeletal tuberculosis of the wrist and hand (27 cases). J Hand Surg 7: 594

Bourguignon RL (1981) Recurrent turret exostoses. J Hand Surg 6: 578

Bouvier M, Le Jeune E, Queneau P (1971) Les localisations distales du cancer secondaire des os. Rhumatologie 23: 353

Brette MR, Grandmottet P (1961) Métastase métacarpienne isolée d'un épithélioma colique. Travaux originaux, pp 733-739

Bryan RS, Soule EH (1974) Metastatic lesions of the hand and forearm. Clin Orthop 101: 167-170

Camiel MR, Aron BS (1969) Metastases to palm, sole, nailbed, nose, face and scalp from unsuspected carcinoma of the lung. Cancer 23: 214-220

Coley BL, Higinbotham NL (1939) Tumors primary in bones of hands and feet. Surgery 5: 112-128

De Massary E, Weil P (1907) Carcinose généralisée. Cancer des doigts simulant des troubles trophiques. Bull Soc Méd Hôpitaux Paris 24: 1456-1462

Depass S, Roswitt B (1958) Metastatic carcinoma in the bones of the hands. AJR 79: 643

Dickson RA (1976) Thorn-induced granulomata of bone. Hand 8/1: 69-71

Dolich BH, Spinner M, Kaufman G (1970) Isolated metastasis to the carpal bones. Report of a case. Bull Hosp Jt Dis Orthop Inst 31: 78

Finsterbush A, Husseini N, Rousso M (1981) Multifocal hemangioendothelioma of bones in the hand. J Hand Surg 6: 353

Fitzgerald P, Meenan F (1958) Sarcoidosis of hands. J Bone Joint Surg [Br] 40: 256-261

Fragiadakis EG, Panayotopoulos G (1972) Metastatic carcinoma of the hand. Hand 4: 268

Gelberman RH, Stewart WR (1978) Hand metastasis from melanoma, a case study. Clin Orthop 136: 264-265

Gold GL, William F (1963) Carcinoma and metastases to the bones of the hand. JAMA 184/3: 237-239

Graham WP, Kilgore ES (1973) Metastatic tumors of the hand. Hand 5/2: 177-179

Greene MH (1957) Metastasis of pulmonary carcinoma of phalanges of hand. J Bone Joint Surg [Am] 39: 972

Guttman G, Stein I (1968) Metastatic tumor of the thumb from adenocarcinoma of the colon. Int Surg 49/3: 217-221

Hacihanefioglu U (1978) Tumoral calcinosis. A clinical and pathological study of eleven cases in Turkey. J Bone Joint Surg [Am] 60: 1131-1135

Handley WS (1906) Cancer of the breast and its operative treatment. Murray, London, p 32

Harkess J, Peters H (1967) Tumoral calcinosis. A report of six cases. J Bone Joint Surg [Am] 49: 721-731

Haverbush TJ, Wilde AH, Phalen GS (1972) The hand in Paget's disease of bone. J Bone Joint Surg [Am] 54: 173-175

Johnson MK, Lawrence JF (1975) Metaplastic bone formation (myositis ossificans) in the soft tissues of the hand. J Bone Joint Surg [Am] 57: 999-1000

Kerin R (1953) Metastatic tumors of the hand. J Bone Joint Surg [Am] 40: 263

Kerin R (1983) Metastatic tumors of the hand. J Bone Joint Surg [Am] 65: 1331

Kominoth J, Florange W (1977) Métastase cutanée digitale d'un cancer du plancher buccal. Ann Otolaryngol Chir Cervicofac 94/1.2: 53-56

Lee BS, Kaplan R (1974) Turret exostosis of the phalanges. Clin Orthop 100: 186-189

Lerner M, Southwick W (1968) Keratin cysts in phalangeal bones. Report of an unusual case. J Bone Joint Surg [Am] 50: 365-372

Lombard A, Seignon B (1976) Métastase phalangienne simulant une ostéite et révélant un cancer bronchique. Sem Hôp Paris 52/2: 119-121

Marmor L, Horner RL (1959) Metastasis to a phalanx simulating infection in a finger. Am J Surg 97: 236-237

McCarthy EF, Ireland DC et al (1976) Parosteal (nodular) fasciitis of the hand. J Bone Joint Surg [Am] 58: 714-716

Mohanty S, Federiwicz TE, Uehara H (1968) Metastatic lesions of the fingers. Surgery 64: 411
Murphy F, Dahlin DC, Sullivan CR (1962) Articular synovial chondromatosis. J Bone Joint Surg [Am] 44: 77-86
Phalen G (1952) Calcification adjacent to the pisiform bone. J Bone Joint Surg [Am] 34: 579-583
Plonk JW, Feldman JM (1974) Carcinoid arthropathy. Arch Int Med 134: 651-654
Pollen A (1961) Calcareous deposits about the MP joints. J Bone Joint Surg [Br] 43: 250-255
Schajowicz F et al (1979) Juxta articular bone cysts (intraosseous ganglia). J Bone Joint Surg [Br] 61: 107-116
Svenes KJ, Halleraker B (1977) Epidermal bone cyst. Acta Orthop Scand 48: 29-31
Tondeur G, Verschraegen J, Depauw J (1969) Tumeur métastatique de la main. A propos d'une observation. Acta Chir Belg 68: 141
Toubiana GG, Proux C (1965) Les métastases osseuses distales. Ann Radiol 8: 217
Uribururu IJF (1976) Metastases of carcinoma of the larynx and thyroid gland to the phalanges of the hand. Report of two cases. J Bone Joint Surg [Am] 58/1: 134-136
Wissinger H et al (1966) Turret exostosis. Ossifying hematoma of the phalanges. J Bone Joint Surg [Am] 48: 105-110
Wu KK, Guise E (1978) Metastatic tumors of the hand. Report of six cases. J Hand Surg 3/3: 271-276

IV. Ungual Tumours

Introduction

Tumours of the ungual region are relatively common, but their clinical diagnosis is often difficult, the most frequent error being confusion with a mycosis. Biopsy should be routinely performed with any persistent lesion and will correct the diagnosis.

There are two main groups, benign and malignant. Benign tumours are by far the most common: peri- and subungual warts, naevi (black bands), botryomycoma (also found elsewhere), keratoacanthoma, Bowen's disease (at the borderline of malignant tumours) and subungual exostoses, discussed under the heading of osteochondroma (see p. 174). Malignant tumours include: squamous-cell epithelioma (the most common), basal-cell epithelioma (exceptional) and melanotic whitlow, discussed in the chapter on melanoma (see p. 44).

Periungual and Subungual Warts

This is a benign viral lesion, spread by contagion. Lesions of the hand are very common but occur mostly at the nail bed.

Clinical Features

There are two types (Baran 1978), peri- and subungual. The periungual wart (Fig. 124) is embedded in the lateral fold and forms a more or less continuous cushion at the periphery of the nail. It presents as a horny elevation, greyish in colour, with an irregular surface covered with prominences. It is painless unless there is secondary infection. The subungual wart (Fig. 125) differs by being always painful. It elevates the free edge of the nail and is seen through the nail as a yellow translucent spot.

Differential Diagnosis

The periungual wart may be confused with certain other tumours of the ungual region (Bowen's disease, epithelioma of the nail bed, botryomycoma)

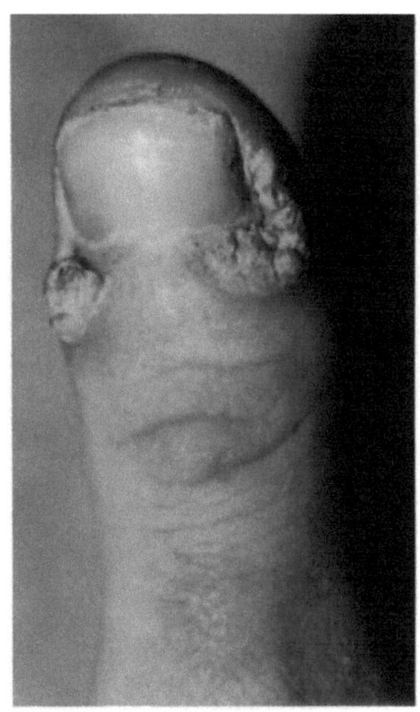

Fig. 124. Periungual warts

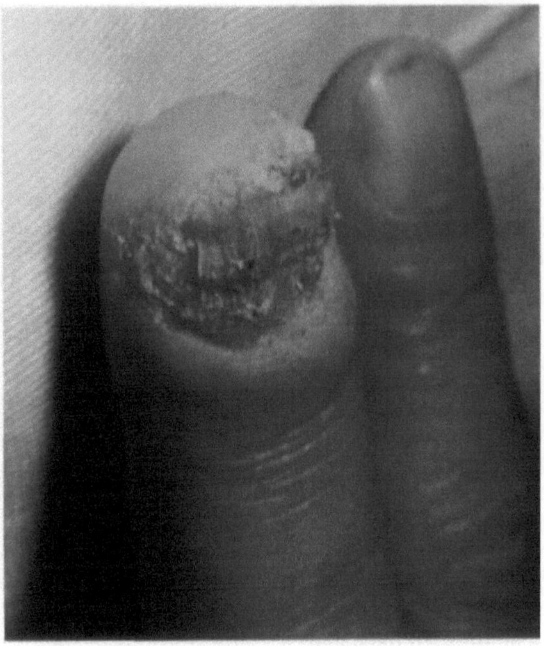

Fig. 125. Subungual wart

but also with the rare verrucous tuberculosis (Baran 1978) and with onychogryphosis (callosity of the nail folds). The subungual wart may be difficult to distinguish from a subungual glomus tumour.

Treatment

The treatment of ungual warts should always aim at conservatism, so as not to result in subsequent onychodystrophy. This is why surgical excision and curettage should be relinquished (Leclercq and Wibault 1972) in favour of local application of liquid nitrogen or carbon dioxide snow.

Bibliography

Baran R (1978) Tumeurs unguéales. In: Pierre M (ed) L'ongle. Monographie du Groupe d'Etude de la Main. L'expansion Scientifique, Paris, pp 76-84

Leclercq R, Wibault C (1972) Tumeurs bénignes infectieuses. In: Encyclopédie médico-chirurgicale: Dermatologie. Paris, pp 12725 - A10: 1-5

White CJ (1966) Diseases of the nails: Importance of proper diagnosis. In: Flynn JE (ed) Hand surgery. Williams & Wilkins, Baltimore, p 873

Naevi of the Nail Matrix

The nail matrix and nail bed are not often the site of naevi. The very existence of melanocytes in the nail bed has been disputed in white persons (Baran 1978b), though Higashi and Saito (1969) have found active melanocytes in the matrix of normal nails of Japanese. However this may be, the appearance of a pigmented subungual band must give rise to thorough investigation. Pigmented subungual bands appear in adults or older children as a brown or black band 1-2 mm wide, not deforming the nail or its setting (Fig. 126). The more diffuse chromonychia of bacterial or fungal dermatological origin can thus excluded (Baran 1978).

The diagnosis can only be affirmed by histological examination after biopsy, done under local anaesthesia and with a digital tourniquet. An incision is made on either side at the lateral margin of the nail matrix and the eponychium is elevated, thus exposing the nail matrix. This may or may not be the site of a pigmented spot. If it is, the naeval zone of the matrix must be excised en bloc, together with the corresponding strip of the nail bed and of the nail itself. Suture of the matrix and nail bed should be made under a lens with a very fine 7/0 or 8/0 material. Nevertheless, regrowth of the nail is usually dystrophic, with a longitudinal cicatricial line. When, as is more often the case (Higashi 1968), no naevus is found at the nail matrix, it is enough to excise the strip of nail bed and nail without touching the matrix; this allows a better quality of regrowth. These pigmented bands or strips without actual naevi are quite common in black and Indian people in the form of diffuse or banded melano-onychia. They are occasionally seen in white-skinned subjects, but the aetiology has not been completely established (Leyden et al. 1972).

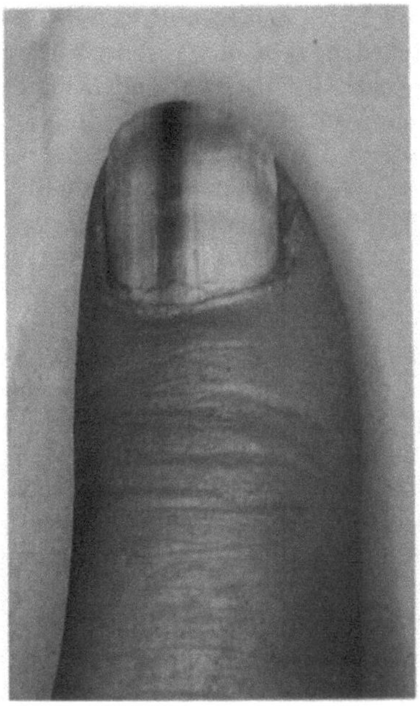

Fig. 126. Naevus of nail matrix: subungual pigmented band

Bibliography

Baran R (1978a) L'ongle dans les affections dermatologiques. In: Pierre M (ed) L'Ongle. Expansion Scientifique, Paris, pp 66-75

Baran R (1978b) Tumeurs unguéales. In: Pierre M (ed) L'Ongle. Expansion Scientifique, Paris, pp 76-83

Higashi N (1968) Melanocytes of nail matrix and nail pigmentation. Arch Dermatol 95: 570-574

Higashi N, Saito T (1969) Horizontal distribution of dopa positive melanocyte in nail matrix. J Invest Dermatol 53: 163-165

Leyden J, Spott DA, Goldschmidt H (1972) Diffuse and banded melanin pigmentation in nails. Arch Dermatol 105: 548-550

Botryomycoma

Botryomycoma is a small, benign angiomatous tumour of connective tissue type. It was described by Kuttner in 1905 (cited by Leclercq and Wibault 1972) under the name of telangiectatic granuloma, but is better known in the English-language literature by the name of pyogenic granuloma (Radcliffe-Crocker, cited in Leclercq and Wibault 1972).

Clinical Features

It is often found in the hand (Fig. 127) but also elsewhere, as on the trunk and face. In the nail, botryomycoma presents more in its "moist" form: a dark-red, fleshy button the size of a pea, bleeding on contact (Fig. 128). Usually it is sessile and seems laid on the skin surface, from which it is separated only by a macerated neck, but sometimes it is completely pedunculated. The usual site of the lesion in the hand is at the nail fold, but it may also be situated on the nail plate (Baran 1978). It should be added that it has a special appearance at the great toe, where it often develops in relation to an ingrowing toenail.

Aetiology

Botryomycoma is normally of microbial origin and forms evidence of chronic secondary infection of a small wound or unnoticed excoriation, as is more often the case at the nail. It may occur apart from any wound, assuming the aspect of an acquired angioma (Leclercq and Wibault 1972). In the absence of any treatment, the course is chronic with progressive increase in size and without any tendency towards cicatrisation. Sometimes, however, the lesion fibroses and gains an epidermal covering (Colomb et al. 1966).

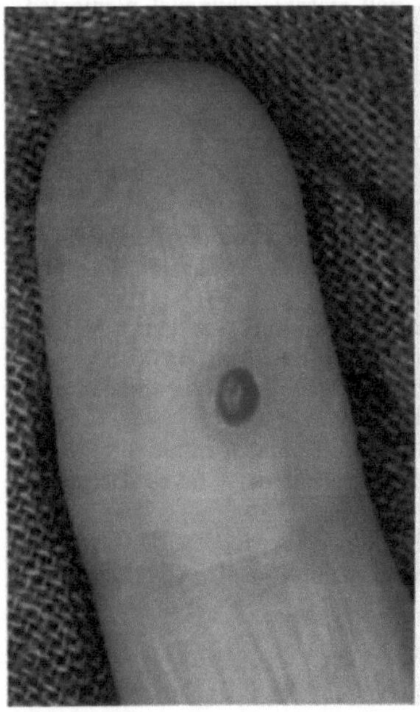

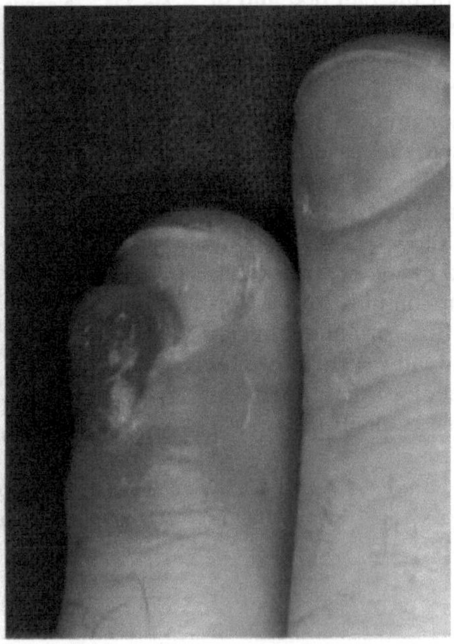

Fig. 127. Botryomycoma of fingertip *(left)*

Fig. 128. Ungual botryomycoma: usual "moist" form *(right)*

Differential Diagnosis

The differential diagnosis relates to the same considerations as with other ungual tumours (see under "Squamous-cell epithelioma of the nail-bed", p. 223), in particular Bowen's disease in its ulcerative form and achromic malignant melanoma. The diagnosis is based on the dogma of the absolute necessity of biopsy for every continuing ungual lesion; histology confirms the appearance of the fleshy nodule, formed of proliferating capillaries with an inflammatory cellular infiltration.

Treatment

Treatment consists of simple excision of the lesion by curette or knife, possibly with added diathermisation or local applications (e.g. permanganate). The resulting gap must not be closed, or there will be a risk of recurrence.

Bibliography

Baran R (1978) Tumeurs unguéales. In: Pierre M (ed) L'ongle. Monographie du Groupe d'Etude de la Main. L'Expansion Scientifique, Paris, pp 76-84

Colomb D, Gonyet C, Durand B (1966) Botriomycome épidermisé et fibrosé. Bull Soc Fr Derm 73: 546

Duperrat B (1961) Note sur l'histologie du botriomycome. Bull Soc Fr Derm 68: 666

Leclercq R, Wibault C (1972) Botriomycome. In: Tumeurs bénignes infectieuses. Encyclopédie médico-chirurgicale: Dermatologie, p 12725 A10-10

Keratoacanthoma of the Nail Bed

This is a benign and ubiquitous lesion characterised by pseudo-epitheliomatous hyperplasia. Though it is only very uncommonly situated at the nail, this site deserves mention because of the resulting diagnostic problems, in particular from squamous-cell epithelioma.

Keratoacanthoma was first described by Sir Jonathan Hutchinson, in 1889, but the first real characterisation was by Levy et al. in 1954. To our knowledge, the first reports of subungual keratoacanthoma were those of Lamp et al. (1964, three cases), followed by Shapiro and Baraf (1970, three cases, Emerson et al. (1971, one case) and Mehregan and Fabian (1973, two cases). Localisation of this lesion in the hand is actually relatively common, representing some 15% of all keratoacanthomata, but involvement of the nail is very rare and thus often passes unrecognised. Lamp et al. (1964) gathered six cases from the literature published under another diagnosis, but whose description is more suggestive of keratoacanthoma.

Clinical Features

Keratoacanthoma has a very characteristic course. At the outset there is the rapid appearance of a small, red, inflammatory papule, soon accompanied (and usually in this order) by pain, first intermittent and then constant and quite intense, swelling and digital erythema. The papule rapidly increases in size to reach a diameter of 1-2 cm. A crusted nodule then appears in its centre, generally at the free edge of the nail, which is elevated and eroded. The lesion is then static for several weeks. It is not altered by any attempt at incision and drainage, which produces not pus, but sometimes a whitish pseudocaseous material. Regression eventually takes place slowly, resulting in a scarred area after a few months.

Radiology

Radiography shows early destruction of the phalangeal tuft, which extends rapidly, thus showing evidence of bony necrosis due to compression without any periosteal reaction. Skeletal reconstruction during the regression phase takes place in parallel with the skin scarring.

Differential Diagnosis

As with other nail bed tumours, keratoacanthoma may also suggest many other possible diagnoses: epidermoid cyst, glomus tumour, osteochondroma, melanoma, metastasis, giant-cell tumour and osteomyelitis. The only difficult distinction arises with squamous-cell epithelioma (see p. 223). The importance of distinguishing the two conditions is obvious, since their treatment varies so radically, and help may be gained from the following points. Keratoacanthoma tends to occur at around 50 years of age, squamous-cell epithelioma at around 60; the former appears very rapidly, the latter much more slowly. Keratoacanthoma really does look like a tumour and alarms the patient from the outset, while the epithelioma assumes very confusing appearances and often resembles a chronic infection. The gross appearance of the central crater differs. In keratoacanthoma it is filled with keratin and occupies the greater part of the lesion, while in squamous-cell carcinoma it is smaller and contains necrotic tissue. Though biopsy is naturally indispensable, histological examination does not always allow a formal decision, and the diagnosis then rests on correlation of the histology with the gross appearances and clinical features mentioned above.

Epidermoid cysts also raise problems of diagnosis from keratoacanthoma, but basically there is a history of trauma, their course is slower, and pain is absent except with fracture. On the other hand, the radiographic features may be very much alike.

Pathology

Histological examination confirms the presence of keratin in the central crust, which is bordered by a pseudo-epitheliomatous hyperplasia. There may be atypical cells and abnormal mitoses, but the basal membrane is essentially intact, even though it may be breached locally due to the rapid growth of the lesion. The inflammatory reaction is always severe.

The general appearance varies in relation to the stage of progress of the lesion.

Development

Regression in several months is the rule, although Carroll and Bowers (1976) seem to have noted in some cases a failure of regression in the hand after a course of 1 year. No similar state of affairs has been reported at the nail, but this is difficult to judge since all reported ungual lesions were surgically excised without awaiting spontaneous progress. Malignant change may occur in some cases.

Aetiopathogenesis

McCormack and Scarff (1936) were the first to postulate that keratoacanthoma is derived from the hair follicle. This has been confirmed, but since then other possible causes, particularly viral, have been reported.

Treatment

For ungual lesions, there seems to be general agreement that keratoacanthoma should be treated by excision (by knife or curette), with subsequent skin grafting if there has been much loss of substance. Some (Shapiro and Baraf 1970) add diathermy coagulation. Recurrence, though rare, is not unheard of (Shapiro and Baraf 1970, one case).

The importance of diagnostic certainty with this lesion is evident; confusion with squamous-cell carcinoma, in the presence of radiological signs, would make amputation of the distal phalanx obligatory.

Bibliography

Baran R (1978) Tumeurs unguéales. In: Pierre M (ed) L'ongle. Monographie du Groupe d'Etude de la Main. Expansion Scientifique, Paris, pp 76-84

Carroll RE, Bowers WH (1976) Keratoacanthoma: An unusual hand tumor. Clin Orthop 118: 173-179

Emerson CW, Hillman JW et al (1971) Keratoacanthoma vs squamous-cell carcinoma. J Bone Joint Surg [Am] 53: 143-146

Hutchinson J (1889) The crateriform ulcer of the face: A form of epithelial cancer. Trans Pathol Soc Lond 40: 275

Lamp JC, Graham JH et al (1964) Keratoacanthoma of the subungual region. J Bone Joint Surg [Am] 46: 1721-1731

Levy E-J, Cahn MM, Shaffer B, Beerman H (1954) Keratoacanthoma. JAMA 155: 552

McCormack H, Scarff RW (1936) Molluscum sebaceum. Br J Dermatol 48: 624

Mehregan AH, Fabian L (1973) Keratoacanthoma of nail bed. Report of two cases. Int J Dermatol 12: 149-151

Shapiro L, Baraf C (1970) Subungual epidermoid carcinoma and keratoacanthoma. Cancer 25: 141

Bowen's Disease of the Nail Bed

Bowen's disease, or intraepidermal (or in situ) squamous-cell epithelioma, is a dyskeratosis with a malignant course. It may occur anywhere on the skin or the genital and buccal mucous membranes. Localisation at the nail is very rare and assumes a particular clinical appearance that explains the frequent failure of recognition.

The lesion was initially described by Bowen in 1912, but it seems that the first report of ungual localisation was that of Pardo-Castello and Pardo in 1960. Some 30 cases have been reported since (Coskey et al. 1972; Dieteman 1973; Mikhail 1974; Larregue et al. 1974; Baran 1978; Defiebre 1978; Stilwell and Maisels 1981; Strong 1983) in which the nail region is affected (Table 6). Thus, this site is extremely rare, lesions of the hand usually involving the dorsum (Pelisse 1981). Letessier, in his thesis (1972), found 4 ungual lesions in 234 cases of Bowen's disease (1.65%).

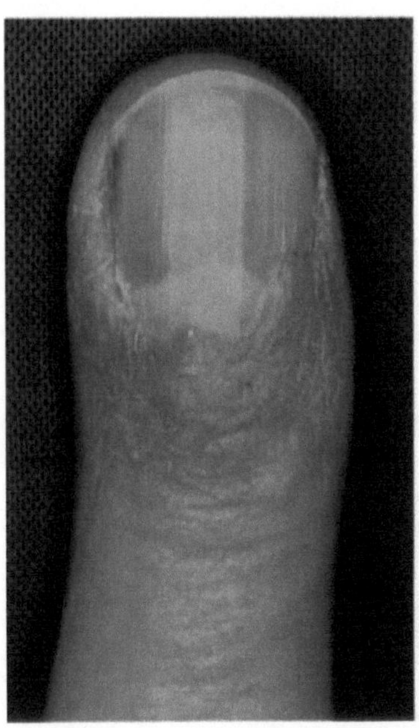

Fig. 129. Bowen's disease of the nail bed

Clinical Features

As at other sites, Bowen's disease of the nail bed preferentially affects patients at around the age of 60 years but sometimes also young adults. The published cases show a slight predominance in the left hand and radial fingers. As far as we know, no lesion of the 5th finger has been reported. Involvement of the toes is exceptional (Coskey et al. 1972). The clinical picture varies, but the lesion always starts at the deep aspect of the lateral or posterior nail fold (Fig. 129). This may appear as a fissure or ulceration, an erythematosquamous lesion or a pseudo-verrucous hyperplasia. The lesion then progressively involves the nail bed, which thickens and becomes hyperkeratotic. The nail is progressively destroyed and disappears in 3–5 years. There is never any bony invasion, and the radiographs of the distal phalanx are normal.

Multiple lesions are rare. Pelisse (1981) finds the lesion solitary in 80% of cases, and, when it is multiple, the lesions usually remain confined to a single area. Coskey et al. (1972) describe a case with a lesion on either hand and Strong (1983), a case with multiple lesions on both hands with four affected fingers.

Differential Diagnosis

This polymorphous lesion may suggest many alternative diagnoses: verruca, onychomycosis, glomus tumour, subungual exostosis, hypochromic form of melanoma, keratoacanthoma and squamous-cell carcinoma. The diagnosis is confirmed by the usual biopsy called for by any chronic lesion of the nail bed; we have seen that close relations exist between Bowen's disease and squamous-cell carcinoma (see p. 223) with the possibility of transition from one to the other.

Histology

Histological examination shows a marked thickening of the epithelium of the nail bed, with diffuse anaplasia. There exist numerous dyskeratotic cells with an increased number of mitoses, overlaid by a

Table 6. Bowen's disease, a review of all the cases from the literature

Reference	Sex	Age	Hand	Finger	Treatment
1 Pardo-Castello and Pardo (1960)	M	75	L	III	Amputation
2 Coskey et al. (1972)	M	41	L	IV	Amputation distal phalanx
3	F	62		Great toe	Excision and graft
4	M	35	L	I	Excision and graft
5			R	II	Excision and graft
6	M	54	R	II	Excision and graft,
7 Dieteman (1973)	M	45	L	II	Amputation distal phalanx
8 Mikhail (1974)	M	54	R	I	Mohs' technique
9	M	62	R	I	Mohs' technique
10	M	47	R	I	Mohs' technique
11	M	75	L	I	Mohs' technique
12 Larrègue et al. (1974)	M	52	L	III	Excision
13	M	76	L	I	Excision-graft; repeated after recurrence
14	F	61	L	IV	Diathermy and curettage
15	M	70	L	II	Diathermy and curettage; recurrence; radiotherapy; recurrence; amputation distal phalanx
16	M	81	L	III	Amputation distal phalanx
17 Baran (1978)	M	50	L	II	Excision
18	M	65	L	I	Mohs' technique
19	M	81	L	II	Diathermy coagulation
20	M	62	R	Great toe	Excision and graft
21	M	31	L	III	Diathermy; recurrence
22 Defiebre (1978)	M	54	L	III	5-Fluorouracil plus keratolytic agent
23 Stilwell and Maisels (1981)	M	59	R	I	Amputation distal half of distal phalanx
24	M	59	R	III	Amputation distal half of distal phalanx
25 Strong (1983)	M	40	Both	Multiple lesions	1. Amputation including 2/3 of P3 2. Excision-graft; recurrence; repeated 3. 5-Fluorouracil; recurrence; repeated

thick parakeratotic crust. In every case the basal membrane is intact, a fundamental feature in confirming the diagnosis of Bowen's disease. If it were otherwise, the lesion would be a squamous-cell (or, very rarely, basal-cell) epithelioma of the nail bed. It is well-known (Mikhail 1974) that biopsy may be inadequate and that examination of the entire lesion may be required to confirm the diagnosis, as the basal membrane may be invaded in places only in squamous-cell carcinoma.

Development

The course of Bowen's disease is very slow and always remains local. The possibility of transformation into squamous-cell epithelioma has been mentioned and, according to Graham and Helwig (1959), histological examination of the excised specimen reveals such an epithelioma in 15% of cases. The same authors performed autopsies on 35 patients affected by Bowen's disease and found a deep-seated cancer in 80% (as against 28% with squamous-cell carcinoma). However, not everyone acknowledges the need to carry out complementary investigations in Bowen's disease (Stilwell and Maisels 1981).

Treatment

Very diverse forms of treatment have been suggested, dermatologists generally tending to be more conservative than surgeons. While some prefer the security of amputation of whole or part of the terminal phalanx (Stillwell and Maisels 1981), others are content with excision and grafting of the nail bed and clinically affected adjacent areas even if, in the light of the histological findings, it is necessary to extend an inadequate excision secondarily (Coskey et al. 1972; Larrègue et al. 1974; Strong 1983). Some (Larrègue et al. 1974) perform diathermy coagulation together with curettage, or use radiotherapy, but these do not always prevent recurrence (Graham and Helwig 1973; Dieteman 1973). Yet others propose successive excisions with histological monitoring according to the technique of Mohs (Mikhail 1974), the wound being left open for monitored spontaneous healing (see p. 224). The local application of 5-fluorouracil has also been suggested, together with a keratolytic agent to facilitate its penetration (Defiebre 1978). If amputation of the phalanx appears excessive, especially in the thumb, it seems nevertheless essential to excise the entirety of the lesion to avoid the risk of development of squamous-cell epithelioma.

Bibliography

Baran R (1978) Tumeurs unguéales. In: Pierre M (ed) L'ongle. Monographie du Groupe d'Etude de la Main. Expansion Scientifique, Paris, pp 76-77

Bowen JT (1912) Precancerous dermatoses. J Cutan Dis 30: 241

Coskey RJ, Mehregan A et al (1972) Bowen's disease of the nail bed. Arch Dermatol 106: 79-80

Defiebre BK (1978) Bowen's disease of the nail bed: A case presentation and review of the literature. J Hand Surg 3: 184-186

Dieteman DF (1973) Bowen disease of the nail bed (Letter to the Editor). Arch Dermatol 108: 577-578

Graham JH, Helwig EB (1959) Bowen's disease and its relationship to systemic cancer. Arch Dermatol 80: 133

Kern AB, Schiff BL (1977) Bowen's disease of the finger. Cutis 19: 229-232

Larregue M, Civatte J et al (1974) Maladie de Bowen digitale. Bull Soc Fr Dermatol Syphiligr 81: 24-26

Letessier S (1972) La maladie de Bowen. A propos de 243 cas. Thèse, Paris, p 76

Mikhail GR (1974) Bowen disease and squamous cell carcinoma of the nail bed. Arch Dermatol 110: 267-270

Pardo-Castello V, Pardo OA (1960) Diseases of the nail. Thomas, Springfield, p 80

Pelisse M (1981) Maladie de Bowen. In: Encyclopedie médico-chirurgicale: Dermatologie. Paris, p 12755-C10

Stilwell JH, Maisels DO (1981) Subungual Bowen's disease. Hand 13: 287-290

Strong ML (1983) Bowen's disease in multiple nails beds. Case report. J Hand Surg 8: 329-330

Wagers LT, Shapiro L et al (1973) Bowen disease of the hand. Arch Dermatol 107: 745-746

Squamous-Cell Epithelioma of the Nail Bed

Squamous-cell epithelioma of the nail bed is a rare, locally malignant lesion mainly affecting the fingers, seldom the toes. It is by far the most common of the malignant tumours of the hand (variously assessed at 70%–90%).

The lesion was first mentioned by Velpeau in 1850 and has since been the subject of many studies. In 1939 Levine and Lise reported 18 cases published elsewhere and added 1 personal case. In a more recent article, Attiyeh et al. (1979) collected 58 cases from the literature and added 12 personal cases. There are also the 27 cases of Carroll (1976), making around 100 cases in all.

Clinical Features

Squamous-cell epithelioma generally affects patients around the age of 60 years, males slightly more often than females, and more frequently on the radial digits (thumb and index finger). The initial features are variable and deceptive, sometimes giving the appearance of a wart (Carroll 1976), of a pseudokeratosis elevating the free edge of the nail (Fig. 130), of a weeping ulcer or of a paronychia with erythema and oedema around the nail (Baran 1978). Three constant features should suggest the diagnosis.

1. Infection is constant, and often masks the tumour with a pyogenic granuloma, so explaining the anti-infective treatment which is often administered to these patients.
2. Pain is constant, localised and usually moderate, associated with the infection and frequently an ingrowing toenail.
3. The long course should immediately arouse suspicion of squamous-cell epithelioma, for this tumour resists all symptomatic treatment and develops very slowly. In the very great majority of published cases, the correct diagnosis was not made until 4–5 years on average after initial presentation.

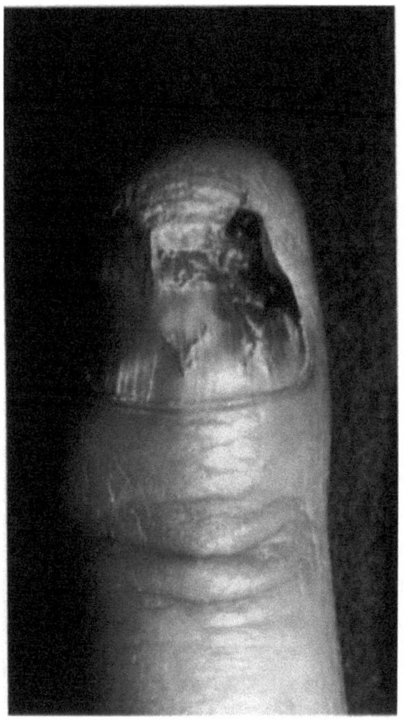

Fig. 130. Squamous-cell epithelioma of the nail bed

Radiology

The radiographic changes, showing invasion of the distal phalanx, are inconstant and delayed. There is a cavity at the phalangeal tuft, with blurred outlines and destruction of the posterior cortex. In very advanced lesions the entire tuft may be destroyed. In the series of Carroll (1976), two-thirds of the cases exhibited radiographic lesions, as against only 2 in the 12 cases of Attiyeh et al. (1979).

Aetiopathogenesis

There are two contrasting theories regarding the origin of squamous-cell epithelioma. Some, like Gellman (1963), ascribe a leading role to injury, and indeed a certain number of cases do appear immediately following injury of the nail region. Still, injuries of this region occur often. Others (Silverman 1935; Levine and Lise 1939; Eichenholz and de Angelis 1965; Mineiro et al. 1970) believe that the lesion originates from chronic infection; it is usually difficult to discover whether the infection really preceded the tumour or merely concealed it.

The debate remains unsettled. There is a third factor, hardly mentioned in the literature, that is certainly responsible for some of these tumours. First described by Pardo-Castello in 1960, *post-irradiation* squamous-cell carcinoma of the nail bed was well documented by Carroll in 1976. He reported 11 cases in doctors and especially in dentists after repeated exposure of the fingers to X-rays. The subungual involvement usually precedes the eventual skin lesions, the early features consisting of changes in the nail, which becomes dry and fissured. Finally, there is the case of multiple squamous-cell epitheliomata arising in epidermal dysplasia reported by Campbell and Keokarn in 1966.

Differential Diagnosis

An impressive number of alternative diagnoses may be suggested by this slowly developing lesion (Eibel 1971): chronic whitlow, chronic osteomyelitis, botryomycoma, glomus tumour, chondroma of the distal phalanx, melanoma, naevus or even subungual haematoma. There are three other lesions that may pose true diagnostic problems.

1. *Digital metastases* present with intense pain, the course is rapid, and the context is often suggestive.
2. *Keratoacanthoma* often initially exhibits the same features as squamous-cell epithelioma. It occurs, however, in rather younger patients (aged 50 years), its course in particular is much more rapid, and bony destruction begins very early on and is limited to the tuft. Biopsy confirms the diagnosis, and treatment is conservative (Shapiro and Baraf 1970; Emerson et al. 1971).
3. *Bowen's disease* poses the greatest problem, for there is a close relationship between the two disorders. The clinical picture is very similar to that of squamous-cell epithelioma, though Bowen's disease generally starts at a lateral fold and then progressively reaches the nail bed and detaches the nail. Here again, it is the histological examination that decides the diagnosis of an intra-epidermal epithelioma respecting the basal membrane. But Mohs' (1956) technique of serial excision has shown that, in fact, squamous-cell degeneration is common at certain sites in lesions labelled as Bowen's disease after biopsy. The diagnosis of Bowen's disease based on biopsy does not therefore exclude the possibility of squamous-cell epithelioma in the midst of the same lesion (Mikhail 1974). In every case of a slowly developing lesion of the nail region, a biopsy is therefore absolutely essential.

Development

Subungual squamous-cell epithelioma is not a very aggressive tumour, and its malignancy is essentially local. Metastases are exceptional. Regional nodal invasion is rare, and, if there is any axillary lymphadenopathy, this is usually an inflammatory sequel to the associated infection. It should be remembered that at other sites in the hand this epithelioma often runs a graver course (Forsythe et al. 1978).

Treatment

The classical treatment for subungual squamous-cell epithelioma is amputation of the distal phalanx, with reshaping of the head of P2 and skin cover by local flaps. If the distal phalanx is massively invaded, the amputation can if necessary be extended into the middle phalanx. In all cases cure is to be expected.

Some authors propose more limited treatment by local excision and skin grafting (Canipe et al. 1964; Eibel 1971). Obviously, this can only be considered in the absence of any bony involvement; it may be justified in limited lesions of the thumb, where amputation of the distal phalanx is mutilating. For some 15 years other surgeons have advised a technique of serial excisions under histological control according to the "chemosurgical" method of Mohs (1956, 1978). After macroscopic curettage of the lesions, a zinc chloride paste is applied to facilitate millimetric excision; then, from the next day, tangential resection is performed and repeated until there is no histological evidence of cancer cells. The remaining tissue is left to monitored, spontaneous healing. Tomsick and Menn (1981) used this technique on eight patients and saw no recurrence. Should the successive slides show any bony invasion, immediate amputation of the distal phalanx would be called for. Nodal eradication is justified only in the presence of palpable axillary adenopathy persisting or appearing after surgical treatment (Forsythe et al. 1978).

To conclude, squamous-cell epithelioma is a rare tumour in which the diagnosis is often not made until very late. In the absence of regional or remote invasion, which is exceptional, it is routinely curable by amputation of the terminal phalanx. Conservative measures may be justified in limited lesions of the thumb.

Bibliography

Albom AJ (1975) Squamous cell carcinoma of the finger and nail bed. A review of the literature and treatment by the Moh's surgical technique. J Dermatol Surg 1/2: 43-48

Attiyeh FF, Shah J et al (1979) Subungual squamous cell carcinoma. JAMA 241: 262-263

Baran R (1978) Tumeurs unguéales. In: Pierre M (ed) L'ongle. Monographie du Groupe d'Etude de la Main. Expansion Scientifique, Paris, pp 76-84

Beasley RW, Ristow BB (1976) Cancers of the hand. In: Andrade et al (eds) Cancers of the hand. Saunders, Philadelphia, pp 1437-1448

Campbell CJ, Keokarn T (1966) Squamous-cell carcinoma of the nail bed in epidermal dysplasia. J Bone Joint Surg [Am] 48: 92-99

Canipe TL, Howell JA, Howell CM (1964) Subungual carcinoma. Plast Reconstr Surg 33: 263

Carroll RE (1976) Squamous cell carcinoma of the nail bed. J Hand Surg 1: 92-97

Duperrat B, Grepinet H, Chambon R (1969) L'épithélioma du lit unguéal. Bull Soc Fr Dermatol Syphiligr 76: 560-562

Eibel P (1971) Squamous cell carcinoma of the nail bed. Clin Orthop 74: 155

Eichenholtz SN, de Angelis C (1965) Squamous cell carcinoma of the nail bed. JAMA 191: 1080

Ellis VH (1948) Squamous-cell carcinoma of the nail bed. J Bone Joint Surg [Br] 30: 656-658

Emerson CW et al (1971) Keratoacanthoma vs squamous cell carcinoma. J Bone Joint Surg [Am] 53: 143-146

Forsythe RL, Bajaj P et al (1978) The treatment of squamous cell carcinoma of the hand. Hand 10: 101-108

Gellman SB (1963) Primary carcinoma of the nail bed. NY J Med 63: 2408

Jaffe RH (1930) Malignant tumors of the nail bed. Surg Gynecol Obstet 50: 847-850

Kaminsky CA et al (1978) Squamous cell carcinoma of the nail bed. Dermatologica 157: 48-53

Levine J, Lise JR (1939) Primary carcinoma of the nail. Arch Surg 38: 107-112

Mikhail GR (1974) Bowen disease and squamous cell carcinoma of the nail bed. Arch Dermatol 110: 267-270

Mineiro LEG, Salter JJ, Orduna CC (1970) Squamous cell carcinoma of the nail bed. Arch Surg 100: 6

Mohs FE (1956) Chemosurgery in cancer, gangrene and infections. Thomas, Springfield

Mohs FE (1978) Chemosurgery: Microscopically controlled surgery for skin cancer. Thomas, Springfield

Nelson LM, Hamilton CF (1970) Primary carcinoma of the nail bed. Arch Dermatol 101: 63-67

Pardo-Castello V (1960) Diseases of the nails. Thomas, Springfield, p 47

Robin P (1976) Moh's surgery in the treatment of basal cell carcinomas of the skin. In: Andrade et al (eds) Cancer of the skin. Saunders, Philadelphia, pp 1537-1550

Schreiber MM (1979) Nail bed carcinoma. Cutis 23: 597-599

Shapiro L, Baraf CS (1970) Subungual epithelial carcinoma and keratoacanthoma. Cancer 25: 141

Silverman I (1935) Epithelioma following chronic paronychia. Am J Surg 29: 141-142

Tomsick R, Menn H (1981) Chemosurgical excision of subungual squamous cell carcinoma. Hand 13: 177-182

Velpeau M (1850) Tumeur cancéreuse de la phalange unguéale de l'indicateur. Gaz Hop 2: 314

Basal-Cell Epithelioma of the Nail Bed

At the nail bed this lesion exhibits the same clinical features and course as squamous-cell epithelioma. It was first reported by Eisenklam in 1934 (see Levine and Lise 1939), and there are less than ten cases in the literature, the two most recent being those of Alpert et al. (1972) and Hoffman (1973). Distinction from a squamous-cell tumour is purely histological, and classical treatment is amputation of the distal phalanx. At this site, radiotherapy seems ineffective (Hoffman 1973).

Bibliography

Alpert LI, Zak FG et al (1972) Subungual basal cell epithelioma. Arch Dermatol 106: 599

Hoffman S (1973) Basal cell carcinoma of the nail bed. Arch Dermatol 108: 828

Levine J, Lisa JR (1939) Primary carcinoma of the nail. Arch Surg 38: 107

Subject Index

U. Heim, Muri; **K. M. Pfeiffer,** Basel, Switzerland

Internal Fixation of Small Fractures

3rd edition of Small Fragment Set Manual

1987. Approx. 259 figures. Approx. 450 pages.
ISBN 3-540-17728-0

The subject of this book is the operative treatment of small-bone fractures which demand a specially refined instrumentarium and small implants. The emphasis is on articular fractures, particularly in the shoulder girdle, elbow, hand, talocalcaneal joint and foot, and minor fractures in the area of the knee. The book is intended as a companion volume to the **Manual of Internal Fixation.**

The bulk of the text is concerned with indications and step-by-step technique illustrated in semidiagrammatic figures. Each chapter contains examples of typical clinical radiographs, accompanied in some cases by details of long-term results.

New additions to the AO instrumentarium and techniques have been integrated into this third edition. Alternative techniques which have proved their worth in recent years are also described and illustrated.

The aim of the book is to promote a greater depth of knowledge among surgeons and orthopaedists who need to familiarize themselves with the operative traumatology and reconstruction of the smaller bones. All material has been completely revised and substantial additions made for the third edition.

Springer-Verlag
Berlin Heidelberg New York
London Paris Tokyo

MIX
Papier aus verantwortungsvollen Quellen
Paper from responsible sources
FSC® C105338

If you have any concerns about our products,
you can contact us on
ProductSafety@springernature.com

In case Publisher is established outside the EU,
the EU authorized representative is:
Springer Nature Customer Service Center GmbH
Europaplatz 3, 69115 Heidelberg, Germany

Printed by Libri Plureos GmbH
in Hamburg, Germany